The Generation and Control of Harmful Products in Food Processing

The Generation and Control of Harmful Products in Food Processing

The Generation and Control of Harmful Products in Food Processing

Guest Editor

Maomao Zeng

MDPI

Basel • Beijing • Wuhan • Barcelona • Belgrade • Novi Sad • Cluj • Manchester

Guest Editor
Maomao Zeng
State Key Laboratory of Food
Science and Resources
Jiangnan University
Wuxi
China

Editorial Office
MDPI AG
Grosspeteranlage 5
4052 Basel, Switzerland

This is a reprint of the Special Issue, published open access by the journal *Foods* (ISSN 2304-8158), freely accessible at: https://www.mdpi.com/journal/foods/special_issues/generation_control_products.

For citation purposes, cite each article independently as indicated on the article page online and as indicated below:

Lastname, A.A.; Lastname, B.B. Article Title. *Journal Name* **Year**, *Volume Number*, Page Range.

ISBN 978-3-7258-5799-9 (Hbk)
ISBN 978-3-7258-5800-2 (PDF)
https://doi.org/10.3390/books978-3-7258-5800-2

Contents

Navneet Kaur and Nigel G. Halford
Reducing the Risk of Acrylamide and Other Processing Contaminant Formation in Wheat Products
Reprinted from: *Foods* **2023**, *12*, 3264, https://doi.org/10.3390/foods12173264 1

Lixian Li, Yingjun Zhuang, Xiuzhi Zou, Maolong Chen, Bo Cui, Ye Jiao and Yunhui Cheng
Advanced Glycation End Products: A Comprehensive Review of Their Detection and Occurrence in Food
Reprinted from: *Foods* **2023**, *12*, 2103, https://doi.org/10.3390/foods12112103 26

Biao Zhang, Yihan Lang, Bowen Guo, Zhengyang Cao, Jin Cheng, Danfeng Cai, et al.
Indirect Competitive Enzyme-Linked Immunosorbent Assay Based on Broad-Spectrum Antibody for Simultaneous Determination of Thirteen Fluoroquinolone Antibiotics in *Rana catesbeianus*
Reprinted from: *Foods* **2023**, *12*, 2530, https://doi.org/10.3390/foods12132530 49

Limin Zhang, Wei Li and Shimin Wu
Rapid Determination of Oxygenated and Parent Polycyclic Aromatic Hydrocarbons in Milk Using Supercritical Fluid Chromatography-Mass Spectrometry
Reprinted from: *Foods* **2022**, *11*, 3980, https://doi.org/10.3390/foods11243980 61

Xing Shen, Yang Chen, Jacob Ojobi Omedi, Emel Oz, Fatih Oz, Chunwang Xiao, et al.
The Effects of Volatile Organic Compounds (VOCs) on the Formation of Heterocyclic Amines (HAs) in Meat Patties, under Different Smoking Temperatures and Durations
Reprinted from: *Foods* **2022**, *11*, 3687, https://doi.org/10.3390/foods11223687 73

Yang Xu, Ye Jiao, Jie Luo, Zhiyong He, Maomao Zeng, Qingwu Shen, et al.
The Influence of Deep Eutectic Solvents Extract from Ginger on the Formation of Heterocyclic Amines and Advanced Glycation End Products in Roast Beef Patties
Reprinted from: *Foods* **2022**, *11*, 3161, https://doi.org/10.3390/foods11203161 89

Yu-Wen Lai, Yu-Tsung Lee, Baskaran Stephen Inbaraj and Bing-Huei Chen
Formation and Inhibition of Heterocyclic Amines and Polycyclic Aromatic Hydrocarbons in Ground Pork during Marinating
Reprinted from: *Foods* **2022**, *11*, 3080, https://doi.org/10.3390/foods11193080 104

Weiwei Tang, Jiancheng Wang, Wei Li, Chaojun Zhang, Ping Li and Jun Chen
Changes in Triacylglycerols Content and Quality Control Implications of Coix Seeds during Processing and Storage
Reprinted from: *Foods* **2022**, *11*, 2462, https://doi.org/10.3390/foods11162462 133

Zhenhui Zhang, Lin Li, Yuting Li, Yi Wu, Xia Zhang, Haiping Qi and Bing Li
Insight on a Competitive Nucleophilic Addition Reaction of Nε-(Carboxymethyl) Lysine or Different Amino Acids with 4-Methylbenzoquinone
Reprinted from: *Foods* **2022**, *11*, 1421, https://doi.org/10.3390/foods11101421 145

foods
MDPI

Review

Reducing the Risk of Acrylamide and Other Processing Contaminant Formation in Wheat Products

Navneet Kaur and Nigel G. Halford *

Rothamsted Research, Harpenden AL5 2JQ, Hertfordshire, UK; navneet.kaur@rothamsted.ac.uk
* Correspondence: nigel.halford@rothamsted.ac.uk; Tel.: +44-(0)1582-763133 (ext. 2203); Fax: +44-(0)785-762-6662

Abstract: Wheat is a staple crop, consumed worldwide as a major source of starch and protein. Global intake of wheat has increased in recent years, and overall, wheat is considered to be a healthy food, particularly when products are made from whole grains. However, wheat is almost invariably processed before it is consumed, usually via baking and/or toasting, and this can lead to the formation of toxic processing contaminants, including acrylamide, 5-hydroxymethylfurfural (HMF) and polycyclic aromatic hydrocarbons (PAHs). Acrylamide is principally formed from free (soluble, non-protein) asparagine and reducing sugars (glucose, fructose and maltose) within the Maillard reaction and is classified as a Group 2A carcinogen (probably carcinogenic to humans). It also has neurotoxic and developmental effects at high doses. HMF is also generated within the Maillard reaction but can also be formed via the dehydration of fructose or caramelisation. It is frequently found in bread, biscuits, cookies, and cakes. Its molecular structure points to genotoxicity and carcinogenic risks. PAHs are a large class of chemical compounds, many of which are genotoxic, mutagenic, teratogenic and carcinogenic. They are mostly formed during frying, baking and grilling due to incomplete combustion of organic matter. Production of these processing contaminants can be reduced with changes in recipe and processing parameters, along with effective quality control measures. However, in the case of acrylamide and HMF, their formation is also highly dependent on the concentrations of precursors in the grain. Here, we review the synthesis of these contaminants, factors impacting their production and the mitigation measures that can be taken to reduce their formation in wheat products, focusing on the role of genetics and agronomy. We also review the risk management measures adopted by food safety authorities around the world.

Keywords: acrylamide; asparagine; food safety; free amino acids; hydroxymethylfurfuryl; Maillard reaction; polycyclic aromatic hydrocarbons; processing contaminants; reducing sugars; wheat

1. Introduction

Wheat is a staple crop, consumed worldwide as a major source of starch and protein, accounting for 20% of the total calories consumed by humans. Hexaploid wheat, *Triticum aestivum*, generally known as "common" or "bread" wheat, is the most widely grown species worldwide and is used to make a variety of products, including bread, biscuits, breakfast cereals, flatbreads, cakes, pies, batter and savoury snacks. The tetraploid species, *T. turgidum* var. *durum*, contributes 6–7% of total wheat production and is mainly used for making pasta (it is often referred to as "pasta" as well as "Durum" wheat). It is cultivated mainly in hot, dry conditions. Some other species, such as tetraploid emmer wheat (*Triticum dicoccoides*, also known as *Triticum turgidum* subsp. *dicoccoides*), hexaploid spelt wheat (*Triticum spelta*) and diploid einkorn wheat (*Triticum monococcum*) are cultivated in relatively small, localised regions [1].

The production of food products from wheat grain requires processing, the first stage of which is the milling of the grain to produce flour. The principal components of the grain are the bran, embryo and endosperm, and these can be separated out in different milling fractions. About 70% of the flour derives from the endosperm, and in refined, white flour, the other components are removed. Wheat varieties are grouped according to their suitability for different end uses. In the UK, for example, a classification defined by UK Flour Millers (UKFM) is used [2], with the following groupings: Group 1, varieties suitable for bread-making, with consistent milling and baking performance; Group 2, varieties with bread-making potential but not suited to all grists; Group 3, soft varieties used for biscuits, breakfast cereals, cakes and similar products; Group 4, sub-grouped into hard and soft types and used mainly for animal feed and bioethanol, although incorporated into some grists for food use. The grain of Groups 1 and 2 varieties has 11–15% protein content, while the grain of soft varieties has 10% protein content or less.

The mixing of flour with water and other ingredients, followed by baking and/or toasting, imparts texture, colour, flavour and aroma. In short, it turns the flour into a palatable food. Many of the compounds that are responsible for the colours, flavours and aromas associated with baked and toasted foods are produced in the Maillard reaction, a complex series of reactions involving free amino acids and reducing sugars. A simple representation of the Maillard reaction is given in Figure 1. However, high-temperature cooking and processing also lead to the production of undesirable processing contaminants via thermally-induced chemical reactions between constituents, as well as dehydration, caramelisation, lipid oxidation and other mechanisms. We define processing contaminants as substances that are produced in a food when it is cooked or processed, are not present or are present at much lower concentrations in the raw, unprocessed food, and are undesirable either because they have adverse effects on product quality or because they are potentially harmful [3]. The major processing contaminants of wheat products are acrylamide, furans and furanic compounds, of which 5-hydroxymethylfurfural (HMF) has received most attention to date, and polycyclic aromatic hydrocarbons (PAHs) [4].

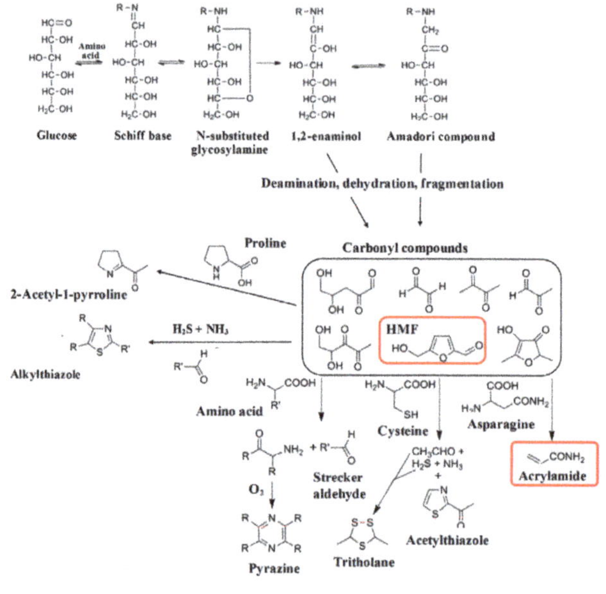

Figure 1. Simplified representation of the Maillard reaction showing the formation of acrylamide, HMF and some of the many other reaction products, many of which contribute to the flavour and aroma of fried, baked, roasted and toasted foods. The pathway shown involves the formation of an Amadori compound from a Schiff base produced by a reaction between glucose and an amino acid. A related Heyns compound is formed from fructose. The deamination, dehydration and fragmentation of these compounds give rise to highly reactive carbonyl compounds, including HMF. These may react with free amino acids again, giving rise to a plethora of products. Acrylamide and HMF are highlighted by red frames. Adapted from [5].

The formation of these contaminants depends upon many factors, from the effects of variety selection and crop management on flour composition through to product recipes, measures taken by processors and food preparers, such as in restaurants, as well as the actions of consumers in the home. In this review, we will discuss the occurrence of these processing contaminants, factors impacting their production, the exposure of consumers and the measures taken by regulators. We will also describe the strategies used to reduce the risk of their formation in wheat products, focusing in detail on the role of genetics and agronomy.

2. Acrylamide

2.1. Acrylamide Toxicity and Dietary Exposure

Acrylamide (Figure 2) is a white, odourless, crystalline, water-soluble solid with the chemical formula C_3H_5NO and relative molecular mass of 71.08. In its monomeric form, it is regarded as a hazardous chemical: in the USA, for example, it is classified as an extremely hazardous substance under the Emergency Planning and Community Right-to-Know Act. Acrylamide is a potent neurotoxin affects male reproduction, causes birth defects and is genotoxic and carcinogenic in laboratory animals [6]. It is classified as a Group 2a carcinogen (probably carcinogenic to humans) by the International Agency for Research on Cancer (IARC) [7].

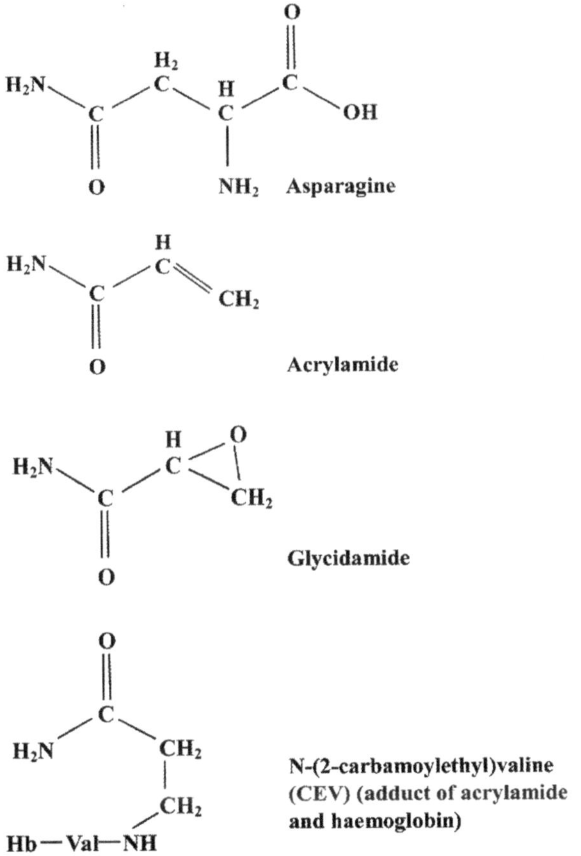

Figure 2. Diagrams representing the chemical structures of asparagine, acrylamide, glycidamide and the adduct, N-(2-carbamoylethyl)valine (CEV), which forms through the reaction of acrylamide with the N-terminal valine residue of a globin chain of haemoglobin (Hb).

Acrylamide is metabolised to glycidamide ($C_3H_5NO_2$) (Figure 2), and it may actually be glycidamide that is responsible for the genotoxic and carcinogenic effects attributed to acrylamide exposure. Both acrylamide and glycidamide form adducts with haemoglobin, and the detection of these adducts is the favoured method for obtaining quantitative measurements of acrylamide exposure [6]. Both also form adducts with DNA in vitro, but only glycidamide-DNA adducts have been found in human or animal tissues. These may also be used as markers for acrylamide exposure. The primary detoxification and excretion route for both acrylamide and glycidamide involves a reaction with glutathione and the conversion of the resulting adducts to mercapturic acids in the liver. These mercapturic acids are excreted, and their presence in urine represents another marker for acrylamide exposure.

The United Nations Food and Agriculture Organization (FAO) and World Health Organization (WHO) Joint Expert Committee on Food Additives (JECFA) issued reports on acrylamide exposure in 2006 and 2011 [8,9], calculating the average dietary exposure for the general population to be 1 μg per kg bodyweight per day, with consumers in the high percentile (i.e., those eating the most acrylamide-containing foods) exposed to 4 μg per kg bodyweight per day. The European Food Safety Authority (EFSA) has also provided estimates of the exposure levels of Europeans to dietary acrylamide [6], calculating a mean of 0.31 to 1.1 μg per kg bodyweight per day for adults, 0.43 to 1.4 μg per kg bodyweight per day for adolescents (11–17 years old), 0.70 to 2.05 μg per kg bodyweight per day for children (3–10 years) and 1.2 to 2.4 μg per kg bodyweight per day for toddlers (1–3 years). These figures are slightly lower than the JECFA estimate.

Toxicological studies with acrylamide have been conducted in a range of animal species, and a meta-analysis of the results of these studies was published by the European Food Safety Authority (EFSA)'s Scientific Panel on Contaminants in the Food Chain (CONTAM) in 2015 [6]. Toxicological studies use high doses of the chemical being tested so that statistically significant effects can be demonstrated, and the doses of acrylamide used are typically measured in mg per kg body weight, i.e., orders of magnitude higher than the estimates of human exposure arising from acrylamide in the diet. Nevertheless, in 2006, JECFA issued an opinion on the risks posed by dietary acrylamide [8], stating that the margins of exposure (MOEs) indicated a health concern due to acrylamide's genotoxic and carcinogenic properties. The CONTAM Panel report of 2015 also concluded that the MOEs for acrylamide indicated 'a concern for neoplastic effects' [6] (the MOE is defined by EFSA as the ratio of the level at which a small but measurable effect is observed to the estimated exposure dose), essentially reaffirming the opinion issued by JECFA almost a decade earlier. However, the CONTAM Panel did not consider exposure from the diet to be sufficient to cause neurological, reproductive or developmental effects.

The CONTAM Panel [6] also considered the results of a wide range of epidemiological studies, including 36 papers investigating associations between dietary acrylamide intake and cancer risk. The Panel concluded that the results of epidemiological studies had not been consistent enough to enable firm conclusions to be drawn on whether acrylamide was a human carcinogen. Hence, the Panel's opinion on the risk posed by dietary acrylamide was based solely on toxicological data.

Since the CONTAM report was published, a unique mutational 'signature' has been linked to acrylamide and glycidamide [10]. This signature was found in approximately one third of the 1600 tumour genomes analysed in the study, derived from 19 human tumour types from 14 organs. This looks alarming, but it should be noted that the study did not provide evidence that the mutations were responsible for the cancers. Nevertheless, the mutations in 184 liver tumour samples and 217 tumours of 15 other cancer types were considered to have arisen from dietary exposure to acrylamide because other mutations associated with the effects of tobacco smoke, the other major source of acrylamide exposure, were not present. On the other hand, Eisenbrand [11] reviewed the evidence for a genotoxic mode of action for acrylamide/glycidamide and found that 'compelling evidence is lacking', concluding that any genotoxicity of acrylamide occurred at doses that were irrelevant to dietary exposure. However, this analysis does not seem to have swayed the opinion of the CONTAM Panel or affected the European Commission's risk management measures (Section 2.3).

2.2. Formation

Acrylamide is formed in foods by the thermal degradation of free (soluble, non-protein) asparagine in the presence of reducing sugars (principally glucose, fructose and maltose) in the Maillard reaction (Figure 1) [12–14]. The Maillard reaction, which is named after the French scientist Louis-Camille Maillard, who described it in 1912, is a complex chain of non-enzymic chemical reactions taking place between free amino acids and reducing sugars. The Maillard reaction can occur at room or even lower temperatures, but the rate is extremely slow and it is only at high temperatures (>120 °C) that it becomes relevant to food processing. It was an important reaction for the food industry long before acrylamide was discovered in food because it is responsible for the colours, flavours and aromas that are associated with fried, baked, roasted and toasted foods and are demanded by consumers. There are numerous reviews available describing the Maillard reaction [5,15–21] and we will not go into detail here. In brief, it involves sugar amine condensation, Amadori or Heyns rearrangement, the degradation and fragmentation of sugars as well as free amino acids and, in the last steps of the reaction, the condensation of aldols and aldehyde-amines and the formation of heterocyclic nitrogen compounds. It is when free asparagine participates in these later stages of the reaction, undergoing decarboxylation and deamination, that acrylamide is formed [12,13], with the entire carbon skeleton of the acrylamide that forms deriving from asparagine (Figure 2).

Another route for acrylamide formation from asparagine involves 3-aminopropionamide (3-APA) as a transient intermediate, with deamination of 3-APA resulting in acrylamide formation even in the absence of reducing sugars/catalysts [22]. Asparagine may also react directly with reducing sugars to form acrylamide [23]. Acrolein, acrylic acid and wheat gluten have also been postulated to be additional precursors for acrylamide [24,25].

The Maillard reaction is complex, multi-step and affected by many factors, including temperature, pH, water activity, and the presence of catalysts. Higher temperatures and alkaline conditions, for example, promote the Maillard reaction, whereas acidic conditions inhibit it, while the presence of metal ions such as copper, iron, or manganese can act as catalysts and accelerate the reaction [19]. Since the discovery of acrylamide in food in 2002, food manufacturers have sought to exploit these factors to reduce the concentration of acrylamide in their products. They have also improved the control of cooking temperature and duration, and of moisture levels in the finished product, and for some products, applied quality control based on colour. Additionally, for some products, food manufacturers have used asparaginase to reduce asparagine content as far as possible before baking [26]. These methods are more effective in some products than in others and also risk having a negative impact on product quality or the characteristics that define brands and set them apart from the competing products on the shop shelf. They have been compiled in a 'Toolbox' produced by FoodDrinkEurope, the latest update of which was published in 2019 [27], and are reviewed in [28].

2.3. Risk Management Measures (Regulations) and the Challenge of Regulatory Compliance

The amount of acrylamide present in a food sample can be measured using either gas or liquid chromatography coupled with mass spectrometry (GC-MS or LC-MS/MS). In the GC-MS method, the acrylamide has to be brominated to produce 2,3-dibromopropionamide. This is converted to 2-bromopropenamide by dehydrobromination with triethylamine, and 2-bromopropenamide is the compound that is actually loaded into the GC-MS. This makes the GC-MS method considerably more laborious than the LC-MS/MS method, which requires no prior derivatisation, and the LC-MS/MS method has become more popular in the last decade or so as the equipment has become more widely available. However, the

Comité Européen de Normalisation (European Committee for Standardisation) (CEN) has developed standard methods for both (EN 16618:2015 and FprCEN/TS 17083).

The European Union set Indicative Values for different food types in 2011 [29] and reduced them in 2013 [30] (Table 1). Indicative Values were described as 'triggers for investigation' by the European Commission, although they were often interpreted as safety limits by consumer groups and the general media. In 2017, Regulation (EU) 2017/2158 [31] replaced Indicative Values with Benchmark Levels (Table 1). These were lower than the corresponding Indicative Values for most products and were described as 'performance indicators'. It is unlikely that consumers understand the difference between 'triggers for investigation' and 'performance indicators', and EFSA's data on acrylamide concentrations in food showed that the levels had been flat for most products at least since 2011, so setting the Benchmark Levels lower than the Indicative Values just meant that there were more apparent 'failures'. Data in the public domain on acrylamide levels in cereal products are relatively sparse, but a recent study in Spain found that 15% of breakfast cereals contained acrylamide above the Benchmark Level [32]. It is alarming for the food industry and its supply chain, therefore, that the European Commission is currently considering reducing Benchmark Levels further and imposing Maximum Levels for some foods, including all of the major cereal products [33]. It would, of course, be illegal to sell a product containing acrylamide above the Maximum Level. The Benchmark and Maximum Levels that are under discussion are given in Table 1, but it should be noted that no final decision has been taken on these at the time of writing.

Table 1. Indicative Values and Benchmark Levels (μg kg^{-1}; parts per billion (ppb)) for acrylamide in food, set by the European Commission, and the possible new Benchmark and Maximum Levels that are under discussion.

Food	Indicative Value 2011	Indicative Value 2013	Benchmark Level 2017	Under Discussion	
				Benchmark Level	Maximum Level
Soft bread	150	80	50	50	75
Soft bread (other)		150	100		
Breakfast cereals: bran products, whole grain cereals, gun puffed grain		400	300	300	500
Breakfast cereals: wheat and rye-based	400	300	300	300	400
Breakfast cereals: maize, oat, spelt, barley and rice-based		200	150		
Biscuits	500	500	350	300	500
Crackers	500	500	400		
Crispbread	500	450	350	300	400
Gingerbread	-	1000	800		
Cereal-based baby foods	100	50	40		
Biscuits and rusks for infants and young children	250	200	150		

At the time that Regulation (EU) 2017/2158 came into force, the UK was part of the EU, and the Regulation rolled over into UK law when the UK left the EU in 2020. The EU is the UK's largest trading partner for both wheat grain and wheat-based food products, so any developments in the EU will affect UK farmers and food businesses. What is not clear is whether the UK will follow the EU's lead in setting Maximum Levels for acrylamide. UK food businesses and their suppliers, therefore, face uncertainty over whether the EU is about to impose Maximum Levels for acrylamide, what those levels will be, and whether

Maximum Levels will also be applied in the UK. This uncertainty makes it very difficult to make the preparations necessary to ensure regulatory compliance.

No other regulatory authorities around the world have gone as far as the EU in developing risk management measures for acrylamide. The State of California did file and win a lawsuit in 2005 against a number of potato crisp (US chip) and French fry manufacturers for selling products without a Proposition 65 warning on them. Proposition 65 requires businesses in California to post warnings of any chemical in their products that may cause cancer. The lawsuit resulted in the manufacturers having to pay fines and commit to cutting the level of acrylamide in their products by between 20 and 87%. In 2010, an NGO, the Council for Education and Research on Toxics (CERT), brought a similar lawsuit under Proposition 65 against more than 90 companies producing or selling coffee. At the Federal level in the USA, however, the Food and Drug Administration (FDA) has gone no further than issuing an 'action plan' [34].

In Canada, acrylamide has been added to the list of chemicals in the government's Chemicals Management Plan, and Health Canada has been working with the food industry to develop and share acrylamide reduction methods. It has also implemented an acrylamide monitoring programme and discussed the possibility of setting 'reduction targets'. However, to date, no targets or limits have been set. Food authorities in Japan and Hong Kong and Food Standards Australia New Zealand (FSANZ) have also issued advice on acrylamide reduction without going so far as to introduce targets or limits.

2.4. Acrylamide in Wheat Products and the Contribution of Wheat Products to Dietary Exposure

Data on the presence of acrylamide in food in the European Union have been collected since 2003 and have been extremely important in informing the development of the European Commission's risk management measures. Data on cereal products from the 2015 CONTAM Panel report [6] are shown graphically in Figure 3A (mean) and Figure 3B (95th percentile), with the highest levels in biscuits and breakfast cereals and the lowest in bread. The report also included estimates of the contribution of different foods to dietary intake in different European Union Member States, summarising the data from 16 studies in the form of a table giving the number of studies that had placed the contribution of each food type with a range of 0–5%, 5–10%, 10–25%, 25–50% and >50%. Estimates from those data of the contribution of different cereal food types to dietary acrylamide exposure in the European Union are shown graphically in Figure 3C. Bread is the highest contributor to intake despite containing relatively low concentrations of acrylamide because so much of it is consumed, and a significant proportion of it is toasted before consumption.

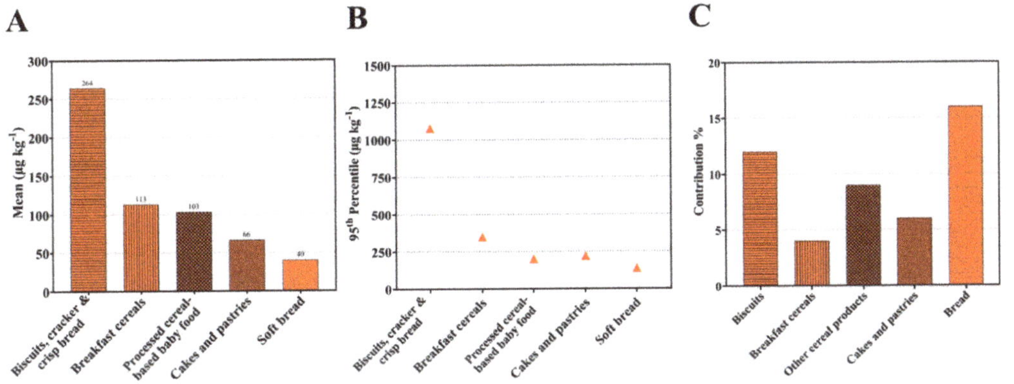

Figure 3. Acrylamide in cereal products in the European Union. (**A**). Mean (μg kg^{-1}) and (**B**). 95th percentile (μg kg^{-1}) acrylamide levels in selected cereal-based food groups or products. (**C**). Contribution of different cereal-based foods to dietary acrylamide exposure for adults. Plotted using data from [6].

2.5. Crop Management Measures to Reduce the Acrylamide-Forming Potential of Wheat Grain

Many studies have shown the level of acrylamide formation in wheat-derived food products to depend upon the concentration of free asparagine in the grain [35–38]. Asparagine is an amino acid amide that has a molecular mass of 132.12 and an N:C ratio of 2:4 (Figure 2). It is also relatively unreactive, is a substrate for only a few enzymatic reactions and has little net charge under physiological conditions. These factors make it an ideal molecule for nitrogen storage and transport. Not surprisingly, therefore, it accumulates in plants at some developmental stages, including senescence and seed germination, and in response to various stresses, such as drought or salt stress (osmotic stress), mineral deficiency, toxic metals or pathogen attack [39]. The use of free asparagine as a nitrogen storage and transport molecule means that an asparaginase activity is required to remobilise the nitrogen when required, and asparaginase is present in many plant tissues, including wheat seeds [40]. Research on methods to reduce the free asparagine content of wheat grain has included the effects of fertilisation and disease control, varietal comparisons and genetic approaches to provide wheat breeders with the resources and knowledge to produce low asparagine varieties.

2.5.1. Fertilisation

Given the potential role of free asparagine in nitrogen transport and storage, it is no surprise that its concentration in wheat grain rises in response to increased nitrogen availability. Increased nitrogen supply was shown to cause a rise in free asparagine concentration in barley grain as long ago as 1980 [41]. This was considered to be of largely academic interest until free asparagine was shown to be a precursor for acrylamide formation. Subsequent studies showed that increased nitrogen fertilisation of wheat promoted free asparagine accumulation, with a concomitant effect on acrylamide formation in bread [42,43]. The effect of nitrogen availability, however, was found to be dwarfed by that of sulphur deficiency. Increases of up to 30-fold in free asparagine concentration were measured in response to severe sulphur deficiency imposed in a pot experiment, for example [35], and a series of pot- and field-based studies have confirmed the effect [36, 37,44–46]. Indeed, sulphur deficiency is by far the most important factor affecting the concentration of free asparagine in wheat grain, and varieties with low free asparagine under sulphur-fed conditions tend to be more affected, meaning that any attempt to rank varieties according to free asparagine concentration in grain breaks down under conditions of sulphur deficiency [45].

Analyses of the contrasting gene expression patterns in the developing seeds of wheat grown with and without sulphur supplied showed gene expression in the embryo to be the determining factor for asparagine accumulation in the grain, with increased expression of asparagine synthetase-2 (*TaASN2*) and glutamine synthetase in the embryo in response to sulphur deficiency and a decrease in asparaginase gene expression [40]. The signalling networks responsible for these changes in gene expression have not yet been elucidated, but genes encoding regulatory protein kinases, sucrose nonfermenting-1-related protein kinase-1 (SnRK1) and general control nonderepressible-2 (GCN2), also responded to sulphur deficiency [40,47]. Both of these have been implicated in regulating asparagine synthetase gene expression, possibly in the case of SnRK1 via bZIP transcription factors that contain SnRK1 target sites, including Opaque2/bZIP9, SPA/bZIP25 and BLZ1/OHP1/bZIP63, all of which are expressed in wheat seeds [40,47].

These results are consistent with the plant using free asparagine as a nitrogen store in the grain and directing more nitrogen into free asparagine accumulation when sulphur is not available. This may, at least in part, be because it has insufficient sulphur to make sulphur-rich seed storage proteins [48]. This hypothesis led to an investigation of the importance of the N:S ratio in fertiliser application [38]. That study found the effect of the N:S ratio to be significant (*F* probability < 0.001) over two years of field trials. There was also an interaction between the N:S ratio and variety, meaning that the varieties in the trials responded differently to the treatment [38]. Figure 4A shows the responses (asparagine content) of four varieties in the trial, and Figure 4B shows the darker colour of biscuits cooked with flour from the low sulphur plots. Biscuits made with flour from plots fertilised with 100 kg ha^{-1} nitrogen and 10 kg ha^{-1} sulphur contained a mean of 488 µg kg^{-1} acrylamide, while biscuits from the 100 kg ha^{-1} nitrogen, no sulphur plots, contained 6114 µg kg^{-1} and the 200 kg ha^{-1} nitrogen, no sulphur plots an astonishing 13,523 µg kg^{-1} acrylamide. Biscuit acrylamide correlated with the free asparagine content of the grain and with biscuit colour (top surface hue angle), with R^2 values from linear models of 89% and 91%, respectively ($p < 0.001$). The acrylamide figures show just how much can form in foods if the raw material has a lot of free asparagine in it, and the importance for manufacturers of not allowing wheat grown with insufficient sulphur availability into their processing lines. They also showed very nicely the interaction between nitrogen and sulphur. It was concluded that a nitrogen to sulphur ratio of 10:1 was required and sufficient to prevent large increases in the accumulation of free asparagine. Ensuring that 'good agricultural practices' are followed on fertilisation, to 'maintain balanced sulphur levels in the soil and to ensure correct nitrogen application' are also included in Regulation (EU) 2017/2158 [31].

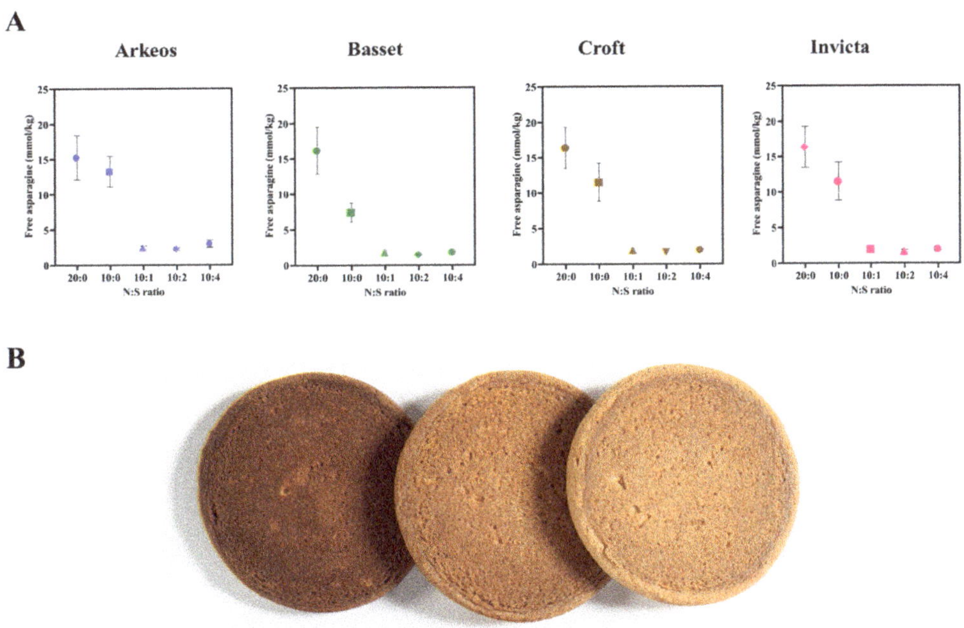

Figure 4. Effects of the nitrogen to sulphur fertilisation ratio on free asparagine concentration. (**A**). Graphs representing the effect of different nitrogen to sulphur ratios on free asparagine content (mmol kg^{-1}) in the grain of four different wheat varieties. Each point represents the mean value from two field trials carried out in years 2019/2020 and 2020/2021 at two different locations. (**B**). Biscuits made from the Basset variety grown with different nitrogen to sulphur ratio treatments. N:S from left to right: 20:0, 10:0 and 10:1. Data from [38]. Image produced by Graham Shephard (Rothamsted Research) and kindly provided by Joseph Oddy.

The study also investigated the effect of withholding two other important minerals for crop growth and health, potassium and phosphorus, but found that deficiencies in these minerals did not cause increases in free asparagine concentration when sulphur was applied, so the response is specific to sulphur availability [38].

2.5.2. Disease Control

Asparagine concentration in plant tissues is also sensitive to biotic stress [39,49], possibly due to asparagine's involvement in the remobilisation of nitrogen and/or ammonia detoxification during pathogen attack [50]. A field-based study in 2009 showed that fungicide treatment of three wheat varieties reduced free asparagine accumulation [43]. A larger study was then performed with 24 varieties of wheat grown in a field trial in the UK in 2011–2012, with adjacent plots treated in an identical fashion, except that fungicide was applied to one plot and not the other, the lack of fungicide resulting in visible infection by *Septoria tritici*, yellow rust and brown rust [51]. Both free asparagine concentration and acrylamide formation in flour (heated for 20 min at 180 °C) reduced in response to fungicide treatment, and the reductions in free asparagine concentration are shown in Figure 5. These studies showed disease control by fungicide application to be an important crop management measure for mitigating the problem of acrylamide formation in wheat products, and EC Regulation (EU) 2017/2158 [31] states that 'application of good practices on crop protection measures to prevent fungal infection' must be ensured. It is ironic, therefore, that the European Commission has recently made crop protection more difficult for farmers by withdrawing or refusing to renew the licences for some fungicides, giving the impression that different offices within the Commission do not communicate with each other.

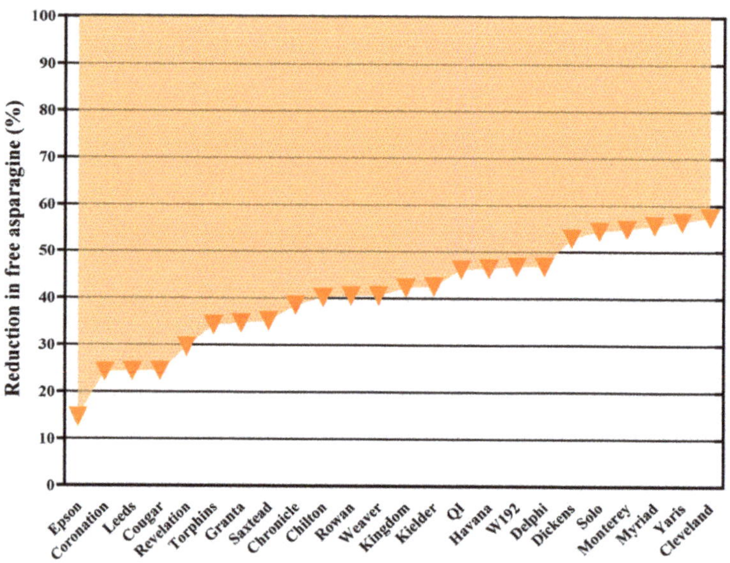

Figure 5. Graph representing the impact of disease control on free asparagine content in the grain of 24 winter wheat varieties. Each triangle represents the percent reduction in free asparagine content in grain from plots treated with fungicide compared with grain from untreated plots. The fungicide treatment was given from growth stages 30 to 65. The main diseases observed in the untreated plots were *Septoria tritici*, yellow rust and brown rust. Drawn using data from [51].

The interplay between the fungal and plant molecular factors involved in this aspect of plant–pathogen interactions is fascinating, even with much of the system still to be elucidated. There is evidence that SnRK1, which was discussed above in the context of the sulphur response, also plays a central role in co-ordinating pathogen defence mechanisms and activation of adaptive responses, including the increase in asparagine synthesis. Perochon et al. [52], for example, reported that SnRK1 is involved in the response of wheat to infection by *Fusarium graminearum* (*Fg*), the main causal agent of Fusarium head blight disease (FHB, also known as 'scab'). This disease reduces yield and grain quality and contaminates grain with mycotoxins, most commonly deoxynivalenol (DON), a toxic sesquiterpenoid. DON treatment of wheat has been shown to increase the levels of free asparagine, glutamine and aspartate in the grain [49,53]. Under *Fg* infection, the expression of a gene encoding a wheat orphan protein, TaFROG, is upregulated, and the interaction of TaFROG with SnRK1 enhances FHB and DON resistance [54]. In direct opposition to the action of TaFROG, an *Fg*-secreted effector protein, OSP24, interacts with SnRK1 and promotes SnRK1 degradation by the ubiquitin-26S proteasome. TaFROG competes with OSP24 for binding with SnRK1, thereby protecting it from degradation [55]. Our current working model of these interactions is given in Figure 6.

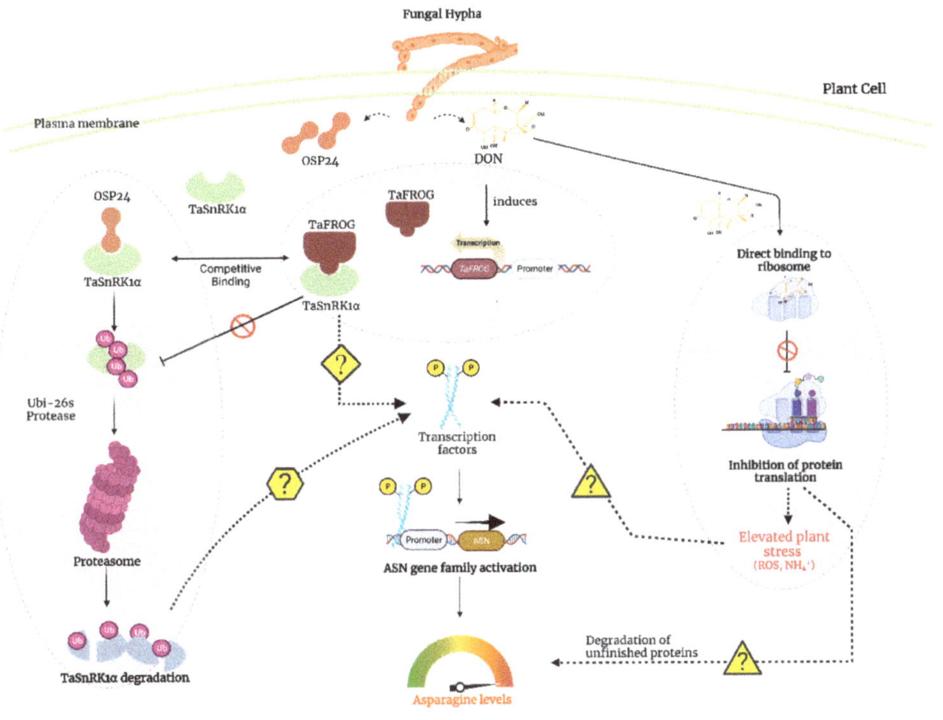

Figure 6. Working model of the signalling hub linking pathogen (*Fusarium graminearum*) infection with asparagine synthetase gene expression in wheat. DON (deoxynivalenol) binds to ribosomes and inhibits protein synthesis, leading to the accumulation of reactive oxygen species (ROS) and free ammonia (NH_4^+). DON also induces expression of the *TaFROG* gene, and TaFROG competes with a fungal protein, OSP24, for binding to TaSnRK1α, thereby safeguarding TaSnRK1α from degradation. However, the precise mechanism through which this signalling pathway ultimately activates the asparagine synthetase gene family and the transcription factors involved remain unknown. This model is based on an adaptation from [52,55]. Created with BioRender.com.

2.6. Genetic Approaches to Reducing the Acrylamide-Forming Potential of Wheat Grain

Free asparagine concentration has been shown to vary among wheat varieties in several studies [35–37,45,51,56,57]. Curtis et al. [45], for example, reported a range of free asparagine concentrations from 0.708 to 11.29 mmol kg^{-1} in different varieties of sulphur-fed wheat in a 2012–2013 field trial in the UK, while Tafuri et al. [57] reported a much tighter range of 0.55 to 2.84 mmol kg^{-1} in 54 Italian bread-making varieties over two years of field trials. It is clear from these studies that genetic factors (G) play a part in determining free asparagine concentration, suggesting that it might be possible to reduce acrylamide formation in wheat products by selecting low asparagine varieties and that free asparagine concentration in wheat grain may be amenable to breeding. However, there are some obstacles that make both of those aspirations difficult to achieve. The first is the impact of environmental factors (E), within which we would include crop management for optimum fertilisation and good disease control. Neither sulphur deficiency nor disease affects different varieties to the same extent; if anything, they cause greater increases in free asparagine in otherwise low asparagine varieties, meaning that varietal rankings break down; i.e., there is a large G × E component to the variance.

Despite the effects of E and G × E, Curtis et al. [45] did identify eight varieties that were consistently low in free asparagine in the grain. However, food businesses who wish to purchase low asparagine wheat, at least in the UK, face another obstacle in that the turnover in wheat varieties is so rapid that by the time the free asparagine concentration of the grain of a new variety has been measured over enough seasons to give an accurate assessment, the variety is no longer on the market. The solution to this problem would be for breeders to measure free asparagine concentration during variety development so that the information could be provided when the variety was launched. To date, however, this has not been done, and in our view, this needs to change.

The food industry would undoubtedly also like to see more progress from breeders in producing low-asparagine wheat varieties. However, to our knowledge, it remains a low priority for breeders, who continue to chase their traditional targets of yield, protein content and disease resistance. We may see more action on asparagine content if the regulatory situation in the EU changes in the expected way (Table 1) and the pressure from the food industry intensifies, or if readily-usable tools and resources can be developed to make the task easier for breeders.

2.6.1. Identifying Quantitative Trait Loci (QTL) That Control Free Asparagine Accumulation in Wheat Grain

Knowledge about the genetic architecture or QTL associated with the free asparagine content of wheat grain could enable the development of genomic tools to assist breeders in the selection of low asparagine varieties in the early generations of a breeding programme. However, to date, the search for consistent QTL has been frustrating. In a study conducted by Emebiri in 2014 [58], for example, 92 different wheat varieties were examined, with a wide range of free asparagine concentrations. Through a genome-wide scan for single-nucleotide polymorphism (SNP) markers associated with this variance, nine SNPs were identified on chromosome 5A. These SNPs accounted for 14–24% of the observed variation in free asparagine concentration and could be potential candidates for further investigations aimed at developing molecular markers. However, the study concluded that breeding cultivars with naturally low levels of grain asparagine would pose a challenge, as the heritability of the trait was only 32%.

Subsequently, in 2018, Peng et al. [59] conducted a study on 182 bread wheat accessions, employing a single-locus genome-wide association study (GWAS) with a mixed linear model (MLM) and six multi-locus GWAS models. They reported that an aminopeptidase on chromosome 4B served as an underlying factor in controlling free asparagine content in wheat. Further, in 2018, Rapp et al. [60] mapped eight putative QTL explaining 78.5% of the variance for free asparagine content among 149 wheat varieties, with the QTL accounting for the highest variance (18.4%) on chromosome 7B. However, the QTL detected in this study were different from those reported by Emebiri [58], presumably arising from the strong impact of environmental factors and signifying that the QTL being identified were based on phenotypic data specific to the local environment.

In 2023, Oddy et al. conducted another study [61], this time to examine the genetic architecture underpinning the variance in levels of free asparagine in a doubled haploid mapping population produced from a cross between varieties Claire and Robigus. These varieties have been extensively used as parent lines in soft wheat breeding in the UK, and have been shown to have contrasting concentrations of free asparagine in the grain, with the concentration in Claire lower and more consistent than that in Robigus [37]. A QTL for free asparagine content in grain was identified on chromosome 4B, located approximately 60 Mbp away from a QTL reported by Peng et al. [59], the first time that two studies had identified QTL that might overlap. However, this QTL was also affected by environmental factors, accounting for only 2.6% of the variance in a 2017–2018 field trial of the population, but 14.8% of the variance in a 2018–2019 field trial when free asparagine concentrations were higher [61]. Even so, the Robigus allele was associated with higher free asparagine concentrations in both years.

The contributions of the different QTL to the variance in free asparagine concentrations between varieties are shown graphically in Figure 7. The overall conclusion to be drawn from these studies is that 'traditional' breeding alone probably has limited scope for reducing free asparagine concentration in wheat grain.

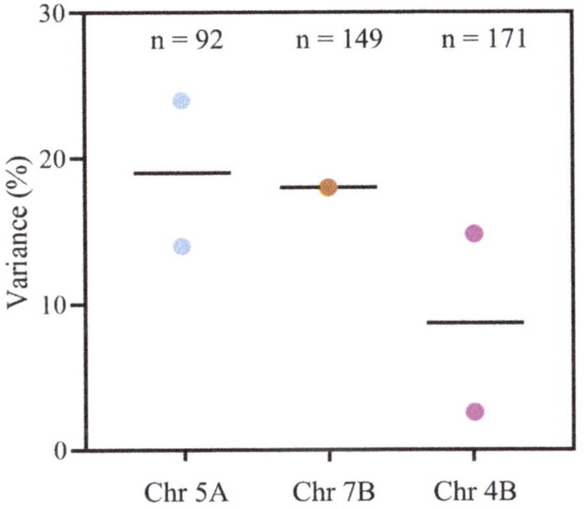

Figure 7. Graph illustrating the maximum amount of variance in free asparagine concentration in the grain that can be attributed to quantitative trait loci (QTL) on chromosomes 5A, 7B and 4B in different varieties of wheat. The QTL on chromosome 5A was responsible for explaining 14–24% of the variance, while those on chromosomes 7B and 4B explained 18.4% and 2.6–14.8% of the variance, respectively. The black lines in the plots indicate the average percentage of variance observed. Plotted using data from [58,59,61].

2.6.2. The Asparagine Synthetase Gene Family of Wheat and Characterisation of a Natural Genetic Event Associated with Reduced Free Asparagine Concentration in Wheat Grain

Asparagine is made via the ATP-dependent transfer of an amino group from glutamine to aspartate to generate glutamate and asparagine. The reaction is catalysed by the enzyme asparagine synthetase and wheat, like all the members of the Triticeae tribe, has five asparagine synthetase genes: *TaASN1*, *TaASN2*, *TaASN3.1*, *TaASN3.2* and *TaASN4* [62,63]. Of these, *TaASN2* is most highly expressed in the grain, while it is not expressed elsewhere in the plant. The closely related gene, *TaASN1*, is the next most highly expressed gene in the grain but is also expressed elsewhere [64]. For both genes, expression in the embryo is much higher (approximately 10-fold) than in the endosperm [64]. This was confirmed in an analysis of RNA-seq data for developing grains, with *TaASN2* gene expression in the embryo increasing in response to sulphur deficiency, in line with the observed increase in free asparagine concentration in the grain [40]. That study also showed the A genome homeologue, *TaASN-A2*, to be much more highly expressed than *TaASN-D2*, while *TaASN-B2* expression was not detectable at all in those genotypes [40].

Analyses of wheat genome data showed all of the *TaASN* genes to be present as a single copy per genome, with *TaASN1* on chromosome 5, *TaASN2* on chromosome 3,

TaASN3.1 and *TaASN3.2* on chromosome 1 and *TaASN4* on chromosome 4 [62,63]. These analyses also showed that variety Chinese Spring lacked a *TaASN2* gene on the B genome (*TaASN-B2*) as a result of a natural deletion of almost 13 kb (Figure 8A). A subsequent study by Oddy et al. [65] determined the presence and absence of the *TaASN-B2* gene across a range of UK and global common wheat varieties and related species. The study showed that the deletion encompassing the *TaASN-B2* gene was present in some wild emmer wheat genotypes and most, but not all, cultivated varieties. Notably, there was a trend for free asparagine concentrations in field-produced grain to be lower in varieties lacking *TaASN-B2* (Figure 8B), although the effect was lost when free asparagine accumulated to very high concentrations as a result of sulphur deficiency [65].

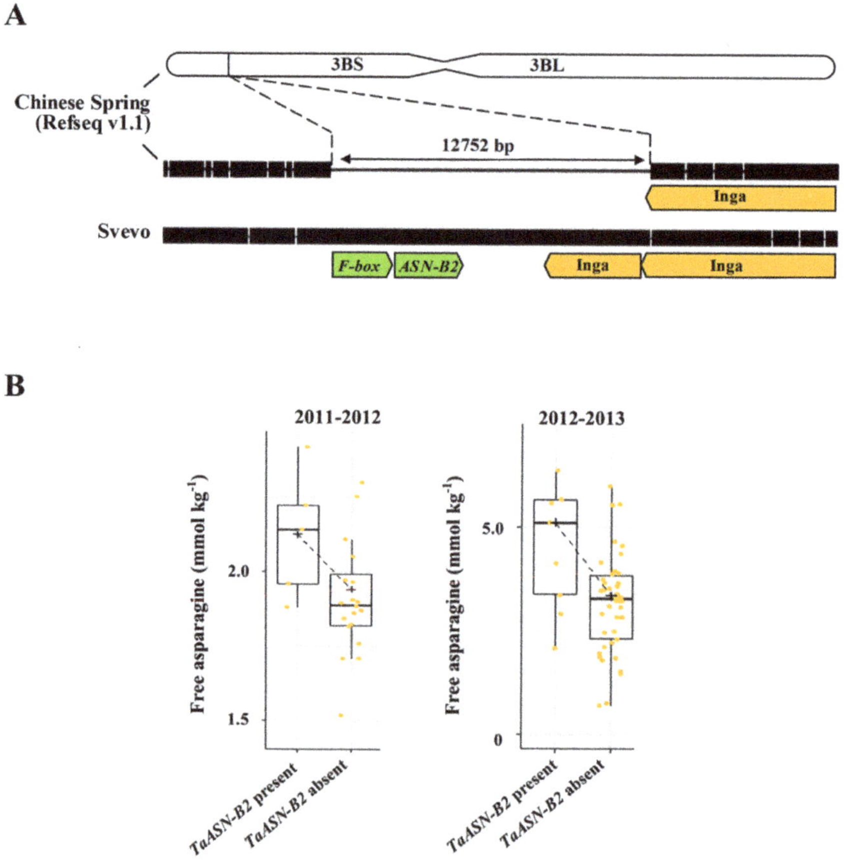

Figure 8. (**A**). Representation of the region of chromosome 3B of wheat variety Chinese Spring in which there is a natural deletion of ~13 kb, and the corresponding region in variety Svevo, in which the deletion has not occurred and the *TaASN-B2* gene is still present. (**B**). Graph showing the impact of the presence/absence of the *TaASN-B2* gene on free asparagine content in the grain of 50 varieties of winter wheat in two consecutive field trials (2011–2012; 2012–2013). Adapted from [65].

These results suggest that selecting wheat genotypes lacking the *TaASN-B2* gene may be a simple and rapid way for breeders to reduce free asparagine concentrations in commercial wheat grain. However, most varieties already lack the gene, across all milling types, which would explain why the null allele has not been identified in QTL studies: both Claire and Robigus, for example, lack a *TaASN-B2* gene, so the effect of the presence/absence of the gene would not have been detected in the analysis of the

Claire × Robigus mapping population [61]. This limits the scope for breeders to exploit this natural genetic event. On the other hand, it is something that breeders could be doing already, but to our knowledge, they are not.

2.6.3. The Application of Genome Editing and Chemical Mutagenesis to Reducing Free Asparagine Concentrations in Wheat Grain

The discovery that *TaASN2* is expressed grain-specifically and is by far the most highly expressed asparagine synthetase gene in the grain, makes it an obvious target for genetic interventions, and in 2021, Raffan et al. [66] reported the generation of bread wheat cv Cadenza genotypes in which *TaASN2* had been 'knocked out' using CRISPR-Cas9. The editing was achieved by introducing a 4-gRNA polycistronic gene into wheat embryos along with a *Cas9* gene and a marker gene (*Bar*). Free asparagine concentration in plants in which all six *TaASN2* alleles were knocked out was 43–57% of the wildtype over two generations [66], and in plants that were subsequently found to be edited in *TaASN1* as well as *TaASN2* (i.e., *TaASN1/TaASN2* total knockouts) [67] was 9–48% of the wildtype [65]. In plants containing edits only in the A genome *TaASN2* alleles, the concentration of free asparagine was 56–68% of wildtype [66].

The fact that *TaASN2* is present as a single copy gene means that it has also been possible to identify plants carrying mutations in the A, B or D genome genes within mutant wheat populations (known as TILLING populations) of cv. Cadenza and cv. Kronos (a tetraploid, genomes AB) produced by ethyl methanesulphonate treatment [68]. Alarcón-Reverte et al. [69] reported reductions in free asparagine concentration of 28% in A genome TILLING nulls in the Cadenza background, and 24–34% in the tetraploid Kronos background in field trials.

In 2021–2022, Raffan et al. carried out field trials of *TaASN2* A genome null and total null gene-edited lines (Figure 9A) [67]. Also included in the field trial were AB genome nulls derived by backcrossing an A genome null TILLING line into the cv. Claire background (cv. Claire lacks a B genome *TaASN2* gene due to the 'natural' deletion described in Section 2.6.2 [65]).

The mean free asparagine concentrations in the total *TaASN2* nulls were just under 50% that of the Cadenza control, while in the A genome nulls the concentration was 86% that of Cadenza [67]. The TILLING lines showed more variable responses: they did show a reduction of 20–40% compared with a TILLING control (a line that had come through the same process but did not carry mutations in *TaASN2*), but Claire was also lower than the TILLING control, so further trials will be required before firm conclusions can be drawn. The lower free asparagine concentration in the grain from the *TaASN2* total null edited lines was reflected in the amount of acrylamide that formed when flour was heated to 160 °C for 20 min, with an almost 50% decrease compared with the control. The data for individual plots of two total null lines (23.60 and 23.75) are shown graphically in Figure 9B; the means for these lines were 427 and 421 $\mu g\ kg^{-1}$, respectively, compared with 761 $\mu g\ kg^{-1}$ in the Cadenza control [67].

This study showed that step reductions in the free asparagine concentration and, therefore, the acrylamide-forming potential of wheat grain could be achieved using genome editing, with the reductions observed under glass maintained in the field. Importantly, while thousand grain weight was lower in the edited lines, yield was not significantly reduced; in other words, the edited lines produced more but smaller grains, although the grain size was well within the normal range for wheat. In contrast, there was a yield drag for the TILLING lines, presumably due to the background mutations introduced by the chemical treatment, making it a nice example of the advantages of the targeted versus random technique.

2.7. The Use of Imaging Techniques to Predict Free Asparagine Concentrations in Wheat Grain and the Anticipated Development of Free Asparagine Sensors

The use of hyperspectral imaging in plants has proven to be an efficient method for measuring plant metabolites [70] and has the potential to assist growers, millers and food processors in evaluating the quality of grains. In a study conducted by Oddy et al. [38], a Tec5 HandySpec Field spectrometer and Videometer SeedLab system were employed in the field to predict the free asparagine concentration of grain that was subsequently harvested with an average accuracy of 71%, demonstrating the effectiveness of this approach. Furthermore, a combination of multispectral, fluorescence and morphological measurements of grains after harvest could distinguish between high and low free asparagine grain with an accuracy of 86%.

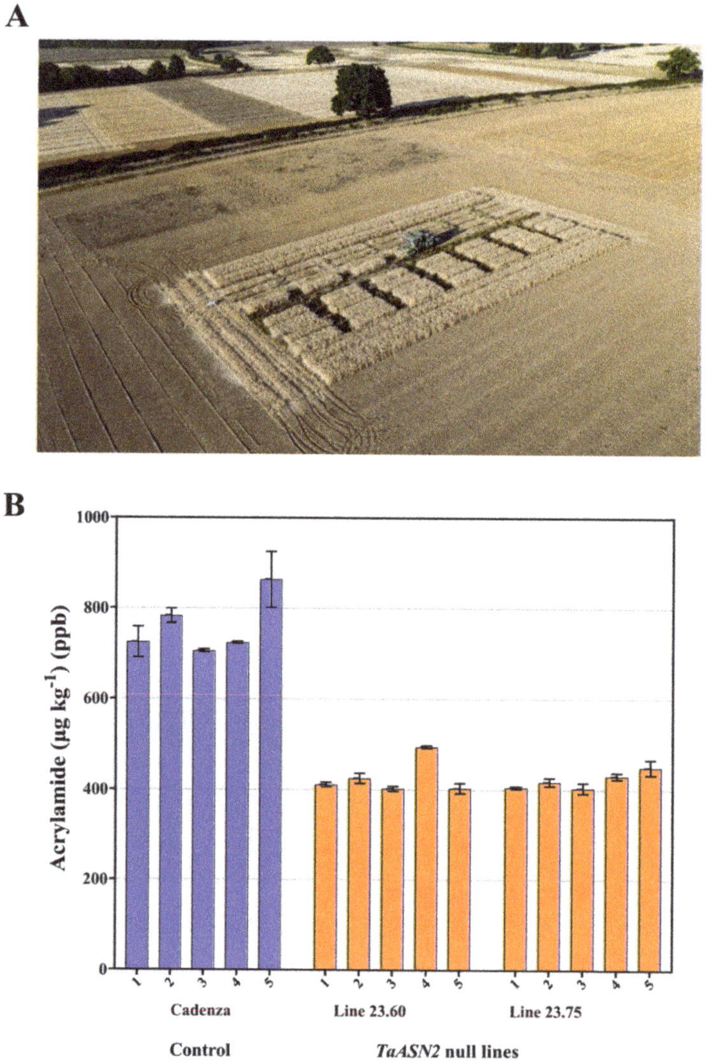

Figure 9. (**A**). Drone picture of Europe's first genome-edited wheat field trial being harvested at Rothamsted, August 2022, showing the layout of 56 plots of genome-edited lines (Cadenza background), TILLING lines (Claire background) and controls in a randomised block design, with a surrounding pollen barrier. (**B**). Graphs representing the acrylamide content of heated flour (160 °C for 20 min) from five individual plots each of genome-edited *TaASN2* null lines (Line 23.60 and line 23.75) and the Cadenza control. Data from [67].

The use of imaging methods to predict free asparagine concentration requires further research to confirm how consistently these prediction figures can be achieved. However, we are also aware of work being carried out by several teams on the development of on-site sensors to enable free asparagine concentration in grain to be measured 'at the factory gate', enabling food businesses to turn away high asparagine grain before it enters their manufacturing lines. Some of these teams claim to be in an advanced stage of development, although we are not aware of publications in the scientific literature that describe such a sensor.

The development of these systems would help food businesses to comply with regulations on acrylamide more consistently, taking away the concern that the use of grain with unexpectedly high concentrations of free asparagine could undermine their mitigation measures. For farmers, on the other hand, it raises the possibility that high-value markets could be shut off from them if the grain they produce contains too much free asparagine, and we expect farmers to look to breeders to help them produce grain that food businesses will use.

3. Other Processing Contaminants in Wheat Products

3.1. 5-Hydroxymethylfurfural (HMF)

Hydroxymethylfurfural (HMF), a furanic compound with both aldehyde and alcohol functional groups (Figure 10), has been known to be a problem in cereal products for longer than acrylamide. Like acrylamide, HMF is formed through the Maillard reaction during thermal processing (Figure 1) [71,72], but it can also be produced by the direct dehydration of fructose or caramelisation [73,74]. It is produced via an intermediate compound, 3-deoxyosone, which undergoes dehydration and cyclisation to produce HMF. HMF is thought to possess some beneficial biological properties, including antioxidant activity, the inhibition of red blood cell sickling, and the improvement of haemorheology [75]. However, in humans, HMF is metabolised to 5-hydroxymethyl-2-furoic acid (HMFA), which is excreted in urine, and 5-sulfoxymethylfurfural (SMF). SMF can form adducts with DNA or proteins, and rodent toxicology studies have indicated potential genotoxicity and carcinogenicity [76,77].

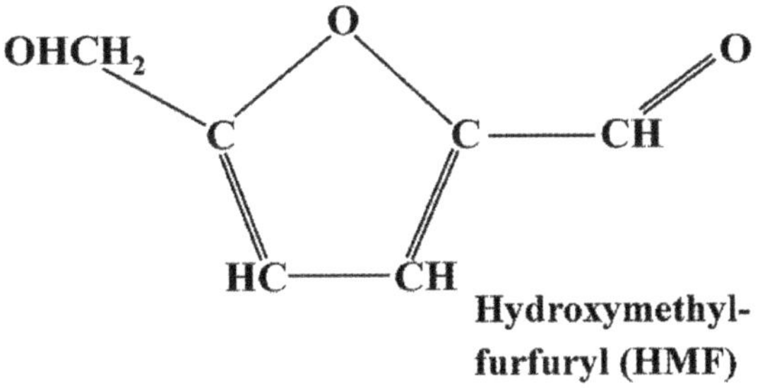

Figure 10. Diagram representing the structure of hydroxymethylfurfuryl (HMF).

HMF is a common contaminant of dark beers [77] and is found in a variety of other food items, such as bakery products, brown sugar, caramels, breakfast cereals, malt, fruit juices, coffee, and vinegar. Among these, bread and coffee are the primary contributors to dietary intake, although dried fruits [78], caramel and vinegar can have particularly high concentrations. The potential toxicity of HMF in humans is still being investigated, but studies have indicated that high concentrations can cause cytotoxic effects and irritation to the eyes, upper respiratory tract, skin, and mucous membranes [79].

HMF can form under acidic conditions at lower temperatures [80]; however, its concentration significantly increases with higher temperatures during thermal processing or storage. Hence, HMF can serve as a valuable indicator for monitoring heating processes in food production, including pasta drying, bread baking and toasting, as well as the extrusion of cereal-based baby foods and breakfast cereals. It is sometimes used as an indicator of excessive heat treatment in biscuit manufacture, and products containing high levels of HMF are likely to contain a lot of acrylamide. HMF concentration is usually measured using high-performance liquid chromatography (HPLC). Mesías et al. [81] reported levels of 14.2–42.0 mg kg^{-1} in breakfast cereals, while Petisca et al. [82] reported levels of 3.0 mg kg^{-1} in cakes and pastries and 7.8 mg kg-1 in biscuits, with high variability between different batches of the same product. However, Ortu and Caboni [83] reported much higher concentrations in bakery products, ranging from 75–395 mg kg^{-1}.

The reduction of HMF levels in food products primarily relies on the chosen processing methods and parameters. There have been no attempts that we are aware of to reduce the concentration of its precursors through genetic or agronomic approaches, for example by assessing the range of fructose and glucose concentrations in different wheat varieties or how these concentrations are affected by crop management. This is surprising because glucose and fructose levels are affected by environmental factors that could be alleviated by crop management, for example increasing in response to drought stress as the plant interconverts simple sugars and complex carbohydrates for osmotic regulation [5].

3.2. Polycyclic Aromatic Hydrocarbons (PAHs)

Polycyclic aromatic hydrocarbons (PAHs) are a group of nearly 10,000 chemical compounds possessing 2–7 fused aromatic rings generated during incomplete combustion of organic compounds. Direct pyrolysis of food constituents, such as proteins and fats, can result in the formation of PAHs, and intense thermal processing methods like toasting, roasting, frying and other high-temperature cooking techniques contribute to PAH generation. PAHs have gained significant attention and recognition as priority pollutants due to their mutagenic and carcinogenic properties [84–87]. They have been included in the priority pollutant lists of several authoritative bodies, such as the Agency of Toxic Substances and Disease Register (ATSDR), the International Agency for Research on Cancer (IARC), the European Union (EU) and the United States Environmental Protection Agency (EPA). The four principal PAHs in cereal products are benzo[a]pyrene, benzo[a]anthracene, benzo[b]fluoranthene and chrysene, the so-called PAH4 (Figure 11), and the European Commission has set a tolerance limit of 1 μg kg^{-1} for PAH4 in cereal products for infants [88]. PAHs present diverse risks to human health, encompassing conditions such as asthma, breast cancer, cardiovascular disease and diminished sperm quality [89]. Additionally, PAHs can impact the prenatal development of foetuses in pregnant women. Exposure to PAHs has also been linked to impaired cognitive function in childhood and decreased birth rates [89].

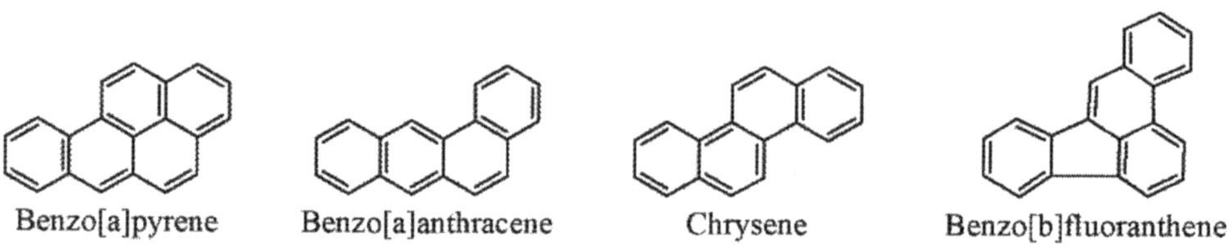

Figure 11. Diagrams representing the structures of the PAHs making up the so-called PAH4, the principal PAHs of cereal products.

Although polycyclic aromatic hydrocarbons (PAHs) are composed of multiple aromatic rings, the initial creation of the primary aromatic ring (benzene or phenyl) is important. Multiple formation mechanisms have been proposed, involving different reactants and reaction pathways. Examples include the interaction between the vinyl radical (C_2H_3) and 1,3-butadiene (1,3-C_4H_6), followed by hydrogen removal, the reaction of diatomic carbon (C_2) with 1,3-butadiene, and the combination of ethylene (C_2H_4) with cyclopentadiene (C_5H_6) [90]. Once the initial aromatic ring forms, subsequent structural expansion happens through diverse mechanisms, resulting in the development of polycyclic aromatic configurations. The processes of PAH creation and enlargement are mainly reliant on the source of carbon and the surrounding environmental conditions. Consequently, a comprehensive approach is imperative to comprehend these mechanisms. Four PAH formation pathways are illustrated in Figure 12.

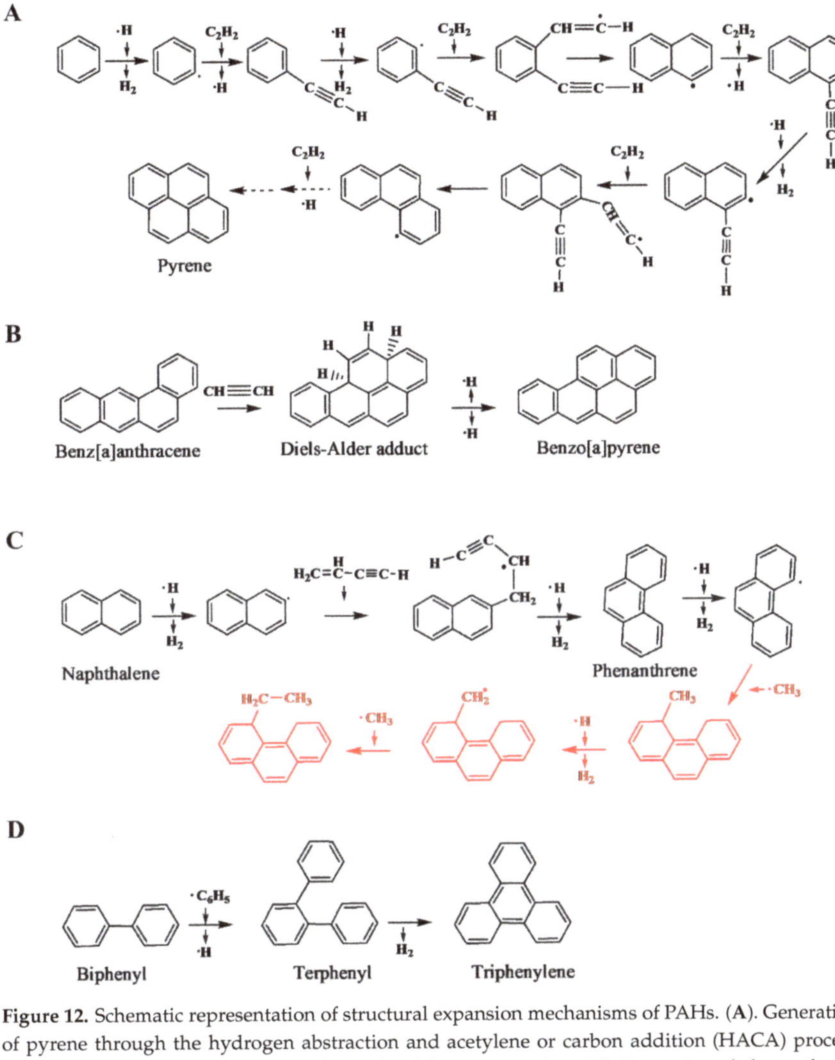

Figure 12. Schematic representation of structural expansion mechanisms of PAHs. (**A**). Generation of pyrene through the hydrogen abstraction and acetylene or carbon addition (HACA) process. (**B**). Creation of benzo[a]pyrene via the Diels-Alder-type reaction. (**C**). Formation of phenanthrene from naphthalene through the hydrogen abstraction vinyl acetylene addition (HAVA) mechanism, followed by the introduction of an ethyl chain onto phenanthrene using the methyl addition cyclisation (MAC) mechanism (shown in red). (**D**). Production of triphenylene from biphenyl through the phenyl-addition cyclisation (PAC) mechanism. Figure adapted from [90].

Research on PAH exposure consistently highlights that foods, particularly those that are cereal-based, serve as the primary source of human exposure to these compounds [91–93]. PAHs can potentially be found in wheat products due to environmental contamination of crops or products arising from atmospheric deposition of pollutants from industrial and vehicle emissions, or formation during processing and cooking. High-temperature cooking techniques such as baking, toasting, roasting and frying can lead to the direct deposition of PAHs onto the surface of the food. Additionally, when food is exposed to smoke for flavouring or preservation purposes, PAHs present in the smoke can be deposited onto the food.

The levels can vary due to several factors, including growing conditions, storage practices and processing methods. Ahmed et al. [94], for example, conducted a study that demonstrated the influence of the fuel type used during baking on the levels of PAHs in bread and observed a range of PAH concentrations in bread baked in different types of bakeries. For instance, bread baked in mazut (heavy fuel oil)-operated bakeries had a PAH concentration of 321 μg kg^{-1}, while bread baked in solid waste, solar, and electricity-operated bakeries had concentrations of 317, 158 and 26 μg kg^{-1}, respectively. Rey-Salgueiro et al. [95] also observed significant effects of the toasting technique on the levels of PAHs in bread, with charcoal and flame grilling resulting in higher levels compared with electric oven and toaster methods. Wood flame toasting produced the highest levels, with up to 350 3.65 μg kg^{-1} total PAHs. In line with this, Rascón et al. [96] determined that home toasting escalated PAH levels, with the degree of increase influenced by the specific raw material used. Among different bread varieties, white bread demonstrated the highest PAH content (17.8 3.65 μg kg^{-1}), closely followed by wholegrain bread (16.9 3.65 μg kg^{-1}), multiseed bread (16.5 3.65 μg kg^{-1}), black bread (13.0 3.65 μg kg^{-1}) and sliced bread (10.3 3.65 μg kg^{-1}). Ciecierska and Obiedziński [97] noted that the collective amount of 19 PAHs present in grain, flour and bran exhibited a range of 1.07 to 3.65 3.65 μg kg^{-1}. The PAH content of bread varied between 1.59 and 13.6 3.65 μg kg^{-1}, depending on both the specific portion of the bread analysed and the temperature employed during baking. Similarly, Chawda et al. [98] investigated total PAH concentration in Indian tandoori and tawa breads, observing a range of 113.36–211.19 3.65 μg kg^{-1} and 59.64–77.12 3.65 μg kg^{-1}, respectively, representing a 71.7% difference between the two types.

Tran-Lam et al. [99] analysed PAH levels in instant noodles, fried noodles and cakes. The mean concentration in noodles was 57.2 μg kg^{-1}, but this increased almost three-fold to 182.8 μg kg^{-1} when the noodles were fried. The concentration in cakes ranged from 2.32–26.92 μg kg^{-1}. There has also been an analysis of PAH concentrations in Nigerian biscuits prepared using different baking techniques, fuel sources, oven conditions and ingredients, revealing a wide range of concentrations in shortcake (35.7 to 645.3 μg kg^{-1}) digestives biscuits (75.9 to 490.7 μg kg^{-1}) cookies (91.5 to 537 μg kg^{-1}), shortbread (18.4 to 522.2 μg kg^{-1}), wafers (123.5 to 393.8 μg kg^{-1}), crackers (167.2 to 880 μg kg^{-1}) and cabin biscuits (135.5 to 241.6 μg kg^{-1}) [100]. Therefore, just like with acrylamide and HMF, the levels of PAHs in food products can be lowered by selecting effective processing methods, and this has been the focus of mitigation methods to date.

4. Concluding Remarks

Cooking or high-temperature processing kills potentially pathogenic micro-organisms, tenderises food, makes it palatable, and produces the colours, flavours and aromas that we associate and demand in fried, baked, roasted and toasted foods. However, food raw materials derived from grains, tubers, storage roots and beans have highly complex compositions, and it is not surprising that heating them to very high temperatures also produces some undesirable contaminants. Here, we have reviewed three contaminants, acrylamide, HMF and PAHs, that are of particular concern in wheat-derived products. In all three cases, food manufacturers can limit the formation of these contaminants through careful control of the manufacturing process and quality control of the final product. However, in the case of acrylamide and HMF, the efforts of manufacturers can be undone by the presence of unexpectedly high concentrations of precursors in the raw material. That might be difficult to address for HMF because reductions in fructose and glucose concentrations would have a broad impact on product flavour, colour and aroma. For acrylamide, however, reducing the concentration of the key precursor, free asparagine, should have less impact on product quality, and free asparagine is amenable to both genetic and agronomic approaches to reducing its concentration.

Acrylamide is a matter of great concern in the realm of food safety due to its classification as a probable carcinogen. With wheat being a staple crop in numerous regions around the world, it serves as a crucial component in many basic food products, and reducing the acrylamide-forming potential of wheat would reduce the exposure of millions of consumers to dietary acrylamide. The presence of acrylamide in food is also a major regulatory-compliance issue for food manufacturers based in or exporting to the European Union, with the prospect of Maximum Levels being imposed in the near future. Effectively tackling this issue is of utmost importance to protect public health and ensure that food businesses can continue to produce wheat-based products that comply with regulations and provide consumers with confidence and peace of mind while enjoying their daily staples.

In the quest to reduce acrylamide content in food products, crop management plays a critical role. By implementing best crop management practices, such as ensuring good disease control and adjusting nitrogen and sulphur fertiliser applications, spikes in free asparagine concentration can be avoided. Furthermore, advancements in biotechnology offer promising solutions for enhancing wheat varieties. The development of low asparagine, genome-edited lines, in particular, shows that step reductions in free asparagine concentration in wheat grain can be achieved. Employing advanced technologies and refining processing conditions can result in significant reductions in acrylamide formation during cooking and manufacturing processes, but any changes to processing methods will be more effective and give more consistent and predictable results with low asparagine raw material. A low asparagine starting point would also mean that manufacturers could reduce the levels of acrylamide in their products without compromising on taste and quality. It is essential, therefore, that the regulatory environment on genome editing and other biotechnological applications gives breeders confidence that they can start using new breeding technologies.

In conclusion, the combination of crop management practices, genomic and biotechnological advancements, and improved processing techniques offer a comprehensive approach to tackling the issue of acrylamide and other processing contaminants in wheat-based products. By proactively addressing this issue, the food industry and its supply chain can enhance food quality and safety, benefiting both producers and consumers alike. The ripple effects of these efforts extend beyond individual health to encompass broader positive impacts on food quality and overall human well-being. A concerted and proactive approach to reducing contaminant levels is an essential step towards building a safer and healthier food landscape for the future.

Author Contributions: N.K. and N.G.H. contributed equally to the writing of this review. All authors have read and agreed to the published version of the manuscript.

Funding: N.K. is supported by a Biotechnology and Biological Sciences Research Council (BBSRC) Responsive Mode grant (BB/W007134/1). BBSRC is part of UK Research and Innovation. N.G.H. is supported at Rothamsted Research by the Lawes Agricultural Trust.

Data Availability Statement: No new data were created or analyzed in this study. Data sharing is not applicable to this article.

Conflicts of Interest: N.G.H. is a Visiting Professor at the Shanghai Academy of Agricultural Sciences and a Special Professor at the University of Nottingham. He also has collaborations with the Universities of Reading and Bristol, as well as a number of plant breeding and food businesses. None of these organisations or the organisations listed in the funding section had any role in the conceptualisation, writing or editing of this review.

References

1. Shewry, P.R.; Hey, S. The contribution of wheat to human diet and health. *Food Energy Secur.* **2015**, *4*, 178–202. [CrossRef]
2. Available online: http://www.nabim.org.uk/wheat/wheat-varieties (accessed on 17 August 2023).
3. Curtis, T.Y.; Postles, J.; Halford, N.G. Reducing the potential for processing contaminant formation in cereal products. *J. Cereal Sci.* **2014**, *59*, 382–392. [CrossRef]

4. Available online: https://www.efsa.europa.eu/en/topics/topic/chemical-contaminants-food-feed (accessed on 17 August 2023).
5. Halford, N.G.; Curtis, T.Y.; Muttucumaru, N.; Postles, J.; Mottram, D.S. Sugars in crop plants. *Ann. Appl. Biol.* **2011**, *158*, 1–25. [CrossRef]
6. European Food Safety Authority Panel on Contaminants in the Food Chain (CONTAM Panel). Scientific opinion on acrylamide in food. *EFSA J.* **2015**, *13*, 4104.
7. International Agency for Research on Cancer (IARC). *IARC Monographs on the Evaluation of Carcinogenic Risks to Humans Volume 60, Some Industrial Chemicals*; International Agency for Research on Cancer (IARC): Lyon, France, 1994.
8. Joint FAO/WHO Expert Committee on Food Additives (JECFA). *Evaluation of Certain Food Contaminants. Sixty-Fourth Report of the Joint FAO/WHO Expert Committee on Food Additives (Rome, 8–17 February 2005)*; WHO Technical Reports Series; WHO: Geneva, Switzerland, 2006; Volume 930.
9. Joint FAO/WHO Expert Committee on Food Additives (JECFA). *Safety Evaluation of Certain Contaminants in Food*; WHO Food Additives Series, Volume 63; FAO JECFA Monographs 2011, Volume 8; World Health Organization: Geneva, Switzerland; Food and Agriculture Organization of the United Nations: Rome, Italy, 2011.
10. Zhivagui, M.; Ng, A.W.T.; Ardin, M.; Churchwell, M.I.; Pandey, M.; Renard, C.; Villar, S.; Cahais, V.; Robitaille, A.; Bouaoun, L.; et al. Experimental and pan-cancer genome analyses reveal widespread contribution of acrylamide exposure to carcinogenesis in humans. *Genome Res.* **2019**, *29*, 521–531. [CrossRef]
11. Eisenbrand, G. Revisiting the evidence for genotoxicity of acrylamide (AA), key to risk assessment of dietary AA exposure. *Arch. Toxicol.* **2020**, *94*, 2939–2950. [CrossRef]
12. Mottram, D.S.; Wedzicha, B.L.; Dodson, A.T. Acrylamide is formed in the Maillard reaction. *Nature* **2002**, *419*, 448–449. [CrossRef]
13. Stadler, R.H.; Blank, I.; Varga, N.; Robert, F.; Hau, J.; Guy, P.A.; Robert, M.-C.; Riediker, S. Acrylamide from Maillard reaction products. *Nature* **2002**, *419*, 449–450. [CrossRef]
14. Zyzak, D.V.; Sanders, R.A.; Stojanovic, M.; Tallmadge, D.H.; Eberhart, B.L.; Ewald, D.K.; Gruber, D.C.; Morsch, T.R.; Strothers, M.A.; Rizzi, G.; et al. Acrylamide formation mechanisms in heated foods. *J. Agric. Food Chem.* **2003**, *51*, 4782–4787. [CrossRef]
15. Hodge, J.E. Chemistry of the browning reaction in model systems. *J. Agric. Food Chem.* **1953**, *1*, 928–943. [CrossRef]
16. Thorpe, S.R.; Baynes, J.W. Maillard reaction products in tissue proteins: New products and new perspectives. *Amino Acids* **2003**, *25*, 275–281. [CrossRef]
17. Mottram, D.S. The Maillard reaction: Source of flavour in thermally processed foods. In *Flavours and Fragrances: Chemistry, Bioprocessing and Sustainability*; Berger, R.G., Ed.; Springer: Berlin/Heidelberg, Germany, 2007; pp. 269–284.
18. Peng, X.; Ma, J.; Chen, F.; Wang, M. Naturally occurring inhibitors against the formation of advanced glycation endproducts. *Food Funct.* **2011**, *2*, 289–301. [CrossRef]
19. Hellwig, M.; Henle, T. Baking, ageing, diabetes: A short history of the Maillard reaction. *Angew. Chem. Int. Ed.* **2014**, *53*, 10316–10329. [CrossRef]
20. Lund, M.N.; Ray, C.A. Control of Maillard Reactions in Foods: Strategies and Chemical Mechanisms. *J. Agric. Food Chem.* **2017**, *65*, 4537–4552. [CrossRef]
21. Koubaa, M.; Roohinejad, S.; Mungure, T.; Alaa El-Din, B.; Greiner, R.; Mallikarjunanet, K. Effect of emerging processing technologies on Maillard reactions. In *Encyclopedia of Food Chemistry*; Elsevier: Amsterdam, The Netherlands, 2019; pp. 76–82.
22. Granvogl, M.; Schieberle, P. Thermally generated 3-aminopropionamide as a transient intermediate in the formation of acrylamide. *J. Agric. Food Chem.* **2006**, *54*, 5933–5938. [CrossRef]
23. Parker, J.K.; Balagiannis, D.P.; Higley, J.; Smith, G.; Wedzicha, B.L.; Mottram, D.S. Kinetic model for the formation of acrylamide during the finish-frying of commercial French fries. *J. Agric. Food Chem.* **2012**, *60*, 9321–9331. [CrossRef]
24. Yasuhara, A.; Tanaka, Y.; Hengel, M.; Shibamoto, T. Gas chromatographic investigation of acrylamide formation in browning model systems. *J. Agric. Food Chem.* **2003**, *51*, 3999–4003. [CrossRef]
25. Claus, A.; Weisz, G.M.; Schieber, A.; Carle, R. Pyrolytic acrylamide formation from purified wheat gluten and gluten-supplemented wheat bread rolls. *Mol. Nutr. Food Res.* **2006**, *50*, 87–93. [CrossRef]
26. Xu, F.; Oruna-Concha, M.J.; Elmore, J.S. The use of asparaginase to reduce acrylamide levels in cooked food. *Food Chem.* **2016**, *210*, 163–171. [CrossRef]
27. FoodDrinkEurope. *Acrylamide Toolbox 2019*; FoodDrinkEurope: Brussels, Belgium, 2019.
28. Boyaci Gunduz, C.P. Formulation and processing strategies to reduce acrylamide in thermally processed cereal-based foods. *Int. J. Environ. Res. Public Health* **2023**, *20*, 6272. [CrossRef]
29. European Commission. *Commission Recommendation of 10 January 2011 on Investigations into the Levels of Acrylamide in Food*; European Commission: Brussels, Belgium, 2011.
30. European Commission. *Commission Recommendation of 8 November 2013 on Investigations into the Levels of Acrylamide in Food*; European Commission: Brussels, Belgium, 2013.
31. European Commission. *Commission Regulation (EU) 2017/2158 of 20 November 2017 Establishing Mitigation Measures and Benchmark Levels for the Reduction of the Presence of Acrylamide in Food*; European Commission: Brussels, Belgium, 2017.
32. Mesías, M.; Sáez-Escudero, L.; Morales, F.J.; Delgado-Andrad, C. Reassessment of acrylamide content in breakfast cereals. Evolution of the Spanish market from 2006 to 2018. *Food Control* **2019**, *105*, 94–101. [CrossRef]

33. European Commission. Food Safety: Acrylamide. Available online: https://ec.europa.eu/food/safety/chemical-safety/contaminants/catalogue/acrylamide_en (accessed on 13 July 2023).
34. Food and Drug Administration. *Guidance for Industry, Acrylamide in Foods*; Food and Drug Administration: College Park, MD, USA, 2016.
35. Muttucumaru, N.; Halford, N.G.; Elmore, J.S.; Dodson, A.T.; Parry, M.; Shewry, P.R.; Mottram, D.S. The formation of high levels of acrylamide during the processing of flour derived from sulfate-deprived wheat. *J. Agric. Food Chem.* **2006**, *54*, 8951–8955. [CrossRef]
36. Granvogl, M.; Wieser, H.; Koehler, P.; Von Tucher, S.; Schieberle, P. Influence of sulphur fertilization on the amounts of free amino acids in wheat. Correlation with baking properties as well as with 3-aminopropionamide and acrylamide generation during baking. *J. Agric. Food Chem.* **2007**, *55*, 4271–4277. [CrossRef]
37. Curtis, T.Y.; Muttucumaru, N.; Shewry, P.R.; Parry, M.A.; Powers, S.J.; Elmore, J.S.; Mottram, D.S.; Hook, S.; Halford, N.G. Effects of genotype and environment on free amino acid levels in wheat grain: Implications for acrylamide formation during processing. *J. Agric. Food Chem.* **2009**, *57*, 1013–1021. [CrossRef]
38. Oddy, J.; Addy, J.; Mead, A.; Hall, C.; Mackay, C.; Ashfield, T.; McDiarmid, F.; Curtis, T.Y.; Raffan, S.; Wilkinson, M.; et al. Reducing dietary acrylamide exposure from wheat products through crop management and imaging. *J. Agric. Food Chem.* **2023**, *71*, 3403–3413. [CrossRef]
39. Lea, P.J.; Sodek, L.; Parry, M.A.; Shewry, P.R.; Halford, N.G. Asparagine in plants. *Ann. Appl. Biol.* **2007**, *150*, 1–26. [CrossRef]
40. Curtis, T.Y.; Raffan, S.; Wan, Y.; King, R.; Gonzalez-Uriarte, A.; Halford, N.G. Contrasting gene expression patterns in grain of high and low asparagine wheat genotypes in response to sulphur supply. *BMC Genom.* **2019**, *20*, 628. [CrossRef]
41. Winkler, U.; Schön, W.J. Amino acid composition of the kernel proteins in barley resulting from nitrogen fertilization at different stages of development. *J. Agron. Crop Sci.* **1980**, *149*, 503–512.
42. Claus, A.; Schreiter, P.; Weber, A.; Graeff, S.; Herrmann, W.; Claupein, W.; Schieber, A.; Carle, R. Influence of agronomic factors and extraction rate on the acrylamide contents in yeast-leavened breads. *J. Agric. Food Chem.* **2006**, *54*, 8968–8976. [CrossRef]
43. Martinek, P.; Klem, K.; Vanova, M.; Bartackova, V.; Vecerkova, L.; Bucher, P.; Hajslova, J. Effects of nitrogen nutrition, fungicide treatment and wheat genotype on free asparagine and reducing sugars content as precursors of acrylamide formation in bread. *Plant Soil Environ.* **2009**, *55*, 187–195. [CrossRef]
44. Curtis, T.; Halford, N.G.; Powers, S.J.; McGrath, S.P.; Zazzeroni, R. *Home Grown Cereals Authority Project Report No. 525*; Agriculture and Horticulture Development Board: Kenilworth, UK, 2014.
45. Curtis, T.Y.; Powers, S.J.; Wang, R.; Halford, N.G. Effects of variety, year of cultivation and sulphur supply on the accumulation of free asparagine in the grain of commercial wheat varieties. *Food Chem.* **2018**, *239*, 304–313. [CrossRef]
46. Wilson, T.L.; Guttieri, M.J.; Nelson, N.O.; Fritz, A.; Tilley, M. Nitrogen and sulfur effects on hard winter wheat quality and asparagine concentration. *J. Cereal Sci.* **2020**, *93*, 102969. [CrossRef]
47. Raffan, S.; Oddy, J.; Halford, N.G. The sulphur response in wheat and its implications for acrylamide formation and food safety. *Int. J. Mol. Sci.* **2020**, *21*, 3876. [CrossRef]
48. Zhao, F.J.; Hawkesford, M.J.; McGrath, S.P. Sulphur assimilation and effects on yield and quality of wheat. *J. Cereal Sci.* **1999**, *30*, 1–17. [CrossRef]
49. Warth, B.; Parich, A.; Bueschl, C.; Schoefbeck, D.; Neumann, N.K.N.; Kluger, B.; Schuster, K.; Krska, R.; Adam, G.; Lemmens, M.; et al. GC-MS based targeted metabolic profiling identifies changes in the wheat metabolome following deoxynivalenol treatment. *Metabolomics* **2015**, *11*, 722–738. [CrossRef]
50. Wang, H.; Liu, D.; Sun, J.; Zhang, A. Asparagine synthetase gene *TaASN1* from wheat is up-regulated by salt stress, osmotic stress and ABA. *J. Plant Physiol.* **2005**, *162*, 81–89. [CrossRef]
51. Curtis, T.Y.; Powers, S.J.; Halford, N.G. Effects of fungicide treatment on free amino acid concentration and acrylamide-forming potential in wheat. *J. Agric. Food Chem.* **2016**, *64*, 9689–9696. [CrossRef]
52. Perochon, A.; Váry, Z.; Malla, K.B.; Halford, N.G.; Paul, M.J.; Doohan, F.M. The wheat SnRK1α family and its contribution to *Fusarium* toxin tolerance. *Plant Sci.* **2019**, *288*, 110217. [CrossRef]
53. Nussbaumer, T.; Warth, B.; Sharma, S.; Ametz, C.; Bueschl, C.; Parich, A.; Pfeifer, M.; Siegwart, G.; Steiner, B.; Lemmens, M.; et al. Joint transcriptomic and metabolomic analyses reveal changes in the primary metabolism and imbalances in the subgenome orchestration in the bread wheat molecular response to *Fusarium graminearum*. *G3* **2015**, *5*, 2579–2592. [CrossRef]
54. Perochon, A.; Jianguang, J.; Kahla, A.; Arunachalam, C.; Scofield, S.R.; Bowden, S.; Wallington, E.; Doohan, F.M. *TaFROG* Encodes a Pooideae orphan protein that interacts with SnRK1 and enhances resistance to the mycotoxigenic fungus *Fusarium graminearum*. *Plant Physiol.* **2015**, *169*, 2895–2906. [CrossRef]
55. Jiang, C.; Hei, R.; Yang, Y.; Zhang, S.; Wang, Q.; Wang, W.; Zhang, Q.; Yan, M.; Zhu, G.; Huang, P.; et al. An orphan protein of *Fusarium graminearum* modulates host immunity by mediating proteasomal degradation of TaSnRK1α. *Nat. Commun.* **2020**, *11*, 4382. [CrossRef]
56. Corol, D.I.; Ravel, C.; Rakszegi, M.; Charmet, G.; Bedo, Z.; Beale, M.H.; Shewry, P.R.; Ward, J.L. ^{1}H-NMR screening for the high-throughput determination of genotype and environmental effects on the content of asparagine in wheat grain. *Plant Biotechnol. J.* **2016**, *14*, 128–139. [CrossRef]

57. Tafuri, A.; Zuccaro, M.; Ravaglia, S.; Pirona, R.; Masci, S.; Sestili, F.; Lafiandra, D.; Ceriotti, A.; Baldoni, E. Exploring variability of free asparagine content in the grain of bread wheat (*Triticum aestivum* L.) varieties cultivated in Italy to reduce acrylamide-forming potential. *Plants* **2023**, *12*, 1349. [CrossRef]
58. Emebiri, L.C. Genetic variation and possible SNP markers for breeding wheat with low-grain asparagine, the major precursor for acrylamide formation in heat-processed products. *J. Sci. Food Agric.* **2014**, *94*, 1422–1429. [CrossRef]
59. Peng, Y.; Liu, H.; Chen, J.; Shi, T.; Zhang, C.; Sun, D.; He, Z.; Hao, Y.; Chen, W. Genome-wide association studies of free amino acid levels by six multi-locus models in bread wheat. *Front. Plant Sci.* **2018**, *9*, 1196. [CrossRef]
60. Rapp, M.; Schwadorf, K.; Leiser, W.L.; Würschum, T.; Longin, C.F. Assessing the variation and genetic architecture of asparagine content in wheat: What can plant breeding contribute to a reduction in the acrylamide precursor? *Theor. Appl. Genet.* **2018**, *131*, 2427–2437. [CrossRef]
61. Oddy, J.; Chhetry, M.; Awal, R.; Addy, J.; Wilkinson, M.; Smith, D.; King, R.; Hall, C.; Testa, R.; Murray, E.; et al. Genetic control of grain amino acid composition in a UK soft wheat mapping population. *Plant Genome* **2023**, *16*, e20335. [CrossRef]
62. Xu, H.; Curtis, T.Y.; Powers, S.J.; Raffan, S.; Gao, R.; Huang, J.; Heiner, M.; Gilbert, D.; Halford, N.G. Genomic, biochemical and modelling analyses of asparagine synthetases from wheat. *Front. Plant Sci.* **2018**, *8*, 2237. [CrossRef]
63. Raffan, S.; Halford, N.G. Phylogenetic analysis of cereal asparagine synthetase genes. *Ann. Appl. Biol.* **2021**, *178*, 6–22. [CrossRef]
64. Gao, R.; Curtis, T.Y.; Powers, S.J.; Xu, H.; Huang, J.; Halford, N.G. Food safety: Structure and expression of the asparagine synthetase gene family of wheat. *J. Cereal Sci.* **2016**, *68*, 122–131. [CrossRef]
65. Oddy, J.; Alárcon-Reverte, R.; Wilkinson, M.; Ravet, K.; Raffan, S.; Minter, A.; Mead, A.; Elmore, J.S.; Moreira de Almeida, I.; Cryer, N.C.; et al. Reduced free asparagine in wheat grain resulting from a natural deletion of *TaASN-B2*: Investigating and exploiting diversity in the asparagine synthetase gene family to improve wheat quality. *BMC Plant Biol.* **2021**, *21*, 302. [CrossRef]
66. Raffan, S.; Sparks, C.; Huttly, A.; Hyde, L.; Martignago, D.; Mead, A.; Hanley, S.J.; Wilkinson, P.A.; Barker, G.; Edwards, K.J.; et al. Wheat with greatly reduced accumulation of free asparagine in the grain, produced by CRISPR/Cas9 editing of asparagine synthetase gene *TaASN2*. *Plant Biotechnol. J.* **2021**, *19*, 1602–1613. [CrossRef]
67. Raffan, S.; Oddy, J.; Meade, A.; Barker, G.; Curtis, T.; Usher, S.; Burt, C.; Halford, N.G. Field assessment of genome edited, low asparagine wheat: Europe's first CRISPR wheat field trial. *Plant Biotechnol. J.* **2023**, *21*, 1097–1099. [CrossRef]
68. Rakszegi, M.; Kisgyörgy, B.N.; Tearall, K.; Shewry, P.R.; Láng, L.; Phillips, A.; Bedő, Z. Diversity of agronomic and morphological traits in a mutant population of bread wheat studied in the Healthgrain program. *Euphytica* **2010**, *174*, 409–421. [CrossRef]
69. Alarcón-Reverte, R.; Xie, Y.; Stromberger, J.; Cotter, J.D.; Mason, R.E.; Pearce, S. Induced mutations in *ASPARAGINE SYNTHETASE-A2* reduce free asparagine concentration in the wheat grain. *Crop Sci.* **2022**, *62*, 1484–1496. [CrossRef]
70. Burnett, A.C.; Anderson, J.; Davidson, K.J.; Ely, K.S.; Lamour, J.; Li, Q.; Morrison, B.D.; Yang, D.; Rogers, A.; Serbin, S.P. A best-practice guide to predicting plant traits from leaf-level hyperspectral data using partial least squares regression. *J. Exp. Bot.* **2021**, *72*, 6175–6189. [CrossRef]
71. Helou, C.; Gadonna-Widehem, P.; Robert, N.; Branlard, G.; Thebault, J.; Librere, S.; Jacquot, S.; Mardon, J.; Piquet-Pissaloux, A.; Chapron, S.; et al. The impact of raw materials and baking conditions on Maillard reaction products, thiamine, folate, phytic acid and minerals in white bread. *Food Funct.* **2016**, *7*, 2498–2507. [CrossRef]
72. Leiva, G.E.; Naranjo, G.B.; Malec, L.S. A study of different indicators of Maillard reaction with whey proteins and different carbohydrates under adverse storage conditions. *Food Chem.* **2017**, *215*, 410–416. [CrossRef]
73. Román-Leshkov, Y.; Chheda, J.N.; Dumesic, J.A. Phase modifiers promote efficient production of hydroxymethylfurfural from fructose. *Science* **2006**, *312*, 1933–1937. [CrossRef]
74. Thakur, J.S. HMF as a quality indicator in *Garcinia indica* fruit juice concentrate. *Curr. Res. Nutr. Food Sci.* **2018**, *6*, 227–233. [CrossRef]
75. Li, W.; Qu, X.N.; Han, Y.; Zheng, S.W.; Wang, J.; Wang, Y.P. Ameliorative effects of 5-hydroxymethyl-2-furfural (5-HMF) from *Schisandra chinensis* on alcoholic liver oxidative injury in mice. *Int. J. Mol. Sci.* **2015**, *16*, 2446–2457. [CrossRef] [PubMed]
76. Husøy, T.; Haugen, M.; Murkovic, M.; Jöbstl, D.; Stølen, L.H.; Bjellaas, T.; Rønningborg, C.; Glatt, H.; Alexander, J. Dietary exposure to 5-hydroxymethylfurfural from Norwegian food and correlations with urine metabolites of short-term exposure. *Food Chem. Toxicol.* **2008**, *46*, 3697–3702. [CrossRef] [PubMed]
77. Durling, L.J.K.; Busk, L.; Hellman, B.E. Evaluation of the DNA damaging effect of the heat-induced food toxicant 5-hydroxymethylfurfural (HMF) in various cell lines with different activities of sulfotransferases. *Food Chem. Toxicol.* **2009**, *47*, 880–884. [CrossRef]
78. Murkovic, M.; Pichler, N. Analysis of 5-hydroxymethylfurfural in coffee, dried fruits and urine. *Mol. Nutr. Food Res.* **2006**, *50*, 842–846. [CrossRef] [PubMed]
79. Choudhary, A.; Kumar, V.; Kumar, S.; Majid, I.; Aggarwal, P.; Suri, S. 5-Hydroxymethylfurfural (HMF) formation, occurrence and potential health concerns: Recent developments. *Toxin Rev.* **2021**, *40*, 545–561. [CrossRef]
80. Lee, H.S.; Nagy, S. Relative reactivities of sugars in the formation of 5-hydroxymethyl furfural in sugar-catalyst model systems. *J. Food Process. Preserv.* **1990**, *14*, 171–178. [CrossRef]
81. Mesías, M.; Sáez-Escudero, L.; Morales, F.J.; Delgado-Andrade, C. Occurrence of furosine and hydroxymethylfurfural in breakfast cereals. Evolution of the Spanish market from 2006 to 2018. *Foods* **2019**, *8*, 158. [CrossRef]
82. Petisca, C.; Henriques, A.R.; Pérez-Palacios, T.; Pinho, O.; Ferreira, I.M.P.L.V.O. Assessment of hydroxymethylfurfural and furfural in commercial bakery products. *J. Food Compos. Anal.* **2014**, *33*, 20–25. [CrossRef]

83. Ortu, E.; Caboni, P. Levels of 5-hydroxymethylfurfural, furfural, 2-furoic acid in sapa syrup, Marsala wine and bakery products. *Int. J. Food Prop.* **2017**, *20* (Suppl. 3), S2543–S2551. [CrossRef]
84. Diggs, D.L.; Huderson, A.C.; Harris, K.L.; Myers, J.N.; Banks, L.D.; Rekhadevi, P.V.; Niaz, M.S.; Ramesh, A. Polycyclic aromatic hydrocarbons and digestive tract cancers: A perspective. *J. Environ. Sci. Health C Environ. Carcinog. Ecotoxicol. Rev.* **2011**, *29*, 324–357. [CrossRef]
85. Hessel, S.; Lampen, A.; Seidel, A. Polycyclic aromatic hydrocarbons in food—Efflux of the conjugated biomarker 1-hydroxypyrene is mediated by Breast Cancer Resistance Protein (ABCG2) in human intestinal Caco-2 cells. *Food Chem. Toxicol.* **2013**, *62*, 797–804. [CrossRef] [PubMed]
86. Crowell, S.R.; Hanson-Drury, S.; Williams, D.E.; Corley, R.A. In vitro metabolism of benzo[a]pyrene and dibenzo[def,p]chrysene in rodent and human hepatic microsomes. *Toxicol. Lett.* **2014**, *228*, 48–55. [CrossRef] [PubMed]
87. Zelinkova, Z.; Wenzl, T. The Occurrence of 16 EPA PAHs in food—A review. *Polycycl. Aromat. Compd.* **2015**, *35*, 248–284. [CrossRef] [PubMed]
88. European Food Safety Authority Panel on Contaminants in the Food Chain (CONTAM Panel). Polycyclic aromatic hydrocarbons in food. *EFSA J.* **2008**, *724*, 1–114.
89. Rengarajan, T.; Rajendran, P.; Nandakumar, N.; Lokeshkumar, B.; Rajendran, P.; Nishigaki, I. Exposure to polycyclic aromatic hydrocarbons with special focus on cancer. *Asian Pac. J. Trop. Biomed.* **2015**, *5*, 182–189. [CrossRef]
90. Reizer, E.; Viskolcz, B.; Fiser, B. Formation and growth mechanisms of polycyclic aromatic hydrocarbons: A mini-review. *Chemosphere* **2022**, *291*, 132793. [CrossRef]
91. Ibáñez, R.; Agudo, A.; Berenguer, A.; Jakszyn, P.; Tormo, M.J.; Sanchéz, M.J.; Quirós, J.R.; Pera, G.; Navarro, C.; Martinez, C.; et al. Dietary intake of polycyclic aromatic hydrocarbons in a Spanish population. *J. Food Prot.* **2005**, *68*, 2190–2195. [CrossRef]
92. Martí-Cid, R.; Llobet, J.M.; Castell, V.; Domingo, J.L. Evolution of the dietary exposure to polycyclic aromatic hydrocarbons in Catalonia, Spain. *Food Chem. Toxicol.* **2008**, *46*, 3163–3171. [CrossRef]
93. Orecchio, S.; Papuzza, V. Levels, fingerprint and daily intake of polycyclicaromatic hydrocarbons (PAHs) in bread baked using wood as fuel. *J. Hazard. Mater.* **2009**, *164*, 876–883. [CrossRef]
94. Ahmed, M.T.; Hadi, E.A.; Samahy, S.E.; Youssof, K. The influence of baking fuel on residues of polycyclic aromatic hydrocarbons and heavy metals in bread. *J. Hazard. Mater.* **2000**, *80*, 1–8. [CrossRef]
95. Rey-Salgueiro, L.; García-Falcón, M.S.; Martínez-Carballo, E.; Simal-Gándara, J. Effects of toasting procedures on the levels of polycyclic aromatic hydrocarbons in toasted bread. *Food Chem.* **2008**, *108*, 607–615. [CrossRef] [PubMed]
96. Rascón, A.J.; Azzouz, A.; Ballesteros, E. Assessing polycyclic aromatic hydrocarbons in cereal-based foodstuffs by using a continuous solid-phase extraction system and gas chromatography–mass spectrometry. *Food Control* **2018**, *92*, 92–100. [CrossRef]
97. Ciecierska, M.; Obiedziński, M.W. Polycyclic aromatic hydrocarbons in the bakery chain. *Food Chem.* **2013**, *141*, 1–9. [CrossRef] [PubMed]
98. Chawda, S.; Tarafdar, A.; Sinha, A.; Mishra, B.K. Profiling and health risk assessment of PAHs content in tandoori and tawa bread from India. *Polycycl. Aromat. Compd.* **2020**, *40*, 21–32. [CrossRef]
99. Tran-Lam, T.T.; Hai Dao, Y.; Kim Thi Nguyen, L.; Kim Ma, H.; Nguyen Tran, H.; Truong Le, G. Simultaneous determination of 18 polycyclic aromatic hydrocarbons in daily foods (Hanoi Metropolitan Area) by gas chromatography–tandem mass spectrometry. *Foods* **2018**, *7*, 201. [CrossRef]
100. Iwegbue, C.M.; Onyonyewoma, U.A.; Bassey, F.I.; Nwajei, G.E.; Martincigh, B.S. Concentrations and health risk of polycyclic aromatic hydrocarbons in some brands of biscuits in the Nigerian market. *Hum. Ecol. Risk Assess.* **2015**, *21*, 338–357. [CrossRef]

foods

Review

Advanced Glycation End Products: A Comprehensive Review of Their Detection and Occurrence in Food

Lixian Li [1], Yingjun Zhuang [1], Xiuzhi Zou [1], Maolong Chen [1], Bo Cui [2], Ye Jiao [1,*] and Yunhui Cheng [1,2]

1 School of Food Science and Bioengineering, Changsha University of Science & Technology, Changsha 410114, China; cyh@csust.edu.cn (Y.C.)
2 School of Food Science and Engineering, Qilu University of Technology, Shandong Academy of Sciences, Jinan 250353, China
* Correspondence: jiaoye@csust.edu.cn; Tel.: +86-15700710776

Abstract: The Maillard reaction (MR) is a complicated chemical process that has been extensively studied. Harmful chemicals known as advanced glycation end products (AGEs), with complex structures and stable chemical characteristics, are created during the final stage of the MR. AGEs can be formed both during the thermal processing of food and in the human body. The number of AGEs formed in food is much higher compared to endogenous AGEs. A direct connection exists between human health and the build-up of AGEs in the body, which can result in diseases. Therefore, it is essential to understand the content of AGEs in the food we consume. The detection methods of AGEs in food are expounded upon in this review, and the advantages, disadvantages, and application fields of these detection methods are discussed in depth. Additionally, the production of AGEs in food, their content in typical foods, and the mechanisms influencing their formation are summarized. Since AGEs are closely related to the food industry and human health, it is hoped that this review will further the detection of AGEs in food so that their content can be evaluated more conveniently and accurately.

Keywords: advanced glycation end products; food; detection methods; content

1. Introduction

Modern food preparation includes hot processing, which can increase food flavor, extend shelf life, and lower the risk of foodborne illnesses [1]. During thermal processing, most commercially treated foods experience the Maillard reaction (MR), a chemical reaction consisting of a complex interaction of free amino groups (derived from amino acids or proteins) and carbonyl groups (derived from sugars) under the influence of heating [2]. A category of hazardous chemically stable substances known as advanced glycation end products (AGEs), formed in the last step of the MR, have complicated structures [3]. Since AGEs are related to a number of diseases, their formation process, types, and methods of quantification should be understood.

The ionic pathway, oxidation pathway, and free radical pathway are the possible means of AGE production in food [1]. These reaction processes can occur inside proteins or in tiny food molecules, with the consequent creation of AGEs of various sizes as the outcome. The complexity of the AGE formation mechanism, their wide range, and the multitude of nutritional components make it difficult to detect these harmful substances in food. Most current studies have measured a few representative AGEs, such as N^{ε}-carboxymethyl-lysine (CML) and N^{ε}-carboxyethyl-lysine (CEL) [4–7]. Instruments are typically used to analyze AGEs in food, with the most commonly used method being high-performance liquid chromatography–tandem mass spectrometry (HPLC-MS/MS) [8]. Although the detection accuracy is high, the sample pre-treatment process is very cumbersome and complicated [8].

This review discusses the formation and detection of AGEs in food and their distribution in different foods. A discussion and comparison of detection methods and techniques for analyzing AGEs in food and future research directions are proposed. This review is intended to provide guidance and incentive for further research into AGEs in food and encourage the development of simple, convenient, and accurate methods for their detection.

2. Types of AGEs

AGEs are a complex and heterogeneous group of compounds that can be formed from a variety of precursors as a result of the MR [9]. This chapter focuses on the types of AGEs present in foods, including reagents and formation mechanisms, as this is closely linked to AGE detection.

The reaction pathway for the amine group of protein and glucose to generate AGEs was summarized by Lund and Ray, as shown in Figure 1 [10]. Reducing sugars (glucose, aldose) and protein amine groups form Schiff bases, which can be rearranged by 1, 2-amino alcohols to Amadori products (or Haynes products, in the case of ketoses). The Amadori products can then undergo enolization to yield 1, 2-enamelanol or 2, 3-enamelanol and to form deoxyketones (α-dicarbonyl compounds), which rapidly react with other nucleophiles to form aldehydes of the Strecker group. AGEs are obtained following a series of such reactions.

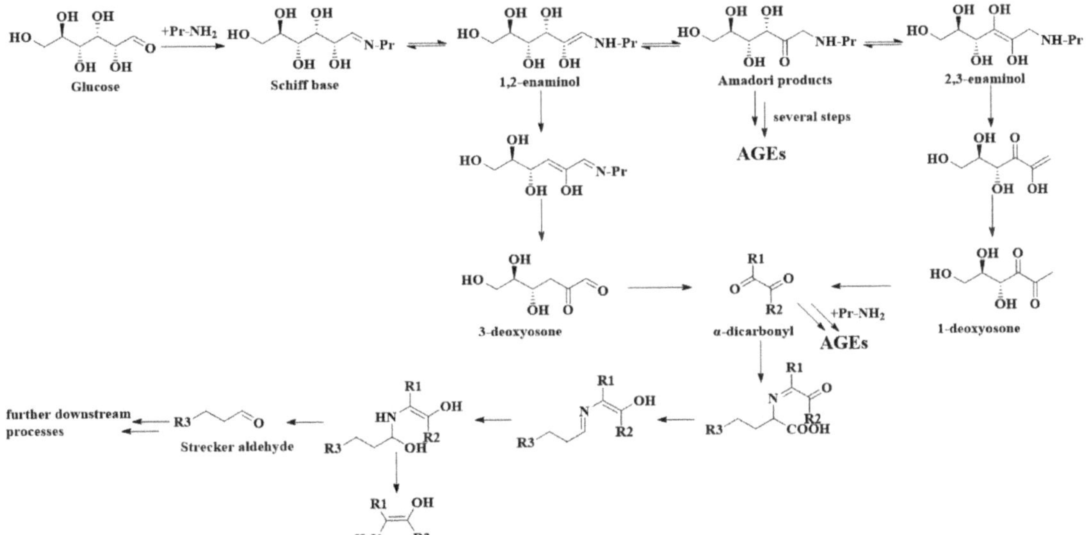

Figure 1. Reaction pathways for AGE formation in foods [10]. Quoted with permission from (Lund, M.N.; Ray, C.A. Control of Maillard Reactions in Foods: Strategies and Chemical Mechanisms. J. Agric. Food Chem. 2017, 65, 4537–4552. https://doi.org/10.1021/acs.jafc.7b00882 (accessed on 12 May 2023)). Copyright (2017) American Chemical Society.

Regarding the classification of AGEs, there are currently three common classification methods. One is to divide AGEs based on their molecular weight as either low molecular weight (LMW) or high molecular weight (HMW), but the exact boundary between LMW and HMW is not clear [1]. The other classification method divides products into fluorescent and crosslinking AGEs and non-fluorescent and non-crosslinking AGEs [11]. Fluorescent and crosslinking AGEs include pentosidine, crossline, 2-(2-furoyl)-4(5)-(2-furanyl)-1 *H*-imidazole, fluorolink, glyoxal-lysine dimer (GOLD), and methylglyoxal-lysine dimer (MOLD). Non-fluorescent and non-crosslinking AGEs include CML, CEL, pyrraline, argpyrimidine, 3-deoxyglucosone (3-DG)-imidazolones, and methylglyoxal (MGO)-imidazolones. In addition, there are methods to classify AGEs according to the types of dicarbonyl compounds, such as glyoxal (GO)-AGEs and MGO-AGEs [12]. CML, N[6]-glycolyl lysine (GALA), glyoxal-lysine-amide (GOLA), and glyoxal-lysine-amidine (GLA) belong to GO-AGEs, while CEL, methylglyoxal-lysine amide (MOLA), N[6]-lactoyl lysine, and methylglyoxal-lysine amidine (MGLA) belong to MGO-AGEs. The chemical structure of different types of AGEs is shown in Figure 2. Common AGEs are explained in the following sections. The possible formation pathways of CML, CEL, pentosidine, and pyrraline are summarized in Figure 3 [13–23].

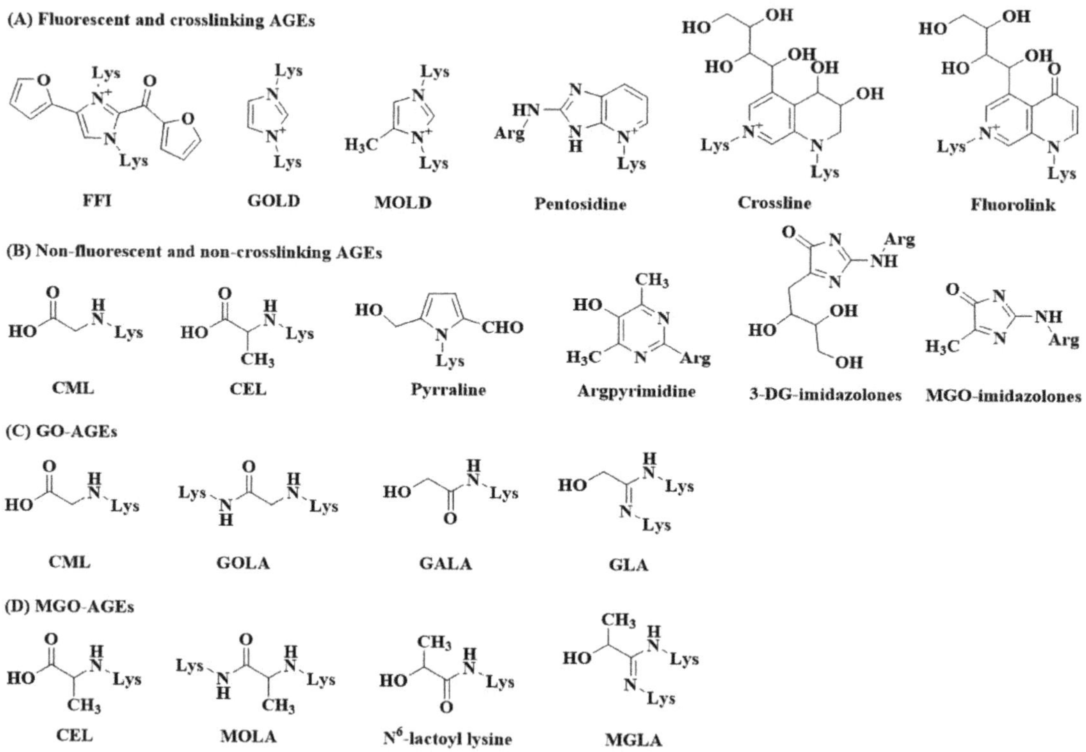

Figure 2. Chemical structural formulas of different types of AGEs.

2.1. CML and CEL

By the middle of the 1980s, scholars believed that all AGE precursor substances were Amadori products and that the outcome of metal-catalyzed oxidative cleavage of the Amadori product to protein was CML [24]. The first AGE, CML, was identified by Ahmed in 1986 [25]. CEL is a CML homologue with a formation pathway that is similar to that of CML [24]. CML/CEL can be formed in several ways, through a reductive sugar and protein reaction, lipid oxidation, and ascorbic acid oxidation. Schiff bases are early MR products (aldose and ketose) formed by the condensation of reducing sugars and amino groups on proteins. Alternatively, they can be formed by cleaving dehydroascorbic acid to form L-threose and then reacting with the lysine residues, followed by a rearrangement of Amadori/Heyns to generate products of Amadori/Heyns, including fructoselysine, lactulosyllysine, and tagatoselysine. Further oxidation of these intermediates leads to CML/CEL formation. Moreover, GO/MGO is a major intermediate in the CML/CEL formation pathway generated by both lipid peroxidation and auto-oxidation of reducing sugars. This reactive dicarbonyl compound reacts further with the lysine residues to form the CML/CEL complex.

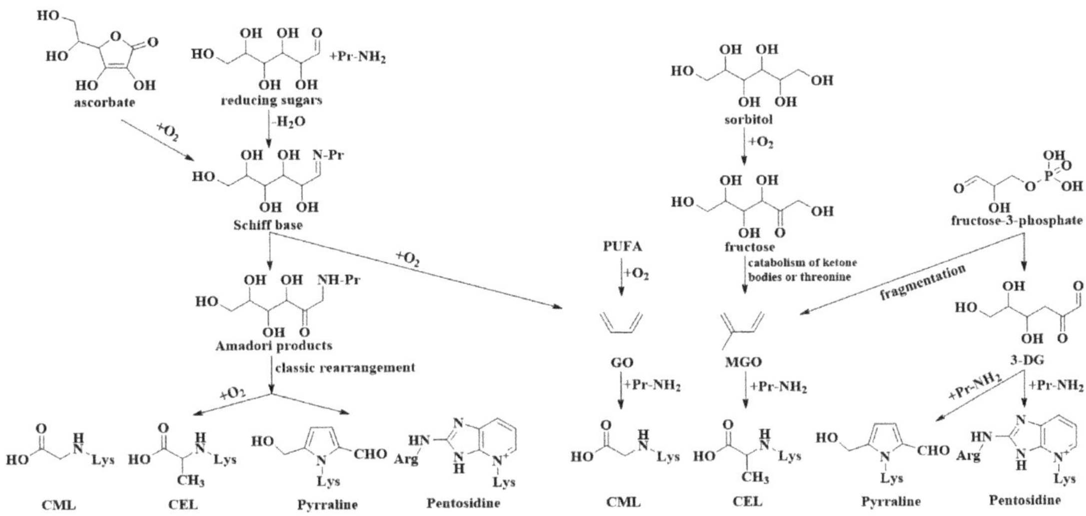

Figure 3. Reaction pathways for common AGE formation in foods.

2.2. Pentosidine

Pentosidine was first discovered as a fluorescent crosslink derived from pentoses formed by arginine and lysine reactions [26]. Next, Grandhee and Monnier [27] explored the mechanism of pentoside, believing that pentosidine can be synthesized in vitro through the reaction of ribose, arginine, and lysine. The precursors of pentosidine include glucose, fructose, and ascorbic acid. [27]. Pentosidine is also thought to be the major fluorophore formed during the non-enzymatic browning of the ribonuclease and lysozyme by glucose [28]. This study showed that pentoside formation occurs most efficiently in the reaction of pentose with lysine and arginine in the model system but is also formed by glucose, fructose, ascorbic acid, Amadori products, and 3-DG as well as other sugars [28]. Notably, pentosidine does not form from peroxidized polyunsaturated fatty acid or malondialdehyde [28].

2.3. Pyrraline

The glycosylated lysine is initially obtained by the reaction of glucose with lysine and then degraded to produce the 3-DG. 3-DG reacts with free amino groups again to form pyrraline, which is a rate-limiting step in the pathway of pyrraline formation, and the lower pH promotes the occurrence of this pathway [29].

The formation of pyrraline was studied with a kinetic model of the MR model system of glucose and lysine and postulated to be a combination of 3-DG or MGO and lysine. The results showed that the activation energy of the 3-DG pathway was significantly lower than that of the MG pathway. Kinetic studies have shown that 1, 4-dicarbonyl compounds were cyclized to form pyrrole, and pyrraline was more amenable to generation via the 3-DG pathway [30].

3. Measurement of AGEs in Food

There are many kinds of AGEs with complex structures and different contents. This chapter focuses on the detection methods of AGEs, including detection principles, application areas, advantages, and limitations. Current methods for the detecting of AGEs in foods fall mainly into two categories: instrument analysis and immunoassay analysis. Instruments used in these processes include HPLC-DAD (diode array detector) [31], HPLC-

FLD (fluorescence detector) [32], HPLC-UV (ultraviolet detector) [33], HPLC-MS/MS [4], UPLC (ultra-high-performance liquid chromatography)-MS/MS [34–36], and GC (gas chromatography)-MS [37–39]. The most commonly used instrumental analysis method is HPLC-MS/MS, while the primary immunoassay method is ELISA [40–45].

3.1. Instrument Analysis

3.1.1. HPLC-Detectors

HPLC is usually used in series with a detector to determine AGEs. A list of commonly used HPLC methods, application areas, and detection markers is detailed in Table 1.

Table 1. AGEs detected by different instruments.

Method	Column	Application Field	Marker	Limit of Detection	References
HPLC-DAD	Reverse-phase (RP)-C18 column (4.6 mm × 250 mm, 5 μm)	cookies	fluorescent AGEs		[31]
HPLC-PDA	Sunfire C18 column (4.6 mm × 250 mm, 5 μm)	baked milk and baked yogurt	pyrraline		[46]
UPLC-DAD	ACQUITY HSS T3 column (100 mm × 2.1 mm, 1.8 μm)	milk	pyrraline	0.50 μg/g protein	[47]
HPLC-FLD	C18 reverse-phase column (250 × 4.6 mm, 5 μm)	human urine and plasma	pentosidine	1 nM	[32]
HPLC-MS	Hydro-RP 80 Å LC column (150 × 4.6 mm)	dicarbonyl-BSA model system	CEL, CML, and MG-H1		[48]
HPLC-MS/MS	Hypercarb guard column (2.1 mm × 10 mm, 5 μm) tandem a C18 column (4.5 mm × 150 mm, 5 μm)	dairy products	CML	0.1 μg/kg	[4]
HPLC-MS/MS	Atlantis silica HILIC column (150 mm × 2.1 mm, 3 μm)	pork tenderloin and offal	CML and CEL		[7]
HPLC-MS/MS	Intrada amino acid column (2.0 mm × 150 mm, 3 μm)	model systems with whey protein	CML, G-H3, and MG-H3		[49]
HPLC-MS/MS	Phenomenex BioZen Peptide XB-C18 column (2.1 × 150 mm, 1.7 μm)	black tea	MG-Hs, GO-Hs, CML, CEL, and argpyrimidine		[50]
HPLC-MS/MS	Atlantis silica HILIC column (150 × 2.1 mm, 3 μm)	minced pork	CML and CEL	CML: 4–5 μg/L; CEL: 12–15 μg/L	[51]
HPLC-MS/MS	Atlantis silica HILIC column (150 mm × 2.1 mm, 3 μm)	silver carp surimi sausages	CML and CEL		[52]
HPLC-MS/MS	Sequant-ZIC-HILIC column (250 mm × 4.6 mm, 5 μm)	roasted nuts and seeds	CML and CEL		[53]
HPLC-MS/MS	Phenomenex Kinetex C-18 column (50 × 2.1 mm, 1.7 μm)	human saliva	CML, CEL, MG-H1, and pyrraline		[54]
HPLC-MS/MS	Phenomenex Synergi Hydro-RP column (150 mm × 2 mm, 4 μm)	MR model systems containing ascorbic acid	pyrraline, CML, and CEL		[55]
HPLC-MS/MS	RP-18 (Xselect HSS T3, 250 mm × 3.0 mm, 5 μm)	bread	CEL, N^6-lactoyl lysine, MG -H1, MG-H3, CEA, CML, GALA, CMA, G-H3, N^6-formyl lysine, N^6-acetyl lysine, pyrraline	CEL: 0.05 mg/kg; N^6-lactoyl lysine: 0.02 mg/kg; MG-H1: 0.07 mg/kg; MG-H3: 0.05 mg/kg; CEA: 0.06 mg/kg; CML: 0.06 mg/kg; GALA: 0.06 mg/kg; CMA: 0.17 mg/kg; G-H3: 0.11 mg/kg; N^6-formyl lysine: 0.03 mg/kg; N^6-acetyl lysine: 0.01 mg/kg; pyrraline: 0.12 mg/kg	[56]
LC-MS/MS	Hypercarb column (100 × 2.1 mm, 5 μm)	human milk	CML and CEL		[57]
LC-MS/MS	BEH Amide column (2.1 × 100 mm, 1.7 μm)	fried sturgeon fillets	CML and CEL		[58]
LC-MS/MS	Hydro-RP 80A LC column (2 × 150 mm, 4 μm)	cheeses	CML and CEL		[59]

Table 1. *Cont.*

Method	Column	Application Field	Marker	Limit of Detection	References
LC-MS/MS	Hydro-RP80A column (250 mm × 2 mm, 4 μm)	mutton	CML and CEL	CML: 3.6 ng/mL; CEL: 1.9 ng/mL	[60]
LC-MS/MS	Phenomenex Synergi 4 μ Hydro-RP80A column (250 mm × 2 mm, 4 μm)	beef	CML and CEL		[61]
UPLC-MS/MS	X-Bridge C18 (4.6 mm × 150 mm, 5 μm)	baked products	CML and CEL		[34]
UPLC-MS/MS	BEH Amide column (100 × 2.1 mm, 1.7 μm)	fish cakes	CML, CEL, and MG-H1		[35]
UPLC-MS/MS	SB-C18 column (2.1 × 50 mm, 1.8 μm)	soy sauce, corn juice	CML		[36]
UPLC-MS/MS	T3 column (150 mm × 4.6 mm × 3 μm)	bread	CML and CEL	CML: 0.75 μg/kg; CEL: 2.5 μg/kg	[62]
UPLC-MS/MS	T3 column (150 mm× 2.1 mm, 3.5 μm)	roasted beef patties	CML and CEL	CML: 0.052 μg/g; CEL: 0.098 μg/g	[63]
UPLC-MS/MS	BEH Amide column (100 × 2.1 mm, 1.7 μm)	sterilized milk	CML	CML: 0.05 mg/kg	[64]
UPLC-MS/MS	HILIC column	antler velvet	CML and CEL	CML: 4.1 ng/g; CEL: 4.3 ng/g	[65]

CML: N^{ε}-(carboxymethyl)lysine. CEL: N^{ε}-(carboxyethyl)lysine. GLAP: glyceraldehyde-derived pyridinium compound. MG-Hs: methylglyoxal-hydroimidazolones. GO-Hs: glyoxal-hydroimidazolones. CEA: N^7-carboxyethyl arginine. CMA: N^7-carboxymethyl arginine. GALA: N^6-glycoloyl lysine.

HPLC-DAD is used to determine the content of pyrraline [46,47] in food because the DAD detector can detect substances with ultraviolet absorption. HPLC-DAD can accurately detect the content of pyrraline with good linearity, but the sample needs to be pre-treated, generally by enzymatic hydrolysis and acid hydrolysis [46,47]. Pepsin is usually used for enzymatic hydrolysis [47], and hydrochloric acid is mainly used for acid hydrolysis [46,47]. In recent years, HPLC-DAD has been regularly used with a C18 column to determine pyrraline in dairy products [31,46]. Poojary et al. [47] developed a method to detect pyrraline with the ACQUITY HSS T3 Column (100 mm × 2.1 mm, 1.8 μm) configured by UPLC-DAD. This method's detection limit was found to be as high as 0.50 μg pyrraline/g protein. A UV detector with the advantages of low noise, wide linear range, and good selectivity is also used. While not very sensitive to ambient temperature, changes in mobile phase composition, and flow rate fluctuations, the detection limit is high. Hellwig and Henle [66] used HPLC-UV to determine formyline and pyrraline quantitatively and found that the UV absorption mode differs greatly from the standard. Further research is needed to confirm whether solid phase extraction is suitable for the samples' purification in this case.

HPLC-FLD is commonly used to detect fluorescent and crosslinking AGEs, such as pentosidine. The method shows high selectivity, high sensitivity, and good reproducibility but only responds to fluorescent AGEs. Non-fluorescent AGEs need to be derived into a fluorescent substance. The use of FLD to detect the content of CML requires hydrolysis and derivatization of the sample. Milk samples should be degreased with n-hexane and reduced with sodium borate before hydrolysis [4,6]. Meat products could be defatted with the degreasing solvents chloroform and methanol before reduction [67]. Hydrolysis is usually carried out with 6 M hydrochloric acid at 110 °C, and derivatization can be achieved by various reagents. The most commonly used reagents are o-phthalaldehyde and 2-mercaptoethanol [68].

HPLC-MS is the most widely used method to measure CML and CEL. However, it is subject to strong interference peaks from the matrix and ion suppression effects, and standard addition and isotope dilution methods are often used to compensate [69]. Ochi [70] used nonafluoropentanoic acid (NFPA) as an ion pair reagent to successfully achieve good retention of CML on a reversed-phase C18 column with good compatibility with HPLC-MS. Sun et al. [71] used water extraction to prepare samples for the analysis of free AGEs. Trichloroacetic acid was used to precipitate proteins to prevent column damage.

CML and CEL content in minced pork was determined by Zhang et al. [51] using the Atlantis silica HILIC column (150 × 2.1 mm, 3 μm) in an HPLC-MS/MS configuration with a low limit of detection of 4 to 5 μg/L for CML and 12 to 15 μg/L for CEL. Subsequent applications were also applied to silver carp surimi sausages [52] and pork tenderloin and offal [7]. When Troise et al. [33] used an HILIC column (75 mm × 2.1 mm, 2.6 μm) to measure CML and CEL in cookies, CML and CEL values > 20 ng/mg could be detected. Xiao et al. [4] used HPLC-MS/MS to configure a Hypercarb guard column (2.1 mm × 10 mm, 5 μm) tandem with a C18 column (4.5 mm × 150 mm, 5 μm) in order to detect the presence of CML in dairy products with a limit of detection of 0.1 μg/kg. Bai et al. [60] developed a method for determining the CML and CEL content in mutton using a Hydro-RP80A column (250 mm × 2 mm, 4 μm). Detection limits were achieved at 3.6 ng/mL for CML and 1.9 ng/mL for CEL. UPLC-MS/MS is also used to measure AGEs in food. The UPLC system can withstand higher pressures and maintain good separation results at lower flow rates [8].

3.1.2. GC

GC is often connected in tandem with MS to detect AGEs. GC requires the pretreatment of samples, usually degreasing, reduction, hydrolysis, and derivatization, as shown in Figure 4 [37,72]. A mixed solution of water/chloroform/methanol is often used to achieve the purpose of degreasing by centrifugation. The reduction often uses a boric acid buffer containing sodium borohydride, and the sample is often placed in 6 M HCl for acid hydrolysis. Derivatization usually involves esterifying the carboxyl group with methanol under strong acid conditions and then acylating the amine group with trifluoroacetic anhydride [72]. Charissou et al. [73] used GC-MS to determine the CML content in commercial milk powder, biscuits, salmon, and beef. The detection limit was 0.1 ng/mg protein. The quantification limit was 1 ng/mg protein, with no matrix effect. However, the derivatization process before the GC-MS measurement is time-consuming. Malgorzata et al. [74] used GC equipped with a flame ionization detector to measure CML in buckwheat samples, and this method required only acid hydrolysis of the samples.

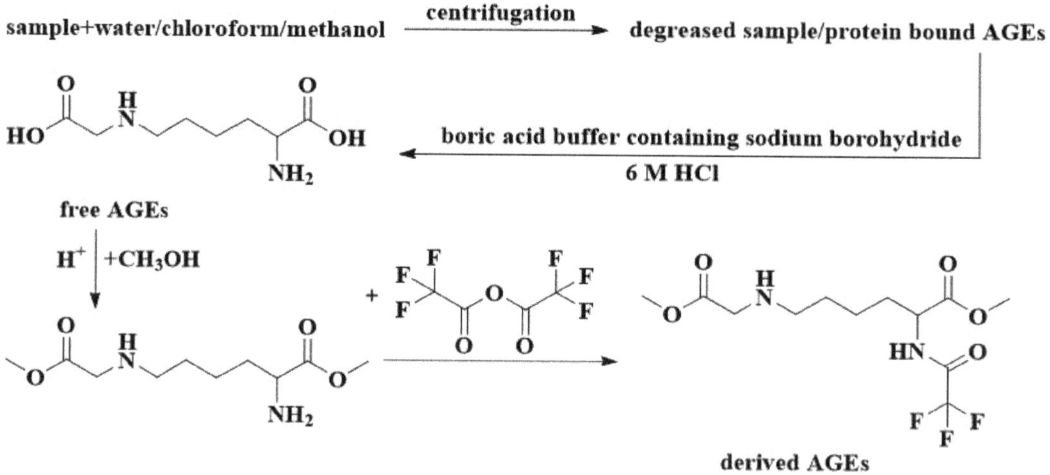

Figure 4. Sample pretreatment for the GC-MS method.

3.2. Immunoassay Analysis

ELISA

ELISA can be used as a rapid means of detecting AGEs and can simplify the process of testing samples. ELISA is currently widely used in the medical field to measure AGEs [75], and there are also examples of the application of ELISA in the food field. Fluorescent AGEs such as pentoside and non-fluorescent AGEs such as CML can be detected by ELISA. The principle of ELISA and the advantages and disadvantages of ELISA in the food field will be discussed in the following.

The basic principle of ELISA involves the immobilization of the antigen or antibody and the enzyme labeling of the antigen and antibody [76]. First, the antibody is bound to the surface of a solid phase carrier. The test antigen is added, and then the antigen–antibody-specific immune reaction is carried out with the corresponding enzyme-labeled antibody, generating an antibody-test antigen-enzyme-labeled antibody complex. The enzyme's substrate reacts to produce a colored product. The quantity of antigen to be tested is directly proportional to the quantity of stained product; therefore, the amount of antigen can be calculated based on absorbance. The principle of antibody measurement is similar to this. With the development of ELISA, commercial ELISA kits have become common. Specific operation of the ELISA kits can be performed as per the manufacturer's instructions. Commonly used antibodies and their application fields are listed in Table 2.

Table 2. Types of antibodies used in ELISA.

Antibodies	Application Field	Limit of Detection	References
4G9	dairy products	CML: 5 ng/mL	[73]
6D12	AGEs-protein		[77]
4G9	food		[78]
4G9 and 3D11	food		[79]
4G9	porridge samples		[80]

In 2004, Goldberg et al. [78] used ELISA to detect the CML content in 250 types of foodstuffs, with a well-characterized anti-CML monoclonal antibody (4G9). Results were expressed as units of AGE per mg of protein or lipid. The author stated that the results of the ELISA assay are preliminary and provide a starting point for developing a research database. The systematic food analysis method can obtain more accurate measurements. Charissou et al. [73] compared ELISA and GC-MS in the determination of dairy products. The anti-CML monoclonal antibody 4G9 was used, and the competitive ELISA had a sensitivity of 5 ng/mL CML. When using ELISA rather than GC-MS, the CML concentration in the liquid sample was almost 10 times higher. The difference between the two methods is greatest for hydrolyzed liquid samples. While ELISA lacks specificity compared with GC-MS, it is rapid, low-cost, and has low detection limits [73]. Uribarri et al. [79] used ELISA to determine the content of AGEs in 549 foods using a monoclonal anti-CML antibody (4G9) and an anti-MG monoclonal antibody (3D11 mAb). Results were expressed as AGE kilounits/100 g of food and nmol/100 g or nmol/100 mL of food. Compared to the anti-MG monoclonal antibody, the sensitivity of the method using the monoclonal anti-CML antibody was better [79]. Tareke et al. [80] used an AGE-specific antibody (mouse monoclonal 4G9) conjugated to horseradish peroxidase. The CML level in the three porridge samples analyzed by ELISA was about 56% of the level detected by LC-MS. The relative deviation of ELISA is fairly large [80]. Three different commercially available ELISA kits were used by Gómez-Ojeda et al. [45]. ELISA-1 was a human CML-AGE ELISA kit, while ELISA-2 was an assay kit for competitive inhibition of total AGEs in human serum. The authors believed that ELISA appears to be particularly suitable for the analysis of products with a relatively high concentration of CML/AGEs, but these methods are less precise, and the difference in the concentration of CML in the same food prepared by different cooking treatments cannot be detected at a statistical level by ELISA. These methods are obviously affected by the matrix effect [45]. Prosser et al. [44] used a commercial CML ELISA kit to measure CML in infant formula. The results obtained in this experiment were consistent with the results of many studies using the ELISA method.

It is recommended that nearly all current commercial ELISA kits be used for clinical sample analysis, but few are used for food analysis. Antibodies such as anti-pentosidine, anti-3-DG-imidazolone, anti-CML, anti-CEL mAb, anti-pyrraline, anti-CMA, and antihuman RAGE (receptor for advanced glycosylation end products) mAbs have been widely used in the medical field [75]. In the future, these antibodies can be considered for use in the food field for ELISA.

3.3. Other Methods

In addition to the standard methods above, others can be used to determine the content of AGEs. Klostermeyer and Henle [81] used ion exchange chromatography in conjunction with a photodiode array to determine pyrraline. This method allows pyrraline to be quantified at levels < 500 μg/kg protein. Capillary electrophoresis was used by Dalin and Dutta [82] to analyze the in vitro MR of glyceraldehyde, glucose, and fructose and the inhibition of glyceraldehyde self-condensation under physiological conditions using naked capillary and 214 nm UV detection. Birlouez-Aragon et al. [83] used front-face fluorescence to determine the CML content in infant milk powder.

3.4. Comparison of the Detection Techniques

There are many methods currently used to detect AGEs. Compared with other instrument analysis methods, HPLC-MS/MS can identify and quantify target analytes with high accuracy and high resolution and does not require sample derivatization [69]. HPLC-MS/MS is suitable for the determination of AGE content in various foods. However, HPLC-MS/MS is a bit more expensive than other methods.

GC-MS is chiefly used for the determination of AGE content in meat, dairy, and cereal products. However, when using HPLC-MS and GC-MS to determine the CML content of porridge, it was found that the standard deviation of GC-MS was ± 28, and the relative standard deviation was 10%, while the standard deviation measured by HPLC-MS was $\pm$ 6 and the relative standard deviation was 5%. Therefore, the precision and stability of GC-MS are lower than those of HPLC-MS [80]. In addition, the application of GC-MS is less extensive than that of HPLC-MS.

Compared to instrumental analysis, ELISA appears to be particularly suitable for studying products with relatively high concentrations of CML/AGEs. ELISA has been used to measure the AGE content of various foods, which facilitates the establishment of a preliminary database. This method, however, is less accurate than LC-MS and is significantly impacted by matrix effects. When using the ELISA method, the number of AGEs in fatty foods may be overestimated and the level of AGEs in starchy foods may be underestimated [45,84]. Nevertheless, ELISA is a good choice for the determination of AGE content in a large number of samples in a short time. Furthermore, it should be emphasized that commercial kits currently used in the study are recommended for clinical sample analysis, while there are actually no commercial kits for food analysis.

4. Levels of AGEs in Common Foods

According to the formation mechanism of AGEs summarized above, AGEs mainly exist in foods rich in proteins, sugars, and fats. Many articles focus on AGEs in foods, and reports of AGE content in hundreds of foods exist. This section focuses on the recent research progress on AGEs in common foods.

4.1. Cereal Products

Cereal products are a common food type consumed in daily life. The MR is an important chemical reaction in bakery products, and the safety of cereal products obtained through baking has received much attention. Cereal products reported to be linked to AGEs in recent years mainly include breads and cookies. There have also been studies on tortillas, sweetheart pastries, and instant noodles. At present, research on AGEs in cereal products primarily focuses on CML, CEL, and methylglyoxal-derived hydroimidazolone-1 (MG-H1), and LC-MS/MS is generally used for measurement (Table 3). Raw materials, ingredients, processing time, and processing temperature are the main factors that affect the content of AGEs in cereal products.

Table 3. AGE levels in different cereal products.

Influencing Factors	Detection Methods	Cereal Products	AGE Content	References
Raw materials	LC-MS	flour tortilla, corn tortilla	CML: 2.09–2.51 μg/g	[45]
	HPLC-FLD	bread (wheat, brown bread, rye bread, pumpernickel, and crispbreads)	CEL: 2.1−8.1 mg/kg; N^6-lactoyl lysine: 0.15−0.72 mg/kg; CEA: 7−19 mg/kg; MG-H1: 13−27 mg/kg; MG-H3: 0.28−0.72 mg/kg; CML: 4.5−10.4 mg/kg; GALA: 0.23−0.59 mg/kg; CMA: 1.1−1.6 mg/kg; G-H3: 3.8−5.5 mg/kg; N^6-formyl lysine: 2.8−5.6 mg/kg; N^6-acetyl lysine: 1.3−3.6 mg/kg; pyrraline: 28−78 mg/kg	[56]
	UPLC-MS	bread crust (whole wheat, refined wheat, whole einkorn, whole corn, whole rye, whole oat)	CML: 15–140 mg/kg; CEL: 30–200 mg/kg	[62]
	LC-MS/MS	cookies (bread wheat, durum wheat, soft wheat, hard wheat, triticale, rye, hulless barley and hulless oat, white-, yellow- and red-colored standard seeded corn and blue-colored popping corn)	CML: 6.3–47.4 mg/kg; CEL: 2.3–38.9 mg/kg	[85]
	HPLC-MS/MS	cookies (okara cookies, cellulose cookies, pea fiber cookies, chitosan cookies)	CML: 6.32–22.84 mg/kg	[86]
	UPLC-QqQ-MS/MS	chicken cookies, baby biscuit, white lotus paste mooncakes, almond biscuit, cookies, bread, sweetheart pastry, fried breadstick, instant noodles	CML: 4.48–35.88 mg/kg; CEL: 1.99–14.49 mg/kg	[87]
Ingredients	UPLC-MS/MS	sponge cakes	protein-bound CML: 80 mg/kg; free CML: 160 mg/kg; protein-bound-CEL: 15 mg/kg; free CEL: 40 mg/kg	[34]
	HPLC-MS/MS	cookies (sucrose, glucose, fructose, honey, banana, invert sugar syrup)	CML: 1.71–42 mg/kg; CEL: 1.541–53 mg/kg; MG-H1: 8.8–218 mg/kg	[88]
	LC-MS/MS	cookies (sucrose, butter, egg liquid)	free CML: 50–250 μg/kg; free CEL: 80–220 μg/kg; protein-bound-CML: 4–6 mg/kg; protein-bound-CEL: 20–70 mg/kg	[89]
Processing time	UPLC-MS	bread crust (5–30 min)	CML: 15–140 mg/kg; CEL: 30–200 mg/kg	[62]
	LC-MS/MS	cookies (7–13 min)	CML: 6.3–47.4 mg/kg; CEL: 2.3–38.9 mg/kg	[85]
	LC-MS/MS	cookies (8–16 min)	free CML: 0–60 μg/kg; free CEL: 100–400 μg/kg free; protein-bound-CML: 0–20 μg/kg; protein-bound-CEL: 100–450 μg/kg	[89]
	UPLC-MS	cookies (0–10 min)	CML: 0–350 mg/kg	[90]
Processing temperature	LC-MS/MS	cookies (130–180 °C)	free CML: 0–100 μg/kg; free CEL: 50–700 μg/kg; protein-bound-CML: 0–100 μg/kg; protein-bound-CEL: 100–700 μg/kg	[89]
	UPLC-MS	cookies (155–230 °C)	CML: 0–350 mg/kg	[90]

4.1.1. Raw Materials

The AGE content in cereal products is affected by the raw materials, in connection with the content of protein, amino acids, lipids, and dietary fiber. Zilic et al. [85] reported the CML and CEL contents of sweet cookies made with whole grain flour of eight wheat genotypes (bread wheat, durum wheat, soft wheat, durum wheat, triticale, rye, hulless barley, and oats) and four maize genotypes (white, yellow, and red standard seed corn, and blue popcorn). CML and CEL contents were found to be positively correlated with the total protein and total amino acid contents of the flour. Hence, the CML and CEL contents of the cookies produced from whole wheat oat flour were the highest because of their high total lysine and protein content [85]. Since okara is known to have various health benefits, it was added to cookies to replace 15% wheat flour; however, the degree of browning and the CML content in these cookies significantly increased with a stronger MR [86]. This increase was related to the protein and lipid of the okara, as unsaturated fats were thermally oxidized during roasting, promoting the formation of CML. Moreover, the authors claimed that this phenomenon was primarily related to the presence of approximately 50% of the insoluble dietary fiber in okara, which reduced the aqueous activity of the biscuits and promoted MR during cooking [86]. Gómez-Ojeda et al. [45] found that the content of CML in flour tortilla and corn tortilla was significantly lower than that of whole wheat bread, possibly because the impact of corn and wheat processing during the tortilla processing on the CML formation was lower than that of the production of other foods based on these grains. Cheng et al. [87] measured CML and CEL content in nine cereal products (chicken cookies, baby biscuits, white lotus paste mooncakes, almond biscuits, cookies, bread, sweetheart pastry, fried breadsticks, and instant noodles). The content of CML and CEL in the bread and biscuits category was significantly higher than that in sweetheart pastry and instant noodles. This may be a result of the high processing temperature and low moisture content of the bread and biscuits, which facilitated the MR and formation of AGEs [87]. In addition, CML and CEL were the highest in baby biscuits. Thus, attention should be paid to controlling the content of AGEs during the processing of baby food.

4.1.2. Ingredients

In order to enrich the taste and improve the flavor of cereal products, various kinds of ingredients are used frequently. The effects of fructose, glucose, honey, and bananas on the number of AGEs in cookies have been explored by Treibmann et al. [88]. Under the same conditions, there was no significant difference in the formation of CML when glucose and fructose were added, although the highest content of MG-H1 and CEL was observed in the biscuits with the highest fructose content, possibly because fructose was more likely to produce MGO [88]. In addition, the effects of the addition of sucrose, butter, and egg fluid on the formation of AGEs in butter cookies were investigated [89]. Sucrose supplementation did not significantly impact free CML and CEL content, which may be a result of the low free lysine content in the biscuits. On the other hand, when the amount of added sucrose was greater than 10 g (25% of the flour weight), both protein-bound CML and protein-bound CEL showed a significant increase with the increase in sucrose addition. The increase in sucrose content promoted the production of α-dicarbonyl compounds (GO, MGO, and 3-DG), leading to the production of more protein-bound AGEs. As for butter, when its addition was greater than 15 g (37.5% of the flour mass), there was a significant increase in the content of protein-bound AGEs. A greater amount of butter contributed to the formation of α-dicarbonyl compounds, which was associated with the formation of short-chain α-dicarbonyl compounds through lipid oxidation reactions, leading to an increase in AGEs bound to proteins. As the addition of whole egg fluid increased, the content of free CML and CEL initially increased and then decreased, suggesting that free CML and CEL were involved in the subsequent reactions as the protein-bound CML and CEL gradually increased. The egg fluid provided sufficient protein and free amino acids for butter biscuit baking. When the amount of amino supplementation was large during cooking, the MR was dominant and may have taken precedence over the caramelization reaction.

4.1.3. Processing Time and Temperature

Processing time and temperature have been shown to have significant influences on the AGE content of cereal products. The effect of baking times (7, 10, and 13 min, at 180 °C) on the AGE content of sweet cookies was investigated [85]. CML and CEL content increased with increasing heating duration, which was linked to a significant increase in the content of the dicarbonyl compounds after baking. The influences of baking time (8–15 min) and temperature (130–180 °C) on the formation of AGEs in butter cookies were investigated by Hu et al. [89]. As cooking temperature and cooking time increased, CML and CEL levels initially increased and then decreased, and changes in protein-bound AGEs lagged behind those in free AGEs. Most α-dicarbonyl compounds (GO, MGO, and 3-DG) positively correlated with baking temperature and baking time. As intermediates, α-dicarbonyl compounds led to the formation of AGEs; however, the pattern of GO and MGO formation was inconsistent with AGEs, indicating that the MR is continuously releasing active dicarbonyl compounds, and that AGEs could be involved in the downstream reaction. Jin et al. [90] studied the changes in the CML content of cookies over a broader range of baking times (1.5–31 min) and temperatures (155–230 °C). CML was found to be more likely to form at a range of relatively elevated temperatures (205–230 °C), and an environment of higher temperature and lower humidity strongly promoted CML formation. CML content was highest in cookies baked at 230 °C for 1.5 min. After an extended cooking period in a bakery, the CML content decreased owing to degradation or further reactions. Deep baking, extended baking times coupled with high baking temperatures, seems to be a potential option for significantly reducing the CML content in cookies but would result in unacceptable organoleptic properties. Çelik et al. [62] reported the content of CML and CEL in bread crust heated for a different amount of times (5, 15, and 30 min) at 200 °C. The concentration of CML in the crust did not change significantly, while the CEL content showed an overall downward trend after heating for more than 5 min. On the other hand, the content of the HMW fractions of AGEs (MW > 10,000 Da) was significantly higher than that of CML and CEL in the bread crust type samples, which was related to the HMW fraction of the melanoidin-containing samples formed at the end of the MR.

4.2. Meat Products

At present, AGE content is reported in fish, pork, beef, and chicken, most of which include CML, CEL, MOLD, MOLA, and MGLA as listed in Table 4. Currently, the content of AGEs in meat products is chiefly determined by GC-MS, ELISA, and HPLC-MS/MS. Studies have shown that the content of AGEs in meat products is mainly related to the heat treatment process, the type of meat and food additives.

Table 4. AGE levels in different meat products.

Influencing Factors	Detection Method	Meat Products	AGE Content	References
Heat treatment process	HPLC-MS/MS	raw pork	CML: 3.0 mg/kg; CEL: 0.9 mg/kg; GALA: 0.9 mg/kg	[12]
		roast pork	CML: 9.2 mg/kg; CEL: 40 mg/kg; GALA: 5 mg/kg; MOLA: 6 mg/kg; GOLA: 1.2 mg/kg; MOLD: 0.5 mg/kg; GOLD: 0.25 mg/kg; MGLA: 0.2 mg/kg	
	HPLC-FLD	meat (fried, boiled and baked)	CML: 1.07–21.84 μg/g	[67]
	GC-MS	fried fish nuggets (180 °C, 4–6 min)	CML: 29.26–59.17 mg/kg	[72]
	UPLC-MS	beef (60–300 °C); (grilled, fried, boiled and baked)	CML: 0.8–13.4 mg/kg	[91]
	UPLC-MS/MS	sausage (70–130 °C; 1–4 h)	CML: 2–6.58μg/g; CEL: 6–16.32 μg/g	[92]

Table 4. *Cont.*

Influencing Factors	Detection Method	Meat Products	AGE Content	References
Type of meat	HPLC-MS/MS	pork tenderloin, hearts, livers and kidneys	CML: 0.41–2.56 mg/kg; CEL: 0.22–2.56 mg/kg; protein-bound CML: 2.53–7.06 mg/kg; protein-bound CEL: 1.60–4.83 mg/kg	[7]
	HPLC-FLD	beef steak, pork top loin, chicken breast, salmon, tilapia	CML: 1.07–2.05 μg/g	[67]
Food additives	UPLC-MS/MS	sausage (0–5 g/100g salt; 0–15 mg/100g $NaNO_2$)	CML: 2.37–6.52 μg/g; CEL: 5.54–6.56 μg/g	[92]
	HPLC-MS/MS	commercially sterilized pork (acetic acid, ethanol, and NaCl)	CML: 2.29–15.35 mg/kg; CEL: 2.75–39.33 mg/kg	[93]

4.2.1. Heat Treatment Process

Common cooking methods for meat include frying, roasting, and stir-frying, accompanied by high-temperature conditions [94,95]. Eggen and Glomb [12] reported that AGEs in raw pork were mainly GO-derived CML. Grilling led to increases in MGO-derived AGEs such as CEL, N^6-lactose lysine, MOLD, and MOLA, while GO-derived AGEs such as CML, GOLD and GOLA had fewer increases. This suggested that the grilling process resulted in a significant rise in MGO quantity. Even though the content of GO in raw meat was approximately 4-fold higher than that of MGO, after grilling, the MGO content was 2-fold higher than GO content. On the contrary, some scholars have found that the level of CML in processed pork was 2–3 times than that of CEL [96], which may be a result of sample type and shape as well as overall processing conditions including temperature and time [12]. Studies have shown that when the internal temperature of roast beef reached 90 °C and 100 °C, the content of fluorescent compounds increased significantly. Nevertheless, the formation of CML was not detected until the internal temperature reached 300 °C [91]. When the effects of temperature and time on the formation of CML and CEL during the thermal processing of sausage were investigated [92], their levels did not change at 70 °C and 90 °C, but rose rapidly at 110 °C, and increased about two times at 130 °C. The accumulation of α-dicarbonyl compounds was thought to accelerate CML and CEL formation. The rapid increase in α-dicarbonyl content at elevated temperatures might result from the combined effect of the MR and lipid oxidation. At the same time, a high-temperature caused the peptide bond to break, expanding the protein structure and exposing active amino acids, which favored CML and CEL production [92]. Even at the same temperature, different heat treatments would also lead to varying levels of AGEs in meat products. More CML was produced by grilling and frying than baking [67]. Furthermore, the CML content in the outer layer of the fried meat was significantly greater than that in the middle layer, which may be related to more water-soluble precursors transferred to the meat surface during the firing process. In addition, the frequency of turning during frying also affected CML formation in meat products. The CML content of the single-turn meat was significantly higher than that of the multiple-turn meat when the internal temperature of the samples was the same. This was possibly due to the accelerated loss of water-soluble precursors by multiple turning.

4.2.2. Type of Meat

CML was reported to form in beef, pork, chicken, and fish during grilling [67]. Although no significant differences between beef, pork, and chicken were found, grilled salmon and tilapia contained less CML than the other muscle samples. Salmon and tilapia had higher moisture content and pH and lower lipid and protein content than beef, pork, and chicken, which may contribute to some extent to differences in CML levels reported. Niu et al. [7] investigated the changes in CML and CEL levels in 10 commercial brands of pork tenderloin, heart, liver, and kidney before and after the heat treatment. The mean values of free CML and CEL in the macroscopic kidneys were higher than those in the other organs, which could be explained by the fact that the kidneys are the primary excretory organs of vertebrates, making them important sites for the clearance and metabolism of AGEs. For example, in the case of free AGEs, AGEs bound to small peptides or amino acids could be rapidly filtered and cleared by the kidneys, leading to significantly higher levels of free AGEs in the kidney compared with other organs. In contrast, the content of heat-induced AGEs bound to proteins in the liver and kidney was lower than that in the spine and heart, which may be attributed to the greater accumulation of antioxidants in the kidneys and in the liver inhibiting or slowing the formation of heat-induced AGEs. At the same time, more pro-oxidants, such as metal ions, may accumulate in the heart and promote heat-induced AGE formation.

4.2.3. Ingredients and Additives

The effect of salt on the formation of AGEs in meat is complicated. Some investigators have proposed that adding salt could make muscle fibers expand with water, improving the water-locking ability of meat, making it less prone to water loss during heating, thereby inhibiting the generation of both CML and CEL [92]. Others have suggested that salt may accelerate lipid oxidation in pork and beef, promote iron ion release, and have pro-oxidative activity [97]. Meanwhile, high concentrations of salt made the lysine residue ε-NH_2 on the protein surface more vulnerable to attack, resulting in an increase in protein carbonyl content [98]. Li et al. [63] found that the CML and CEL content in roast beef patties increased with the addition of salt and polyphosphate (PP) caused by the rising surface temperature of the patties. The addition of polyphosphate could improve the hardness of the roast beef patties by forming a hard shell on the surface of the patty during baking, obstructing the evaporation of surface water and making the surface temperature rise rapidly.

Lu et al. [92] explored the effects of nitrite on AGEs in sausages and found that nitrite supplementation had a strong inhibitory effect on CML production in the samples. One possible reason for this is that nitrite can inhibit protein oxidation and lipid oxidation in meat, thus reducing AGE production. In addition, nitrite potentially forms a large amount of NO and NO_2 in sausages, which is more readily involved in oxidation reactions than lipid and protein. Furthermore, nitrite was able to form stable complexes with iron ions, reduce heme iron release, and then inhibit free iron oxidative catalysis [97].

The effect of acetic acid on the production of CML and CEL in commercially sterilized pork was investigated [93]. In pork heated to 121 °C for 10 min, the addition of 0.5% acetic acid was able to promote CML formation but slowed the formation of CML and CEL in pork heated for 30 min relative to the sample in the absence of acetic acid. The authors interpreted this as the phenomenon by which low pH could also enhance the Amadori rearrangement and the degradation of Amadori rearrangement products catalyzed by acid. On the other hand, the decrease in CML and CEL content may be due to decreased nucleophilicity of the amino groups under acidic conditions, thereby reducing the rate of reaction of the initial step in the MR between the amino group of the amino acid or protein and the carbonyl group of the reducing sugar.

4.3. Dairy Products

The dairy products reported in recent years have mostly included milk powder and liquid milk. Among them, AGEs chiefly include CML, CEL, pyrraline, MG-Hs, glyoxal-hydroimidazolone (GO-Hs), GOLD, MOLD, GOLA, and GALA, which were determined by various methods as shown in Table 5. Studies have shown that the content of AGEs in dairy products is mainly related to the type, sterilization process, and storage conditions of dairy products.

Table 5. AGE levels in different dairy products.

Influencing Factors	Detection Method	Dairy Products	AGE Content	References
Sterilization process	LC-MS	pasteurized milk and sterilized milk	free CML: 8.23–14.02 μg/kg; protein-bound CML: 2.58–5.56 mg/kg	[4]
	LC-MS/MS	whole milk powder (low-temperature long-time (LTLT), high-temperature short-time pasteurization (HTST)) and high-heat process (ultra-pasteurization (ESL), ultra-high-temperature (UHT) treatments, and in-bottle sterilization (BS))	CML: 20–40 mg/kg; CEL: 5–15 mg/kg; pyrraline: 0.15–0.20 mg/kg	[6]
	LC-MS/MS	milk (low/high-temperature)	CML: 2–40 nmol/mg; CEL: 0.2–1.1 nmol/mg; pentosidine: 16–21 pmol/mg	[99]
	UPLC-MS/MS	milk (LTLT, HTST, UHT and BS)	CML: 2–4 mg/kg milk; CEL: 0.4–0.8 mg/kg milk	[100]
	UPLC-MS/MS	milk powder and reconstituted milk (LT, HT, ESL, UHT, CS)	CML: 3–10 mg/kg; CEL: 0.75–2.75 mg/kg	[101]
	UPLC-MS/MS	milk powder	free CML: 0.36–5.22 mg/kg; free CEL: 0.12–1.80 mg/kg; free pyrraline: 0–0.37 mg/kg; protein-bound CML: 21.07–128.34 mg/kg; protein-bound CEL: 13.78–56.62 mg/kg; protein-bound pyrraline: 2.85–81.48 mg/kg	[102]
Storage conditions	LC-MS/MS	whole milk powder (0–18 months)	CML: 20–150 mg/kg; CEL: 5–55 mg/kg; pyrraline: 0.15–0.55 mg/kg	[6]
	ELISA	infant formulas (25 °C, 37 °C; 0–28 days)	CML: 500–600 ng/mL	[40]
	UPLC-MS/MS; UPLC-DAD	UHT milk (22 °C; 12 months)	CML: 25–50 μM; CEL: 2.5–13 μM; G-H1/3: 22–35 μM; G-H2: 4–8 μM; MG-H1/3: 15–35 μM; MG-H2: 0–0.6 μM	[47]
	ELISA	infant formula (7–28 d; 32% RH, 57% RH, 75% RH; 25 °C, 37 °C)	CML: 550–600 ng/mL	[103]
	LC-MS	UHT milk (6 months; 20 °C)	CML: 1.794 ng/mL; CEL: 7.29 ng/mL; G-H1/3: 1.63 ng/mL; G-H2: 2.49 ng/mL; MG-H1/3: 3.82 ng/mL; GOLD: 32.8 ng/mL	[104]
	UPLC-MS; HPLC-UV	infant formulas (65 °C; 48 days)	CML: 280 mg/kg; CEL: 180 mg/kg; MG-H: 190 mg/kg; GO-H: 280 mg/kg; GOLD: 40 mg/kg; MOLD: 20 mg/kg; pyrraline: 60 mg/kg	[105]
	UHPLC-MS/MS	whey proteins (35 °C; 7, 14, 21 days)	CML: 25–150 μM; CEL: 1–2 μM; MG-H1: 4–15 μM	[106]

Table 5. *Cont.*

Influencing Factors	Detection Method	Dairy Products	AGE Content	References
Types of dairy products	LC-MS	liquid milk, milk powder, condensed milk, milk fats, cheese, ice cream, and whey protein	free CML: 0.11–32.71 μg/kg; protein-bound CML: 0.01–134.28 mg/kg	[4]
	UPLC-MS/MS; HPLC-PDA	baked milk and baked yogurt	CML: 0.2–1.7 mg/kg; CEL: 0.06–0.1 mg/kg	[46]

4.3.1. Type of Dairy Products

Different dairy products contain different amounts of AGEs. Xiao et al. [4] determined the content of free and protein-bound CML in seven types of dairy products (28 liquid milk, 26 milk powder, 6 condensed milk, 4 milk fat, 9 cheese, 6 ice cream, and 3 whey proteins). Free CML in liquid milk was lower than protein-bound CML, and correlation analysis showed that protein in liquid milk did not correlate with CML. The concentration of CML in unsweetened condensed milk was higher than that of sweetened condensed milk, which may be related to the processing of condensed milk. Free CML and protein-bound CML concentrations were highest in whey protein samples, possibly owing to the abundance of protein in powdered protein drinks, which made CML form more readily. There was no significant difference in the CML content between natural cheese and processed cheese (several natural cheeses, melted and processed). In addition, CML was found in ice cream (low-temperature production), and the authors believed that the CML might have come from the raw material of the ice cream.

4.3.2. Sterilization Methods

There are different opinions regarding the effect of sterilization on the content of AGEs in liquid milk. Xiao et al. [4] measured the CML content of 28 liquid milk products (9 pasteurized milk, 18 sterilized milk) commercially available in China. The average contents of free and protein-bound CML in the five pasteurized milks were higher than those of all the sterilized milk, which may be related to the longer time of pasteurization treatment than that of sterilization. Some studies determined that high-temperature treatment formed more AGEs. Li et al. [100] quantitated the CML and CEL pasteurization-treated milk (low-temperature long-time (LTLT), high-temperature short-time (HTST)) and sterilization-milk (ultra-high-temperature (UHT), in-bottle sterilization (BS)) that are applicable in China. CML and CEL levels in UHT and BS milk were significantly greater than those in LTLT and HTST milk samples. It was believed that temperature played a more important role in the formation of CML and CEL than the duration of heat treatment.

The increase in CML, CEL, and pyrraline in high-temperature-treated whole milk powder was greater than that in low-temperature-treated samples. It was supposed that the high-temperature preheating of raw milk further promoted the MR, resulting in more AGEs being formed. In addition, the changes in CML and CEL in whole milk powder and milk reconstituted under different preheating conditions during milk powder processing (low-temperature pasteurization (LT), high-temperature pasteurization, ESL, UHT, and in-container sterilization (CS)) were studied by Li et al. [101]. The CML and CEL contents of the ESL, UHT, and CS samples were higher than that of the milk powder samples at low preheating levels. This result can be explained by the fact that preheating at low temperatures (<100 °C) causes a dynamic equilibrium between furfural and AGEs, leading both to accumulate slowly. High-temperature (>100 °C) preheating treatments would promote the loss of 5-methyl-furfural with active-reaction properties, leading to the generation of either aldehyde and ketone compounds or AGEs with significantly increased CML and CEL contents. Furthermore, the greatest increase in CML and CEL content was observed in CS preheated milk powder and reconstituted milk. It was found that as the temperatures of heat treatment increased, the rate of AGE formation increased.

4.3.3. Storage Conditions

Dairy storage conditions also influence AGEs content, and adverse storage conditions further promote the occurrence of the MR. The reactivity between the sugar and the amine moiety increases with the temperature increases. As the reactivity between the sugar and the amine moiety increases, the MR reaction becomes more intense, leading to an increase in CML concentration in infant formula [40]. The change in the CEL content was similar to the trends mentioned previously. MR products in milk stored at 20 °C, 30 °C, and 40 °C for 1 year were determined by Zhang et al. [103], including CML, CEL, MG-H1/3, G-H1/3, and pyrraline. The concentrations of CML and CEL increased with increasing storage temperatures (40 °C). At the end of storage, however, CML and CEL concentrations in milk stored at 30 °C and 40 °C decreased compared to pre-storage levels, which may be because of the further conversion of CML and CEL to other MR products. Pyrraline was more easily formed at 40 °C, and its concentration continuously increased during storage at all storage temperatures, which correlated positively with the loss of the α-dicarbonyl components. At the 40 °C storage condition, MG-H1/H3 had a content that was 2–5 times greater than that of G-H1/H3, indicating that MGO could react more readily with Arg residues than GO [103]. Humidity (52%) was also found to facilitate the formation of CML in whole milk powder over a 200-day analysis period [40]. The storage time had a significant effect on CML and CEL content as well [6]. For whole milk powder, the CML content slowly increased by 59–94% in the first 12 months of storage, and the rate of formation accelerated from the twelfth month. Compared to month 12, the CML content at month 18 increased by 0.9–1.5 times, possibly owing to more α-dicarbonyl compounds formed by the accelerated lipid oxidation in later storage [6,40].

4.4. Coffee, Cocoa and Tea Products

Cocoa, coffee, and tea are three major non-alcoholic beverages enjoyed worldwide in countries with different cultural backgrounds. In these beverages, the AGEs recently studied are CML and CEL, determined by LC-MS/MS.

4.4.1. Coffee Products

Coffee is usually processed by roasting. Liu et al. [107] investigated the effects of roasting temperature (230, 235, 240, and 245 °C) and time (10–18 min) on the CML content of green coffee beans. The CML content of coffee during roasting slowly rose within the first 10 min, then dropped sharply in the next 2 min, then increased again. Baking at 235 °C for 12 min resulted in the lowest CML content. Loaec et al. [108] measured CML in 24 commercial coffee substitutes and 12 instant coffees. CML content varied between 0.17 and 47 mg/kg and increased in proportion to the protein content of the sample used in this study. The CML content in the instant coffee samples was significantly higher than that in the commercial coffee substitute samples because of the relatively high protein content in the instant coffee.

4.4.2. Cocoa Products

Taş and Gokmen [109] conducted a study on the effects of alkali treatment and baking conditions on the CML content of cocoa. The concentration of α-dicarbonyl compounds in the alkali-treated cocoa was higher than in the water-immersed cocoa and untreated cocoa. During the subsequent roasting process, the CML concentration of the alkali-treated cocoa greatly increased to 125 μg/kg of defatted cocoa, almost 2-fold higher than that of the water-immersed and untreated cocoa samples. Under the same pre-treatment, after baking for 30 and 60 min at 135 °C and 150 °C, the concentration of CML in the cocoa samples hardly changed.

4.4.3. Tea Products

Jiao et al. [110] determined the CML and CEL content in 99 tea samples (44 green tea, 7 oolong tea, 41 black tea, and 7 dark tea) from different origins. Black tea and dark tea had higher CML and CEL content, whereas green tea and oolong tea had lower CML and CEL content. The results suggested that the processing of tea significantly influences the CML and CEL content. During the withering process of black tea, the hydrolytic activity of proteases, amylase, and invertases increased with the loss of water, and then amino acids and simple sugars increased through enzymatic reactions. As a result, the levels of active precursors for CML and CEL increased, leading to more production of CML and CEL. For dark tea, the thermal effect of pile fermentation and heat treatment resulted in a high content of CML and CEL. On the contrary, green tea had less CML and CEL because of the low degree of processing involved. Although the CML and CEL content of oolong was relatively low, the degree of withering and fermentation of oolong tea was positively correlated with them. Additionally, the catechin content in green and oolong tea was higher than that of black and dark tea, and catechins were verified to trap GO and MGO.

5. Conclusions and Perspectives

In the past few years, there has been substantial progress in research involving AGE detection methods. Instrumental analysis methods have matured, and commercial ELISA kits are more widely used. However, some challenges remain. First, there are various AGE determination methods, but the results obtained by different methods are difficult to compare. Second, the kinds of AGEs measured by ELISA kits are limited, most of which are used to measure CML, though this method is convenient and fast. Finally, changes in processing methods can appropriately reduce the levels of AGEs, but their effects are limited, considering the flavor of food.

Research on AGEs is crucial for human health, so a rapid and accurate detection method of AGEs in food is necessary. The production of AGEs during food processing, transportation, and storage can be better controlled based on the timely data of AGEs. In the meantime, a corresponding database of AGEs in food can be established according to the dietary needs of different groups, and corresponding dietary recommendations can be given to improve human health.

Author Contributions: Conceptualization, L.L. and Y.J.; resources, L.L., Y.Z. and X.Z.; writing—original draft preparation, L.L.; writing—review and editing, M.C., B.C., Y.J. and Y.C.; funding acquisition, Y.J. and Y.C. All authors have read and agreed to the published version of the manuscript.

Funding: This paper was supported by the National Natural Science Foundation of China (32102114), the Natural Science Foundation of Hunan Province (2021JJ40566), the Postgraduate Scientific Research Innovation Project of Hunan Province (CX20210817), and the Taishan Industry Leading Talents Project (LJNY202004).

Data Availability Statement: Not applicable.

Conflicts of Interest: The authors declare no conflict of interest.

References

1. Poulsen, M.W.; Hedegaard, R.V.; Andersen, J.M.; de Courten, B.; Bugel, S.; Nielsen, J.; Skibsted, L.H.; Dragsted, L.O. Advanced glycation endproducts in food and their effects on health. *Food Chem. Toxicol.* **2013**, *60*, 10–37. [CrossRef] [PubMed]
2. Nesreen, A.; Franck, C. Impact of Maillard reaction products on nutrition and health: Current knowledge and need to understand their fate in the human digestive system. *Crit. Rev. Food Sci. Nutr.* **2019**, *59*, 474–487. [CrossRef]
3. Singh, R.; Barden, A.; Mori, T.; Beilin, L. Advanced glycation end-products: A review. *Diabetologia* **2001**, *44*, 129–146. [CrossRef] [PubMed]
4. Xiao, S.-S.; Shi, L.; Wang, P.-C.; Liu, X.; Fang, M.; Wu, Y.-N.; Gong, Z.-Y. Determination of Nε-(carboxymethyl)lysine in commercial dairy products in China with liquid chromatography tandem mass spectroscopy. *J. Food Meas. Charact.* **2022**, *16*, 714–721. [CrossRef]
5. Prestel, S.; de Falco, B.; Blidi, S.; Fiore, A.; Sturrock, K. Evaluation of the effect of berry extracts on carboxymethyllysine and lysine in ultra-high temperature treated milk. *Food Res. Int.* **2020**, *130*, 108923. [CrossRef] [PubMed]
6. Li, Y.; Quan, W.; Jia, X.; He, Z.; Wang, Z.; Zeng, M.; Chen, J. Profiles of initial, intermediate, and advanced stages of harmful Maillard reaction products in whole-milk powders pre-treated with different heat loads during 18 months of storage. *Food Chem.* **2021**, *351*, 129361. [CrossRef] [PubMed]

7. Niu, L.; Yu, H.; Zhang, L.; Zhao, Q.; Lai, K.; Liu, Y.; Huang, Y. Advanced glycation end-products in raw and commercially sterilized pork tenderloin and offal. *J. Food Compos. Anal.* **2022**, *112*, 104681. [CrossRef]
8. Wei, Q.; Liu, T.; Sun, D.-W. Advanced glycation end-products (AGEs) in foods and their detecting techniques and methods: A review. *Trends Food Sci. Technol.* **2018**, *82*, 32–45. [CrossRef]
9. Luevano-Contreras, C.; Chapman-Novakofski, K. Dietary advanced glycation end products and aging. *Nutrients* **2010**, *2*, 1247–1265. [CrossRef]
10. Lund, M.N.; Ray, C.A. Control of Maillard Reactions in Foods: Strategies and Chemical Mechanisms. *J. Agric. Food Chem.* **2017**, *65*, 4537–4552. [CrossRef]
11. Wu, C.H.; Huang, S.M.; Lin, J.A.; Yen, G.C. Inhibition of advanced glycation endproduct formation by foodstuffs. *Food Funct.* **2011**, *2*, 224–234. [CrossRef] [PubMed]
12. Eggen, M.D.; Glomb, M.A. Analysis of Glyoxal- and Methylglyoxal-Derived Advanced Glycation End Products during Grilling of Porcine Meat. *J. Agric. Food Chem.* **2021**, *69*, 15374–15383. [CrossRef] [PubMed]
13. Loidl-Stahlhofen, A.; Spiteller, G. α-Hydroxyaldehydes, products of lipid peroxidation. Biochimical et. *Biophys. Acta* **1994**, *1211*, 156–160. [CrossRef] [PubMed]
14. Mlakar, A.; Spiteller, G. Reinvestigation of lipid peroxidation of linolenic acid. *Biochim. Biophys. Acta* **1994**, *1214*, 209–220. [CrossRef]
15. Niyati-Shirkhodaee, F.; Shibamoto, T. Gas Chromatographic Analysis of Glyoxal and Methylglyoxal Formed from Lipids and Related Compounds upon Ultraviolet Irradiation. *J. Agric. Food Chem.* **1993**, *41*, 227–240. [CrossRef]
16. Dum, J.A.; Ahmed, M.U.; Murtiashaw, M.H.; Richardson, J.M.; Walla, M.D.; Thorpe, S.R.; Baynes, J.W. Reaction of Ascorbate with Lysine and Protein under Autoxidizing Conditions: Formation of NE-(Carboxymethy1)lysine by Reaction between Lysine and Products of Autoxidation of Ascorbate. *Biochemistry* **1990**, *29*, 10964–10970. [CrossRef]
17. Wells-Knecht, K.J.; Zyzak, D.V.; Litchfield, J.E.; Thorpe, S.R.; Baynes, J.W. Mechanism of Autoxidative Glycosylation: Identification of Glyoxal and Arabinose as Intermediates in the Autoxidative Modification of Proteins by Glucose. *Biochemistry* **1995**, *34*, 3702–3709. [CrossRef]
18. Fu, M.; Requena, J.R.; Jenkins, A.J.; Lyons, T.J.; Baynes, J.W.; Thorpe, S.R. The Advanced Glycation End Product, N-(Carboxymethyl)lysine, Is a Product of both Lipid Peroxidation and Glycoxidation Reactions. *J. Biol. Chem.* **1996**, *271*, 9982–9986. [CrossRef]
19. Kasper, M.; Schieberle, P. Labeling studies on the formation pathway of Nepsilon-carboxymethyllysine in maillard-type reactions. *Ann. N. Y. Acad. Sci.* **2005**, *1043*, 59–62. [CrossRef]
20. Hayashi, T.; Namiki, M. Role of Sugar Fragmentation in an Early Stage Browning of Amino-carbonyl Reaction of Sugar with Amino Acid. *Agric. Biol. Chem.* **2014**, *50*, 1965–1970. [CrossRef]
21. Nguyen, H.T.; van der Fels-Klerx, H.J.; van Boekel, M.A.J.S. Nε-(carboxymethyl)lysine: A Review on Analytical Methods, Formation, and Occurrence in Processed Food, and Health Impact. *Food Rev. Int.* **2013**, *30*, 36–52. [CrossRef]
22. Zhao, X.; Zhang, X.; Ye, B.; Yan, H.; Zhao, Y.; Liu, L. Effect of unsaturated fatty acids on glycation product formation pathways (I) the role of oleic acid. *Food Res. Int.* **2020**, *136*, 109560. [CrossRef] [PubMed]
23. Ahmed, M.U.; Frye, E.B.; Degenhardt, T.P.; Thorpe, S.R.; Baynes, J.W. Nε-(Carboxyethyl)lysine, a product of the chemical modification of proteins by methylglyoxal, increases with age in human lens proteins. *Biochem. J.* **1997**, *324*, 565–570. [CrossRef]
24. Thorpe, S.R.; Baynes, J.W. CML: A brief history. *Int. Congr. Ser.* **2002**, *1245*, 91–99. [CrossRef]
25. Ahmed, M.U.; Thorpe, S.R.; Baynes, J.W. Identification of N epsilon-carboxymethyllysine as a degradation product of fructoselysine in glycated protein. *J. Biol. Chem.* **1986**, *261*, 4889–4894. [CrossRef] [PubMed]
26. Sell, D.R.; Monnier, V.M. Isolation, Purification and Partial Characterization of Novel Fluorophores from Aging Human Insoluble Collagen-Rich Tissue. *Connect. Tissue Res.* **1989**, *19*, 77–92. [CrossRef]
27. Grandhee, S.K.; Monnier, V.M. Mechanism of formation of the Maillard protein cross-link pentosidine. Glucose, fructose, and ascorbate as pentosidine precursors. *J. Biol. Chem.* **1991**, *266*, 11649–11653. [CrossRef]
28. Dyer, D.G.; Blackledge, J.A.; Thorpe, S.R.; Baynes, J.W. Formation of pentosidine during nonenzymatic browning of proteins by glucose. Identification of glucose and other carbohydrates as possible precursors of pentosidine in vivo. *J. Biol. Chem.* **1991**, *266*, 11654–11660. [CrossRef]
29. Hayase, F.; Nagaraj, R.H.; Miyata, S.; Njoroge, F.G.; Monnier, V.M. Aging of Proteins: Immunological Detection of a Glucose-derived Pyrrole Formed during Maillard Reaction in vivo. *J. Biol. Chem.* **1989**, *264*, 3758–3764. [CrossRef]
30. Yu, H.; Zhong, Q.; Xie, Y.; Guo, Y.; Cheng, Y.; Yao, W. Kinetic study on the generation of furosine and pyrraline in a Maillard reaction model system of d-glucose and l-lysine. *Food Chem.* **2020**, *317*, 126458. [CrossRef]
31. Gao, J.; Sun, Y.; Li, L.; Zhou, Q.; Wang, M. The antiglycative effect of apple flowers in fructose/glucose-BSA models and cookies. *Food Chem.* **2020**, *330*, 127170. [CrossRef] [PubMed]
32. Lee, J.S.; Chung, Y.S.; Chang, S.Y.; Jung, Y.S.; Kim, S.H. Simple Quantification of Pentosidine in Human Urine and Plasma by High-Performance Liquid Chromatography. *Int. J. Anal. Chem.* **2017**, *2017*, 1389807. [CrossRef]
33. Troise, A.D.; Colantuono, A.; Fiore, A. Spray-dried olive mill wastewater reduces Maillard reaction in cookies model system. *Food Chem.* **2020**, *323*, 126793. [CrossRef] [PubMed]
34. Wang, S.; Zheng, L.; Zheng, X.; Yang, Y.; Xiao, D.; Zhang, H.; Ai, B.; Sheng, Z. Chitosan inhibits advanced glycation end products formation in chemical models and bakery food. *Food Hydrocoll.* **2022**, *128*, 107600. [CrossRef]

35. Qin, R.K.; Wu, R.L.; Shi, H.A.; Jia, C.H.; Rong, J.H.; Liu, R. Formation of AGEs in fish cakes during air frying and other traditional heating methods. *Food Chem.* **2022**, *391*, 133213. [CrossRef]
36. Li, Y.; Li, L.; Lund, M.N.; Li, B.; Hu, Y.; Zhang, X. Reduction of Nepsilon-(carboxymethyl) lysine by (-)-epicatechin and (-)-epigallocatechin gallate: The involvement of a possible trapping mechanism by catechin quinones. *Food Chem.* **2018**, *266*, 427–434. [CrossRef]
37. Wu, R.L.; Jiang, Y.; Qin, R.K.; Shi, H.N.; Jia, C.H.; Rong, J.H.; Liu, R. Study of the formation of food hazard factors in fried fish nuggets. *Food Chem.* **2022**, *373*, 131562. [CrossRef] [PubMed]
38. Liu, J.; Yang, Z.; Hao, Y.; Wang, Z.; Han, L.; Li, M.; Zhang, N.; Chen, H.; Liu, Y.; Li, H.; et al. Effect of alkylresorcinols on the formation of Nepsilon-(carboxymethyl)lysine and sensory profile of wheat bread. *Food Sci. Nutr.* **2021**, *9*, 489–498. [CrossRef]
39. Asif, A.; Zeeshan, N.; Mehmood, S. Antioxidant and antiglycation activities of traditional plants and identification of bioactive compounds from extracts of Hordeum vulgare by LC-MS and GC-MS. *J. Food Biochem.* **2020**, *44*, e13381. [CrossRef]
40. Wang, M.Q.; Yuan, X.W.; Zheng, Y.; Wu, Z.Y.; Li, H.B.; Li, H.J.; Yu, J.H. Maillard reaction indicators formation, changes and possible intake in infant formula produced by different thermal treatments during domestic use. *Food Chem.* **2022**, *395*, 133576. [CrossRef]
41. Li, H.B.; Zhang, Y.Y.; Li, C.S.; Xu, S.Y.; Song, W.B.; Li, H.J.; Yu, J.H. Content and evolution of Maillard reaction products in commercial brown fermented milk during storage. *Int. Dairy J.* **2022**, *129*, 105343. [CrossRef]
42. Huang, S.H.; Dong, X.L.; Zhang, Y.L.; Chen, Y.R.; Yu, Y.J.; Huang, M.; Zheng, Y.D. Formation of advanced glycation end products in raw and subsequently boiled broiler muscle: Biological variation and effects of postmortem ageing and storage. *Food Sci. Hum. Wellness* **2022**, *11*, 255–262. [CrossRef]
43. Fang, R.; Zhu, Z.S.; Bassey, A.P.; Khan, I.A.; Huang, M. Glyoxal induced advanced glycation end products formation in chicken meat emulsion instead of oxidation. *Food Sci. Hum. Wellness* **2022**, *11*, 828–836. [CrossRef]
44. Prosser, C.G.; Carpenter, E.A.; Hodgkinson, A.J. N(epsilon)-carboxymethyllysine in nutritional milk formulas for infants. *Food Chem.* **2019**, *274*, 886–890. [CrossRef] [PubMed]
45. Gomez-Ojeda, A.; Jaramillo-Ortiz, S.; Wrobel, K.; Wrobel, K.; Barbosa-Sabanero, G.; Luevano-Contreras, C.; de la Maza, M.P.; Uribarri, J.; Del Castillo, M.D.; Garay-Sevilla, M.E. Comparative evaluation of three different ELISA assays and HPLC-ESI-ITMS/MS for the analysis of N(epsilon)-carboxymethyl lysine in food samples. *Food Chem.* **2018**, *243*, 11–18. [CrossRef]
46. Yu, H.; Zhong, Q.L.; Guo, Y.H.; Xie, Y.F.; Cheng, Y.L.; Yao, W.R. Potential of resveratrol in mitigating advanced glycation end-products formed in baked milk and baked yogurt. *Food Res. Int.* **2020**, *133*, 109191. [CrossRef]
47. Poojary, M.M.; Zhang, W.; Olesen, S.B.; Rauh, V.; Lund, M.N. Green Tea Extract Decreases Arg-Derived Advanced Glycation Endproducts but Not Lys-Derived AGEs in UHT Milk during 1-Year Storage. *J. Agric. Food Chem.* **2020**, *68*, 14261–14273. [CrossRef]
48. Peake, B.; Ghetia, M.; Gerber, C.; Costabile, M.; Deo, P. Role of saturated and unsaturated fatty acids on dicarbonyl-albumin derived advanced glycation end products in vitro. *Amino Acids* **2022**, *54*, 721–732. [CrossRef]
49. Nomi, Y.; Sato, T.; Mori, Y.; Matsumoto, H. Evaluation of Fructo-, Inulin-, and Galacto-Oligosaccharides on the Maillard Reaction Products in Model Systems with Whey Protein. *J. Agric. Food Chem.* **2022**, *70*, 9154–9165. [CrossRef]
50. Zhang, W.; Zhang, B.; Ye, Y.; Zhu, H. Methylglyoxal-hydroimidazolones (MG-Hs) instead of Nvarepsilon-(carboxymethyl)-l-lysine (CML) is the major advanced glycation end-product during drying process in black tea. *Food Chem.* **2020**, *333*, 127499. [CrossRef]
51. Zhang, J.J.; Liu, Z.J.; Huang, Y.Q.; Lai, K.Q.; Lin, H.; Liu, Y.L.; Wang, F.X. Effects of sodium bicarbonate and sodium phosphates on the formation of advanced glycation end-products in minced pork during cold storage. *J. Food Meas. Charact.* **2022**, *16*, 4425–4432. [CrossRef]
52. Shen, Z.W.; Li, S.C.; Wu, J.H.; Wang, F.X.; Li, X.H.; Yu, J.; Liu, Y.L.; Ma, X.Y. Effect of different oil incorporation on gelling properties, flavor and advanced glycation end-products of silver carp surimi sausages. *J. Food Meas. Charact.* **2022**, *16*, 5007–5022. [CrossRef]
53. Sen, D.; Gokmen, V. Kinetic modeling of Maillard and caramelization reactions in sucrose-rich and low moisture foods applied for roasted nuts and seeds. *Food Chem.* **2022**, *395*, 133583. [CrossRef] [PubMed]
54. Manig, F.; Hellwig, M.; Pietz, F.; Henle, T. Studies about the Dietary Impact on "Free" Glycation Compounds in Human Saliva. *Foods* **2022**, *11*, 2112. [CrossRef]
55. Liu, L.C.; Liu, L.; Xie, J.H.; Shen, M.Y. Formation mechanism of AGEs in Maillard reaction model systems containing ascorbic acid. *Food Chem.* **2022**, *378*, 132108. [CrossRef]
56. Jost, T.; Henning, C.; Heymann, T.; Glomb, M.A. Comprehensive Analyses of Carbohydrates, 1,2-Dicarbonyl Compounds, and Advanced Glycation End Products in Industrial Bread Making. *J. Agric. Food Chem.* **2021**, *69*, 3720–3731. [CrossRef]
57. Marousez, L.; Sprenger, N.; De Lamballerie, M.; Jaramillo-Ortiz, S.; Tran, L.; Micours, E.; Gottrand, F.; Howsam, M.; Tessier, F.J.; Ley, D.; et al. High hydrostatic pressure processing of human milk preserves milk oligosaccharides and avoids formation of Maillard reaction products. *Clin Nutr.* **2022**, *41*, 1–8. [CrossRef]
58. Liu, Q.Y.; Wang, S.; Wang, X.Y.; Dong, S.Y.; Zhao, Y.H.; Zeng, M.Y. The relationship between the formation of advanced glycation end products and quality attributes of fried sturgeon fillets. *LWT-Food Sci. Technol.* **2022**, *159*. [CrossRef]
59. Li, M.Y.; Lu, J.N.; Huang, Y.S.; Wang, W.J.; Xie, J.H.; Xie, M.Y.; Shen, M.Y. Quantitative assessment of furosine, furfurals, and advanced glycation end products in different types of commercially available cheeses. *Food Control* **2022**, *136*. [CrossRef]

60. Bai, S.; You, L.Q.; Wang, Y.R.; Luo, R.M. Effect of Traditional Stir-Frying on the Characteristics and Quality of Mutton Sao Zi. *Front Nutr.* **2022**, *9*, 925208. [CrossRef]
61. Bai, S.; You, L.Q.; Ji, C.; Zhang, T.G.; Wang, Y.R.; Geng, D.; Gao, S.; Bi, Y.Z.; Luo, R.M. Formation of volatile flavor compounds, maillard reaction products and potentially hazard substance in China stir-frying beef sao zi. *Food Res. Int.* **2022**, *159*, 111545. [CrossRef]
62. Çelik, E.E.; Gökmen, V. Formation of Maillard reaction products in bread crust-like model system made of different whole cereal flours. *Eur. Food Res. Technol.* **2020**, *246*, 1207–1218. [CrossRef]
63. Li, Y.; Xue, C.; Quan, W.; Qin, F.; Wang, Z.; He, Z.; Zeng, M.; Chen, J. Assessment the influence of salt and polyphosphate on protein oxidation and Nepsilon-(carboxymethyl)lysine and Nepsilon-(carboxyethyl)lysine formation in roasted beef patties. *Meat Sci.* **2021**, *177*, 108489. [CrossRef] [PubMed]
64. Chen, Q.; Li, Y.; Dong, L.; Shi, R.; Wu, Z.; Liu, L.; Zhang, J.; Wu, Z.; Pan, D. Quantitative determination of Nε-(carboxymethyl)lysine in sterilized milk by isotope dilution UPLC-MS/MS method without derivatization and ion pair reagents. *Food Chem.* **2022**, *385*, 132697. [CrossRef] [PubMed]
65. Gong, R.-z.; Wang, Y.-h.; Wang, Y.-f.; Chen, B.; Gao, K.; Sun, Y.-s. Simultaneous Determination of Nε-(carboxymethyl) Lysine and Nε-(carboxyethyl) Lysine in Different Sections of Antler Velvet after Various Processing Methods by UPLC-MS/MS. *Molecules* **2018**, *23*, 3316. [CrossRef]
66. Hellwig, M.; Henle, T. Quantification of the Maillard reaction product 6-(2-formyl-1-pyrrolyl)-l-norleucine (formyline) in food. *Eur. Food Res. Technol.* **2012**, *235*, 99–106. [CrossRef]
67. Chen, G.; Smith, J.S. Determination of advanced glycation endproducts in cooked meat products. *Food Chem.* **2015**, *168*, 190–195. [CrossRef]
68. Drusch, S.; Faist, V.; Erbersdobler, H.F. Determination of NE-carboxymethyllysine in milk products by a modified reversed-phase HPLC method. *Food Chem.* **1999**, *65*, 547–553. [CrossRef]
69. Schettgen, T.; Tings, A.; Brodowsky, C.; Muller-Lux, A.; Musiol, A.; Kraus, T. Simultaneous determination of the advanced glycation end product N (epsilon)-carboxymethyllysine and its precursor, lysine, in exhaled breath condensate using isotope-dilution-hydrophilic-interaction liquid chromatography coupled to tandem mass spectrometry. *Anal. Bioanal. Chem.* **2007**, *387*, 2783–2791. [CrossRef]
70. Ochi, N. Simultaneous determination of eight underivatized biogenic amines in salted mackerel fillet by ion-pair solid-phase extraction and volatile ion-pair reversed-phase liquid chromatography-tandem mass spectrometry. *J. Chromatogr. A* **2019**, *1601*, 115–120. [CrossRef]
71. Sun, X.; Tang, J.; Wang, J.; Rasco, B.A.; Lai, K.; Huang, Y. Formation of free and protein-bound carboxymethyllysine and carboxyethyllysine in meats during commercial sterilization. *Meat Sci.* **2016**, *116*, 1–7. [CrossRef] [PubMed]
72. Jiang, Y.; Qin, R.; Jia, C.; Rong, J.; Hu, Y.; Liu, R. Hydrocolloid effects on Nε-carboxymethyllysine and acrylamide of deep-fried fish nuggets. *Food Biosci.* **2021**, *39*, 100797. [CrossRef]
73. Charissou, A.; Ait-Ameur, L.; Birlouez-Aragon, I. Evaluation of a gas chromatography/mass spectrometry method for the quantification of carboxymethyllysine in food samples. *J. Chromatogr. A* **2007**, *1140*, 189–194. [CrossRef] [PubMed]
74. Malgorzata, W.; Konrad, P.M.; Zielinski, H. Effect of roasting time of buckwheat groats on the formation of Maillard reaction products and antioxidant capacity. *Food Chem.* **2016**, *196*, 355–358. [CrossRef]
75. Ashraf, J.M.; Ahmad, S.; Choi, I.; Ahmad, N.; Farhan, M.; Tatyana, G.; Shahab, U. Recent advances in detection of AGEs: Immunochemical, bioanalytical and biochemical approaches. *IUBMB Life* **2015**, *67*, 897–913. [CrossRef]
76. Reen, D.J. The enzyme-linked immunosorbent assay (ELISA) *. *Bull. World Health Organ.* **1976**, *54*, 129–139.
77. Ikeda, K.; Higashi, T.; Sano, H.; Jinnouchi, Y.; Yoshida, M.; Araki, T.; Ueda, S.; Horiuchi, S. N-(Carboxymethyl)lysine Protein Adduct Is a Major Immunological Epitope in Proteins Modified with Advanced Glycation End Products of the Maillard Reaction. *Biochemistry* **1996**, *35*, 8. [CrossRef]
78. Goldberg, T.; Cai, W.; Peppa, M.; Dardaine, V.; Baliga, B.S.; Uribarri, J.; Vlassara, H. Advanced glycoxidation end products in commonly consumed foods. *J. Am. Diet. Assoc.* **2004**, *104*, 1287–1291. [CrossRef]
79. Uribarri, J.; Woodruff, S.; Goodman, S.; Cai, W.; Chen, X.; Pyzik, R.; Yong, A.; Striker, G.E.; Vlassara, H. Advanced glycation end products in foods and a practical guide to their reduction in the diet. *J. Am. Diet. Assoc.* **2010**, *110*, 911–916. [CrossRef]
80. Tareke, E.; Forslund, A.; Lindh, C.H.; Fahlgren, C.; Östman, E. Isotope dilution ESI-LC-MS/MS for quantification of free and total Ne-(1-Carboxymethyl)-L-Lysine and free Ne-(1-Carboxyethyl)-L-Lysine: Comparison of total Ne-(1-Carboxymethyl)-L-Lysine levels measured with new method to ELISA assay in gruel samples. *Food Chem.* **2013**, *141*, 4253–4259. [CrossRef]
81. Henle, T.; Klostermeyer, H. Determination of protein-bound 2-amino-6-(2-formyl-l-pyrrolyl)-hexanoic acid ("pyrraline") by ion exchange chromatography and photodiode array detection. *Z Leb. Unters* **1993**. [CrossRef]
82. Dutta, U.; Dain, J.A. Capillary electrophoretic analysis of advanced glycation endproducts formed from the reaction of reducing sugars with the amino group of glucosamine. *Anal. Biochem.* **2005**, *343*, 237–243. [CrossRef] [PubMed]
83. Birlouez-Aragon, I.; Locquet, N.; de St Louvent, E.; Bouveresse, D.J.; Stahl, P. Evaluation of the maillard reaction in infant formulas by means of front-face fluorescence. *Ann. N. Y. Acad. Sci.* **2005**, *1043*, 308–318. [CrossRef] [PubMed]
84. Niquet-Leridon, C.; Jacolot, P.; Niamba, C.N.; Grossin, N.; Boulanger, E.; Tessier, F.J. The rehabilitation of raw and brown butters by the measurement of two of the major Maillard products, N(epsilon)-carboxymethyl-lysine and 5-hydroxymethylfurfural, with validated chromatographic methods. *Food Chem.* **2015**, *177*, 361–368. [CrossRef] [PubMed]

85. Zilic, S.; Aktag, I.G.; Dodig, D.; Gokmen, V. Investigations on the formation of Maillard reaction products in sweet cookies made of different cereals. *Food Res. Int.* **2021**, *144*, 110352. [CrossRef]
86. Palermo, M.; Fiore, A.; Fogliano, V. Okara promoted acrylamide and carboxymethyl-lysine formation in bakery products. *J. Agric. Food Chem.* **2012**, *60*, 10141–10146. [CrossRef]
87. Cheng, W.; Wang, X.; Zhang, Z.; Ma, L.; Liu, G.; Wang, Q.; Chen, F.; Cheng, K.W. Development of an Isotope Dilution UHPLC-QqQ-MS/MS-Based Method for Simultaneous Determination of Typical Advanced Glycation End Products and Acrylamide in Baked and Fried Foods. *J. Agric. Food Chem.* **2021**, *69*, 2611–2618. [CrossRef]
88. Treibmann, S.; Hellwig, A.; Hellwig, M.; Henle, T. Lysine-Derived Protein-Bound Heyns Compounds in Bakery Products. *J. Agric. Food Chem.* **2017**, *65*, 10562–10570. [CrossRef]
89. Hu, H.; Wang, Y.; Shen, M.; Huang, Y.; Li, C.; Nie, S.; Xie, M. Effects of baking factors and recipes on the quality of butter cookies and the formation of advanced glycation end products (AGEs) and 5-hydroxymethylfurfural (HMF). *Curr. Res. Food Sci.* **2022**, *5*, 940–948. [CrossRef]
90. Cheng, L.; Jin, C.; Zhang, Y. Investigation of variations in the acrylamide and N(epsilon) -(carboxymethyl) lysine contents in cookies during baking. *J. Food Sci.* **2014**, *79*, T1030–T1038. [CrossRef]
91. Trevisan, A.J.; de Almeida Lima, D.; Sampaio, G.R.; Soares, R.A.; Markowicz Bastos, D.H. Influence of home cooking conditions on Maillard reaction products in beef. *Food Chem.* **2016**, *196*, 161–169. [CrossRef]
92. Lu, J.; Li, M.; Shen, M.; Xie, J.; Xie, M. Advanced Glycation End Products and Nitrosamines in Sausages Influenced by Processing Parameters, Food Additives and Fat during Thermal Processing. *Foods* **2023**, *12*, 394. [CrossRef]
93. Li, L.; Kong, S.; Liu, Y.; Huang, Y.; Li, Y.; Lai, K. Effects of acetic acid, ethanol, and sodium chloride on the formation of Nε-carboxymethyllysine, Nε-carboxyethyllysine and their precursors in commercially sterilized pork. *J. Food Meas. Charact.* **2021**, *15*, 5337–5344. [CrossRef]
94. Treibmann, S.; Spengler, F.; Degen, J.; Lobner, J.; Henle, T. Studies on the Formation of 3-Deoxyglucosone- and Methylglyoxal-Derived Hydroimidazolones of Creatine during Heat Treatment of Meat. *J. Agric. Food Chem.* **2019**, *67*, 5874–5881. [CrossRef]
95. Wang, Z.; Cai, R.; Yang, X.; Gao, Z.; Yuan, Y.; Yue, T. Changes in aroma components and potential Maillard reaction products during the stir-frying of pork slices. *Food Control* **2021**, *123*. [CrossRef]
96. Sun, X.; Li, X.; Tang, J.; Lai, K.; Rasco, B.A.; Huang, Y. Formation of protein-bound N(epsilon)-carboxymethyllysine and N(epsilon)-carboxyethyllysine in ground pork during commercial sterilization as affected by the type and concentration of sugars. *Food Chem.* **2021**, *336*, 127706. [CrossRef] [PubMed]
97. Niu, L.; Sun, X.; Tang, J.; Wang, J.; Wang, J.; Rasco, B.A.; Lai, K.; Fan, Y.; Huang, Y. Combination effects of salts and cold storage on the formation of protein-bound N(varepsilon)-(carboxymethyl)lysine and N(varepsilon)-(carboxyethyl)lysine in raw and subsequently commercially sterilized ground pork. *Food Chem.* **2018**, *264*, 455–461. [CrossRef]
98. Li, F.; Zhuang, H.; Qiao, W.; Zhang, J.; Wang, Y. Effect of partial substitution of NaCl by KCl on physicochemical properties, biogenic amines and N-nitrosamines during ripening and storage of dry-cured bacon. *J. Food Sci. Technol.* **2016**, *53*, 3795–3805. [CrossRef]
99. Zenker, H.E.; Raupbach, J.; Boeren, S.; Wichers, H.J.; Hettinga, K.A. The effect of low vs. high temperature dry heating on solubility and digestibility of cow's milk protein. *Food Hydrocoll.* **2020**, *109*, 106098. [CrossRef]
100. Li, Y.; Wu, Y.; Quan, W.; Jia, X.; He, Z.; Wang, Z.; Adhikari, B.; Chen, J.; Zeng, M. Quantitation of furosine, furfurals, and advanced glycation end products in milk treated with pasteurization and sterilization methods applicable in China. *Food Res. Int.* **2021**, *140*, 110088. [CrossRef]
101. Li, Y.; Jia, X.; Wang, Z.; He, Z.; Zeng, M.; Chen, J. Characterizing changes in Maillard reaction indicators in whole milk powder and reconstituted low-temperature pasteurized milk under different preheating conditions. *J. Food Sci.* **2022**, *87*, 193–205. [CrossRef] [PubMed]
102. Li, C.; Zhang, L.; Gao, W.; Lai, C.; Dong, H. Robust Detection of Advanced Glycation Endproducts in Milk Powder Using Ultrahigh Performance Liquid Chromatography Tandem Mass Spectrometry (UHPLC-MS/MS). *Food Anal. Methods* **2021**, *14*, 1472–1481. [CrossRef]
103. Zhang, W.; Poojary, M.M.; Rauh, V.; Ray, C.A.; Olsen, K.; Lund, M.N. Quantitation of alpha-Dicarbonyls and Advanced Glycation Endproducts in Conventional and Lactose-Hydrolyzed Ultrahigh Temperature Milk during 1 Year of Storage. *J. Agric. Food Chem.* **2019**, *67*, 12863–12874. [CrossRef]
104. Poojary, M.M.; Zhang, W.; Greco, I.; De Gobba, C.; Olsen, K.; Lund, M.N. Liquid chromatography quadrupole-Orbitrap mass spectrometry for the simultaneous analysis of advanced glycation end products and protein-derived cross-links in food and biological matrices. *J. Chromatogr. A* **2020**, *1615*, 460767. [CrossRef] [PubMed]
105. Zhu, R.; Sun, X.; Zhang, Y.; Yang, T.; Wang, C.; Zhang, J.; Duan, Z.; Shang, F.; Fan, J.; Liu, Y.; et al. Effect of pectin oligosaccharides supplementation on infant formulas: The storage stability, formation and intestinal absorption of advanced glycation end products. *Food Chem.* **2022**, *373*, 131571. [CrossRef]
106. Zhang, H.; Troise, A.D.; Zhang, H.; Fogliano, V. Cocoa melanoidins reduce the formation of dietary advanced glycation end-products in dairy mimicking system. *Food Chem.* **2021**, *345*, 128827. [CrossRef] [PubMed]
107. Liu, X.; Chen, X.; Lin, T.; Yin, B.; Li, Q.; Wang, L.; Shao, J.; Yang, J. The level variation of N(epsilon)-(carboxymethyl)lysine is correlated with chlorogenic acids in Arabica L. Coffee beans under different process conditions. *Food Chem.* **2021**, *343*, 128458. [CrossRef]

108. Loaec, G.; Jacolot, P.; Helou, C.; Niquet-Leridon, C.; Tessier, F.J. Acrylamide, 5-hydroxymethylfurfural and N(epsilon)-carboxymethyl-lysine in coffee substitutes and instant coffees. *Food Addit. Contam. Part A Chem. Anal. Control Expo. Risk Assess.* **2014**, *31*, 593–604. [CrossRef]
109. Tas, N.G.; Gokmen, V. Effect of alkalization on the Maillard reaction products formed in cocoa during roasting. *Food Res. Int.* **2016**, *89*, 930–936. [CrossRef]
110. Jiao, Y.; He, J.; Li, F.; Tao, G.; Zhang, S.; Zhang, S.; Qin, F.; Zeng, M.; Chen, J. N(epsilon)-(carboxymethyl)lysine and N(epsilon)-(carboxyethyl)lysine in tea and the factors affecting their formation. *Food Chem.* **2017**, *232*, 683–688. [CrossRef]

foods

MDPI

Article

Indirect Competitive Enzyme-Linked Immunosorbent Assay Based on Broad-Spectrum Antibody for Simultaneous Determination of Thirteen Fluoroquinolone Antibiotics in *Rana catesbeianus*

Biao Zhang †, Yihan Lang †, Bowen Guo, Zhengyang Cao, Jin Cheng, Danfeng Cai, Xuping Shentu * and Xiaoping Yu

Zhejiang Provincial Key Laboratory of Biometrology and Inspection & Quarantine, College of Life Sciences, China Jiliang University, Xueyuan Street, Xiasha Higher Education District, Hangzhou 310018, China; zb@cjlu.edu.cn (B.Z.); langyh@cjlu.edu.cn (Y.L.); p22091055019@cjlu.edu.cn (B.G.); 2200901105@cjlu.edu.cn (Z.C.); 2200901122@cjlu.edu.cn (J.C.); caidanfeng@cjlu.edu.cn (D.C.); yxp@cjlu.edu.cn (X.Y.)
* Correspondence: stxp@cjlu.edu.cn; Tel.: +86-0571-8687-6237
† These authors contributed equally to this work.

Abstract: Fluoroquinolone (FQ) is a type of widely used antibiotic in agriculture and aquaculture, and exposure to low doses of FQs may result in the transfer of resistance between animal and human pathogens. Based on the optimization of the operating parameters, an indirect competitive enzyme-linked immunosorbent assay (ic-ELISA) standard curve was constructed for the simultaneous detection of 13 FQs, including enrofloxacin (ENR), ciprofloxacin (CIP), sarafloxacin (SAR), ofloxacin (OFL), norfloxacin (NOR), pefloxacin mesylate (PM), pefloxacin (PEF), enoxacin (ENX), marbofloxacin (MAR), fleroxacin (FLE), lomefloxacin (LOM), danofloxacin (DAN), and difloxacin (DIF). The limit of detection (LOD, computed as IC_{10}) and sensitivity (IC_{50}) of the ic-ELISA for ENR were 0.59 μg/L and 19.23 μg/L, respectively. The precision and dependability of the detection results of this ic-ELISA were properly verified by HPLC in *Rana catesbeianus* samples. This indicated that the established ic-ELISA approach could be utilized to determine the FQs in *Rana catesbeianus*. In addition, this ic-ELISA, based on a broad-spectrum antibody, provides a technical reference and potential strategy for an immunoassay of hazard factors with similar structure.

Keywords: fluoroquinolones; indirect competitive ELISA; broad-spectrum antibody; *Rana catesbeianus*; multicomponent simultaneous detection

1. Introduction

Fluoroquinolone (FQ) antibiotics are one of the most commonly used veterinary drugs, with a low cost and high broad-spectrum antibacterial activity against Gram-positive bacteria such as *Staphylococcus aureus* and many others [1–3]. However, excessive use breeds bacterial resistance, and FQs' residues may be transported into the human body through the food chain [4–6]. The consumption of animal-derived products that contain excessive amounts of fluoroquinolone for an extended period of time was linked to allergies, hemorrhage, and renal failure, ultimately posing a major risk to human health [7–9]. Therefore, FQs are prohibited or restricted in many countries in food-producing animals due to the development of resistance and the impact on human and animal health. Common fluoroquinolones include ciprofloxacin, norfloxacin, ofloxacin, pefloxacin, etc. [10,11]. *Rana catesbeianus* is a sort of popular edible frog, because of its tasty meat and abundant nutrition, with high protein, low fat, and low cholesterol [12,13]. Large-scale farming of *Rana catesbeianus* has rendered it a significant economic breeding animal in China. However, under intensive high-density culture, *Rana catesbeianus* has a slow growth rate, low feed utilization rate, and weak resistance to bacteria and viruses. *Rana catesbeianus* is prone to bacterial or fungal diseases such as red-leg syndrome and chytridiomycosis [14,15]. On account of the pursuit of economic gains, some breeders utilize an excessive amount of prohibited and restricted veterinary drugs during animal feeding and veterinary treatment, which seriously affects the healthy development of the industry and raises social concerns about food safety [16,17]. It is vital to investigate the efficient detection approach of the FQs' residues in *Rana catesbeianus*, since field testing frequently involves the detection of a large number of samples.

Both domestically and internationally, screening and confirmatory analytical procedures are frequently employed for the detection of veterinary medication residues [18,19]. Due to their remarkable specificity, high accuracy, and outstanding repeatability, liquid chromatography/mass spectrometry (LC/MS) and high-performance liquid chromatography (HPLC) are primarily utilized as confirmatory procedures when analyzing veterinary drug residues in animal-derived foods [20–22]. However, these confirmatory methods require tedious pretreatment steps, expensive instruments, specialized technical personnel, and a long testing cycle, which are inappropriate for point-of-care testing (POCT) [23–26]. In the meantime, due to their rapidity, affordability, mobility, and high sample throughput, screening approaches are increasingly being thought of as alternatives and complementing methods for residue analysis [27,28].

The advantages of the enzyme-linked immunosorbent assay (ELISA) over other screening techniques include its high sensitivity, minimal instrumentation, and straightforward sample pretreatment [29–31]. The kit's short 2-h analysis period makes it ideal for rapidly recognizing large quantities of samples in the field. There are, currently, studies for determining the residues of FQs in pork, chicken, milk, and animal feeds [32–34]. Additionally, an assay utilizing a broad-spectrum-specific antibody to detect multiple FQs' residues in a single assay is optimal because all FQs have comparable core structures (Figure 1). In this study, we further explored the indirect competitive ELISA (ic-ELISA) approach for the rapid and concurrent detection of 13 FQs' residues: enrofloxacin (ENR), ciprofloxacin (CIP), sarafloxacin (SAR), ofloxacin (OFL), norfloxacin (NOR). pefloxacin mesylate (PM), pefloxacin (PEF), enoxacin (ENX), marbofloxacin (MAR), fleroxacin (FLE), lomefloxacin (LOM), danofloxacin (DAN), and difloxacin (DIF), with a minimum detection limit of 0.59 μg/L; ic-ELISA was then applied to the detection of an actual aquatic product, *Rana catesbeianus*.

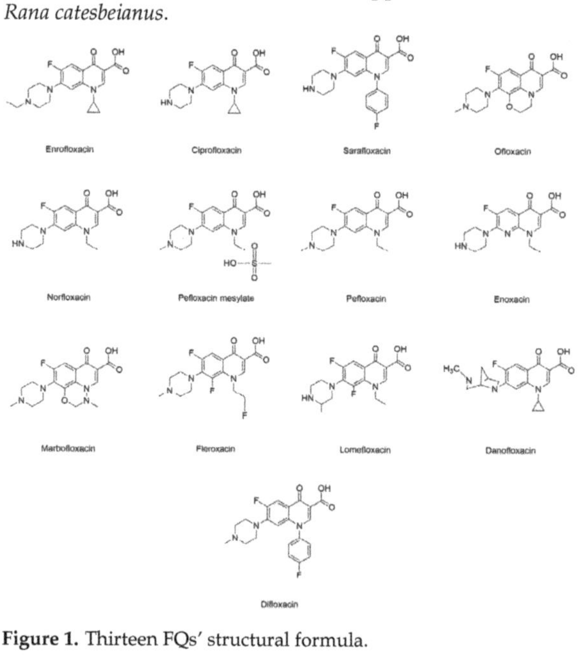

Figure 1. Thirteen FQs' structural formula.

2. Materials and Methods

2.1. Reagents and Instruments

ENR, CIP, SAR, OFL, NOR, PM, ENX, MAR, FLE, LOM, DAN, DIF, and nalidixic acid were obtained from Sigma-Aldrich (Saint Louis, MO, USA). Malachite green, metronidazole, tetracycline, florfenicol, thiamphenicol, sulfamethazine, sulfamethoxazole, and Tween-20 were obtained from Macklin Biochemical Co., Ltd. (Shanghai, China). PEF, 3,3′,5,5′-Tetramethylbenzidine (TMB), dimethyl sulfoxide (DMSO), sodium carbonate (Na_2CO_3), sodium bicarbonate ($NaHCO_3$), sodium chloride (NaCl), skimmed milk powder, acetonitrile, sodium hydroxide (NaOH), dichloromethane, n-hexane, phosphoric acid, and methanol were obtained from Aladdin Reagent Co., Ltd. (Shanghai, China). Goat anti-mouse HRP was obtained from Jackson Immunoresearch Laboratories (West Grove, PA, USA). Chloramphenicol was obtained from Dr. Ehrenstorfer GmbH (Augsburg, Germany). Anti-OFL monoclonal antibody, ENR coating antigen was purchased from Landu Biotechnology Co., Ltd. (Binzhou, China). A phosphate buffer saline (PBS buffer, 0.01 mol/L, pH 7.4) was obtained by accurately weighing 9.00 g NaCl, 13.76 g $Na_2HPO_4{\cdot}12H_2O$, and 1.79 g $NaH_2PO4{\cdot}2H_2O$ and then adding ultra-pure water to 1000 mL. Phosphate buffer solution (PBST) was obtained from 995 mL PBS buffer (10 mM, pH 7.4) containing 10% Tween-20.

Ninety-six-well EIA/RIA plates were obtained from Corning Incorporated (New York, NY, USA). Varioskan LUX multimode microplate reader was obtained from Thermo Fisher Scientific Co. (Waltham, MA, USA) to measure the absorbance. XTerra MS C_{18} column and Arc HPLC System with 2998 PDA Detector were obtained from Waters (Shanghai, China). Rotary evaporator was obtained from BUCHI Corporation (New Castle, DE, USA). Dry nitrogen blower was obtained from Qiuzuo Scientific Instrument Co., LTD. (Shanghai, China). Milli-Q water (Millipore, MA, USA) was used in all the experiments.

2.2. Development of ic-ELISA

With the intention of progressing this supplied ic-ELISA, the critical experimental parametric settings, such as the concentration of coated antigen and dilution factor of antibody were modified. In order to determine the ideal ic-ELISA conditions, different concentrations of coated antigen (0.05, 0.1, 0.2 μg/well) and a series of dilution coefficient of antibody (1:32,000, 1:64,000, 1:128,000, 1:256,000, 1:512,000) were selected by chessboard titration. The ic-ELISA approach could be described as follows: A 96-well plate with the coating antigen solution (100 μL/well) added was incubated at 4 °C overnight. The microwells were blocked through the addition of the blocking solution (200 μL/well) (0.5% skimmed milk powder solution) at 37 °C for 1 h, after being washed three times with PBST (0.01 mol/L PBS with 10% Tween-20). Following the cleaning of the microwells, FQs′ standard or sample solution (50 μL/well) and antibody in PBS (50 μL/well) were added to each well. The microwell plate was then incubated at 37 °C for 1 h. After washing, the 500-fold HRP-labeled secondary antibody solution (100 μL/well) was added to each well, and then the microwell plate was incubated at 37 °C for 30 min. Following cleaning again, TMB substrate solution (100 μL/well) was applied to microwells and then incubated for 15 min at 37 °C. The injection of 50 μL H_2SO_4 (1.25 mol/L) halted the reaction in each well. The absorbance was eventually determined in a multimode microplate reader utilizing a 450 nm test wavelength [35].

2.3. Determination of Sensitivity and Specificity

Sensitivity and specificity are crucial parameters in evaluating ELISA techniques. To determine the inhibition rate (%), several FQs′ standard solution concentrations were utilized (0.01, 0.05, 0.5, 1, 5, 10, 50, 100, 500, 2000, 10,000 μg/L). The following is the formula for estimating the inhibition rate:

$$\text{inhibition rate}(\%) = (A - A_i)/(A - A_0) \times 100$$

where, A_0 is the absorbance value of well with 100 μL PBS solution added, A_i is the absorbance value of the well with 50 μL antibody and 50 μL different doses of FQs' solution added, and A is the absorbance value (at 450 nm) of the well with 50 μL antibody and 50 μL PBS solution added, respectively. An ic-ELISA approach was brought by utilizing the optimal working parameters and gradient concentration standards (0.01, 0.05, 0.5, 1, 5, 10, 50, 100, 500, 2000, 10,000 μg/L). The four-parameter equation was fitted with the average of three distinct experiments performed in triplicate, as reported in Supplementary Table S1. The logistic 5 fitting equation in Origin pro 2021 software was applied to calculate the sensitivity (50% inhibitory concentration, computed as IC_{50}), as well as the limit of detection (LOD, computed as IC_{10}).

Specificity was defined as the capacity of structurally or functionally related compounds to bind to the antibody, and cross-reactivity (CR) was calculated as:

$$\text{Crossreactivity}\ (\%) = 100 \times IC_{50}(\text{ENR})/IC_{50}(\text{other analogues})$$

In the current investigation, CIP, SAR, OFL, NOR, PM, PEF, ENX, MAR, FLE, LOM, DAN, DIF, and nalidixic acid, which were similar to the structure of ENR, and additional 8 common veterinary antibiotics for aquatic products (including malachite green, metronidazole, tetracycline, florfenicol, thiamphenicol, chloramphenicol, sulfamethazine, sulfamethoxazole) were selected for specificity determination. The ratio of IC_{50} of ENR to IC_{50} of other antibiotics was utilized as the cross-reactivity.

2.4. Sample Preparation

Samples of commercial *Rana catesbeianus* were acquired from nearby supermarkets in Hangzhou, Zhejiang, China. Prior to application, all samples were analyzed to ensure they were free of FQs. For edible animal tissues, such as porcine muscle, chicken muscle, fish, shrimp, porcine liver, chicken liver, and other organs, organic solvent extraction was typically utilized. The sample pretreatment method was modified according to the Ministry of China Agriculture Announcement No. 1025-8-2008 [36]: 4.00 g sample and 12 mL acetonitrile-0.1 mol/L sodium hydroxide solution was added into a 50 mL centrifuge tube, shaken for 5 min, and centrifuged at 4000 r/min for 5 min. Then, 6 mL of supernatant was taken, and 6 mL of 0.02 mol/L phosphate buffer solution and 7 mL of dichloromethane was added and then shaken for 5 min. The lower organic phase (6 mL) was dried by nitrogen at 50 °C in a 10 mL glass test tube after centrifugating at 4000 r/min for 5 min. Next, 100 μL of subnatant liquid was removed for further analysis after adding 1 mL of 0.02 mol/L phosphate buffer, eddying for 2 min, then adding 2 mL of n-hexane, and eddying for 30 s. For the purpose of reducing matrix interference, the extracts were redissolved with PBS buffer. The supernatant was diluted by 2 times, 4 times, 6 times, 8 times, and 10 times. By comparing the established matrix curve and standard curve, the dilution multiple, which basically eliminated the matrix effect, was selected for follow-up ic-ELISA experiments.

2.5. Assay Validation

It is vital to confirm the objectivity, accuracy, and feasibility of the established analysis approach, with the goal to demonstrate that it is capable of serving the purpose and needs of detection to evaluate the validity and precision of the supplied ic-ELISA. In this research, negative samples of *Rana catesbeianus* without ENR were spiked with various ENR concentrations (50, 100, 200 μg/kg) and concurrently assessed by HPLC and ic-ELISA. The *Rana catesbeianus* samples that contained PEF, DAN, MAR, FLE, OFL, NOR, CIP, LOM, and ENR (100 μg/kg each) were pitched with the purpose of determining the aggregate amount of FQs by the established ic-ELISA, with the aim of further evaluating the viability of the approach. HPLC was employed to verify the correctness of the results in samples by applying the Chinese National Standard GB 31656.3-2021 with a few modifications [37]. For the chromatographic separation, the XTerra MS C_{18} column (4.6 mm × 250 mm, 5 μm) was utilized. In this research, the Arc HPLC System with 2998 PDA Detector (ultraviolet detector) was utilized, the detection wavelength was 278 nm, the column temperature was 30 °C, the flow rate was 0.8 mL/min, and the injection volume was 10 μL The mobile phases were as follows: A: 0.3% phosphoric acid-5% acetonitrile aqueous solution (pH 4.00 ± 0.01 adjusted with triethylamine); B: methanol [38]. The detecting condition of linear gradient for HPLC analysis was as follows: the volume of mobile phase B was 15% at 0–5 min, increased linearly to 40% at 5–45 min, and then returned to 15% at 45–50 min. The linear gradient parameters for HPLC analysis are reported in Table 1. The sample processing steps are listed in detail in the supporting information.

Table 1. Detecting condition of linear gradient for HPLC analysis.

Time (min)	A: 0.3% Phosphoric Acid-5% Acetonitrile Aqueous Solution (%)	B: Methanol (%)
0	85	15
5	85	15
45	60	40
50	85	15

3. Results and Discussion

3.1. Optimization of Working Parameters

The lowest IC_{50} value with a suitable OD value (ranging from 0.8 to 1.2) is necessary to construct a sensitive ELISA assay. Massive errors may result from OD values that are either too high or too low, which affect the detection sensitivity of the ic-ELISA. In order to achieve the ideal outcomes, the following variables were optimized: the dilution coefficient of the antibody (1:32,000, 1:64,000, 1:128,000, 1:256,000, 1:512,000) and the amount of the coated antigen (0.05, 0.1, 0.2 μg/well). The results are shown in Figure 2. The maximum absorbance (OD_{max}), which indicates the maximum binding between the antibody and coated antigen, was 1.46 when the coating antigen's concentration was 0.05 μg/well and the dilution coefficient of the antibody was 1:32,000, which was beyond the range of the optimal OD value. The OD_{max} value was 0.70 when the dilution coefficient of the antibody was 1:256,000, and the lower OD value would cause the instrument reading to be inaccurate. The detection sensitivity (IC_{50}) was 27.11, 23.83, 21.05, and 19.58 μg/L, when the coating antigen concentration was 0.05 μg/well and the dilution coefficient of the antibody was 1:32,000, 1:64,000, 1:128,000, and 1:256,000, respectively. When the coating antigen concentration was 0.10 μg/well, the half-maximal inhibitory concentration was 36.11, 30.80, 24.03, and 22.97 μg/L, when the dilution coefficient of the antibody was 1:32,000, 1:64,000, 1:128,000, and 1:256,000, respectively. When the coating antigen concentration was 0.20 μg/well, the half-maximal inhibitory concentration was 58.65, 49.24, 47.21, and 34.90 μg/L, when the dilution coefficient of the antibody was 1:64,000, 1:128,000, 1:256,000, and 1:512,000, respectively. The detection sensitivity was not as excellent as 0.05 μg/well when the coated antigen was 0.10 μg/well and 0.20 μg/well; hence, the study did not adopt these amounts. The experimental effect was optimum when the amount of the coated antigen was 0.05 μg/well and the dilution coefficient of the antibody was 1:128,000, since this resulted in an OD_{max} value of 0.91 and a half-maximal inhibitory concentration of 21.09 μg/L. The fitting curve tended to be smooth and a typical S-shaped curve. The OD_{max} value of the operating parameters had high sensitivity and was appropriate for ELISA detection.

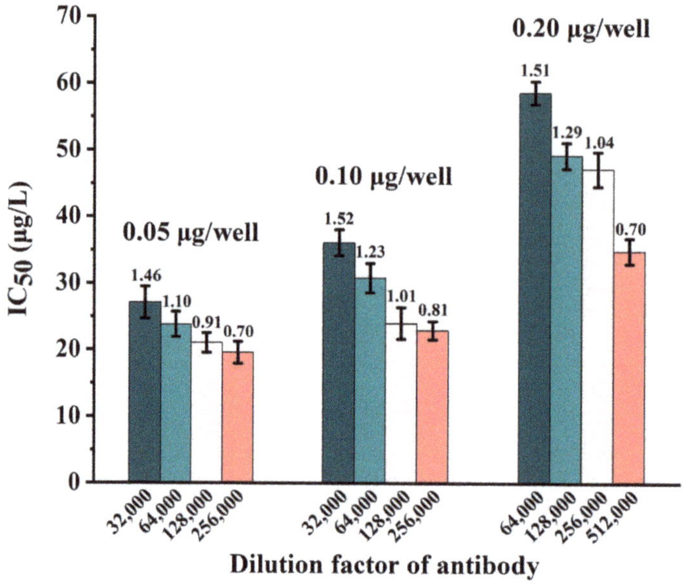

Figure 2. Optimization of antibody dilution (1:32,000, 1:64,000, 1:128,000, 1:256,000, 1:512,000) and coating antigen concentration (0.05, 0.10, 0.20 μg/well) optimization. Each corresponding column bears a mark indicating the average OD value.

3.2. Establishment of icELISA Standard Curve

The IC_{10} values and IC_{50} values were derived from the standard curves, which were employed to evaluate the detection limit and detection sensitivity of the generated ic-ELISA. Under the above-mentioned optimum operating parameters, the standard curves were obtained from various concentrations of FQs' standard solution as the abscissa (0.01, 0.05, 0.5, 1,5, 10, 50, 100, 500, 2000, 10,000 μg/L) and the inhibition rate as the ordinate. The standard inhibition curves of 13 FQs (ENR, CIP, SAR, OFL, NOR, PM, PEF, ENX, MAR, FLE, LOM, DAN, DIF) in PBS by the ic-ELISA were fitted by Origin. The standard curve of ENR is displayed in Figure 3 (R^2 = 0.9995), and the standard curves of CIP, SAR, OFL, NOR, PM, PEF, ENX, MAR, FLE, LOM, DAN, and DIF are shown in Supplementary Figure S1 (R^2 = 0.9984, 0.9964, 0.9986, 0.9923, 0.9994, 0.9991, 0.9990, 0.9976, 0.9949, 0.9981, 0.9974, 0.9978), all of which were typical S-shaped curves. The concentration of FQs in the aforementioned 13 fitting curves all remained stable when they were less than 0.01 μg/L or more than 10,000 μg/L. Utilizing this means, the detection limits (IC_{10}) of ENR, CIP, SAR, OFL, NOR, PM, PEF, ENX, MAR, FLE, LOM, DAN, and DIF were 0.59, 0.97, 1.02, 1.15, 1.37, 1.38, 1.44, 1.66, 1.86, 2.01, 2.04, 2.71, and 3.61 μg/L, respectively, and the half-maximal inhibitory concentrations (IC_{50}) were 19.23, 18.69, 21.55, 22.52, 21.84, 20.93, 21.96, 21.84, 21.69, 21.90, 22.35, 20.61, and 22.52 μg/L, respectively, as reported in Supplementary Table S2. The Ministry of China Agriculture Announcement No. 1025-8-2008 detected ENR, CIP, NOR, OFL, LOM, oxaquic acid, ENX, PEF, DAN, flumequine, MAR, and amifloxacin residues in pig muscle, chicken muscle, chicken liver, honey, eggs, and shrimp, in foods of animal origin, with a detection limit of 3 μg/kg. NOR, CIP, ENR, OFL, oxaquic acid, and flumequine were found in the edible tissues of fish, shrimp, crab, and shellfish, with a detection limit of 2.5 μg/kg, according to the Chinese National Standard GB 31656.3-2021. Due to the lack of antibodies with the appropriate affinity and specificity, previous immunoassay approaches were typically restricted to the detection of a single FQ type or a small number of FQ types. The detection limit of the ic-ELISA, based on a monoclonal antibody against LOM employed by Mukunzi et al. to identify LOM and related drug residues in milk, was 0.04 ng/mL in 0.1 M phosphate-buffered saline (PBS) [39]. The method was highly sensitive to LOM, NOR, and ENX, moderately sensitive to CIP and OFL, and slightly sensitive to FLE and PEF, with a CR < 0.1% compared to the other 12 FQs.

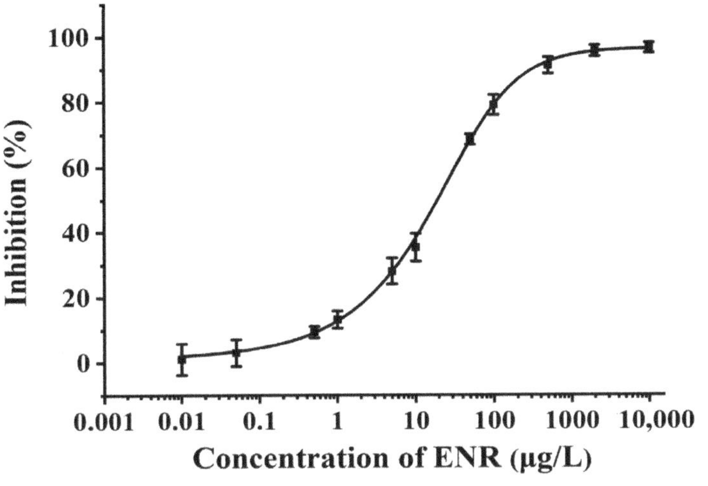

Figure 3. Standard inhibition curve of ic-ELISA for detecting ENR in PBS.

3.3. Determination of Specificity

All immunoassays have the inherent property of specificity. The specificity of the established ic-ELISA was evaluated by calculating the ratio of the IC_{50} of ENR to the IC_{50} of other veterinary antibiotics. In this study, we sought to establish an ic-ELISA method capable of recognizing the majority of common FQ analogs. The IC_{50} values of FQs (CIP, SAR, OFL, NOR, PM, PEF, ENX, MAR, FLE, LOM, DAN, DIF, nalidixic acid) with a similar structure to ENR and an additional eight common veterinary antibiotics (malachite green, fungicide, cephalosporin, florfenicol, thiamphenicol, chloramphenicol, sulfamethazine, sulfamethoxazole) utilized in aquatic products were determined by an ic-ELISA based on an anti-OFL monoclonal antibody and an ENR coating antigen. The results presented in Table 2 demonstrated that the recognition capacity of this ic-ELISA to other antibiotic veterinary drugs was investigated. Table 2 lists the half-maximal inhibitory concentration for the 12 FQs' analogs, including CIP, SAR, OFL, NOR, PM, PEF, ENX, MAR, FLE, LOM, DAN, and DIF, which were 19.23, 18.69, 21.55, 22.52, 21.84, 20.93, 21.96, 21.84, 21.69, 21.90, 22.35, 20.61, and 22.52 μg/L, respectively. Therefore, the cross-reactivities (%) of CIP, SAR, OFL, NOR, PM, PEF, ENX, MAR, FLE, LOM, DAN, and DIF were 102.88%, 89.24%, 85.41%, 88.06%, 91.88%, 87.58%, 88.07%, 88.64%, 87.81%, 86.06%, 93.31%, and 85.38%, respectively. As we expected, these cross-reactivities (%) were all higher than 85%, demonstrating that CIP, SAR, OFL, NOR, PM, PEF, ENX, MAR, FLE, LOM, DAN, and DIF could be recognized by utilizing the established ELISA approach. Meanwhile, the cross-reactivity (%) of this ic-ELISA was 14.97% for nalidixic acid. In addition, the findings demonstrated that the IC_{50} of the above eight antibiotic veterinary drugs was greater than 15,000 μg/L, and the cross-reactivity (%) was less than 0.2%, indicating that the ic-ELISA had no obvious recognition ability for them. As a result, the established ic-ELISA approach has broad-spectrum specificity for the 13 FQs, with low cross-reactivity with nalidixic acid (<15%), and other antibiotic veterinary medications have no impact on the properties of the ic-ELISA (<0.2%).

Table 2. Specificity evaluation of ic-ELISA ($n = 3$).

Name	IC_{50} (μg/L)	Cross-Reactivity (%)
ENR	19.23	100.00
CIP	18.69	102.88
SAR	21.55	89.24
OFL	22.52	85.41
NOR	21.84	88.06
PM	20.93	91.88
PEF	21.96	87.58
ENX	21.84	88.07
MAR	21.69	88.64
FLE	21.90	87.81
LOM	22.35	86.06
DAN	20.61	93.31
DIF	22.52	85.38
nalidixic acid	128.43	14.97
malachite green	>15,000	<0.2
metronidazole	>15,000	<0.2
tetracycline	>15,000	<0.2
florfenicol	>15,000	<0.2
thiamphenicol	>15,000	<0.2
chloramphenicol	>15,000	<0.2
sulfamethazine	>15,000	<0.2
sulfamethoxazole	>15,000	<0.2

3.4. Removal of Matrix Effects

In general, the recovery rate may be affected by the sample type, analyte solubility, sample preparation technique, and antibody effectiveness. In the actual analysis of food samples, compounds such as proteins and fats in complex food matrices may reduce the intensity of color development or interfere with the binding of antibodies and analytes, thus reducing the accuracy of the results. Therefore, the removal of matrix effects is primary in ELISA experiments. Typically, diluting sample extracts and preparing standard curves in blank tissue are two effective methods for minimizing matrix interference. After the sample extracts of *Rana catesbeianus* were pretreated with the modified Ministry of China Agriculture Announcement No. 1025-8-2008, the extracts were redissolved with PBS buffer. Following redissolving, the extracts were diluted with PBS by 2-fold, 4-fold, 6-fold, 8-fold, and 10-fold, and the matrix curve was generated. The matrix curve was compared with the standard curve obtained in the blank tissue. As shown in Figure 4, when the sample extract was diluted by 2-fold, 4-fold, 6-fold, and 8-fold, then the matrix curve of the *Rana catesbeianus* extract sample significantly deviated from the standard curve, indicating that the matrix effect was still clearly visible; when the extract was diluted 10-fold, the matrix curve was basically consistent with the standard curve, and the matrix effect was negligible. As a consequence, this investigation utilized the 10-fold diluted *Rana catesbeianus* extract sample for subsequent ic-ELISA detection.

3.5. Analysis of Spiked Samples

The recoveries of ENR in *Rana catesbeianus* samples were employed by HPLC. Three various spiking concentrations of ENR (50, 100, 200 μg/kg) were set to confirm the viability of the established ELISA approach. The unspiked samples yielded no positive results throughout the spiking and recovery testing, which confirmed that the ELISA method had a low false positive rate and was more suitable for rapid detection in the actual field. *Rana catesbeianus* spiked with FQs exhibited an excellent correlation between the spiking degree and the observed concentration. The results of recovery are shown in Table 3. It can be seen from Table 3 that when the spiking concentrations were 50, 100, and 200 μg/kg, the recoveries of ENR in *Rana catesbeianus* samples by the ic-ELISA were 93.87%, 90.39%, and 90.89%, respectively, while those by HPLC were 85.88%, 102.70%, and 92.18%, respectively.

The coefficients of variation (CVs) of the ic-ELISA were 3.86%, 2.30%, and 5.11%, and those of HPLC were 7.01%, 3.18%, and 1.52%, respectively. The CVs of both approaches were less than 10%. The ic-ELISA outcomes were consistent with those of HPLC with a correlation coefficient (R^2) of 0.9936. For the analysis of the ENR residues in *Rana catesbeianus*, the sample extraction technique and ic-ELISA were satisfactory.

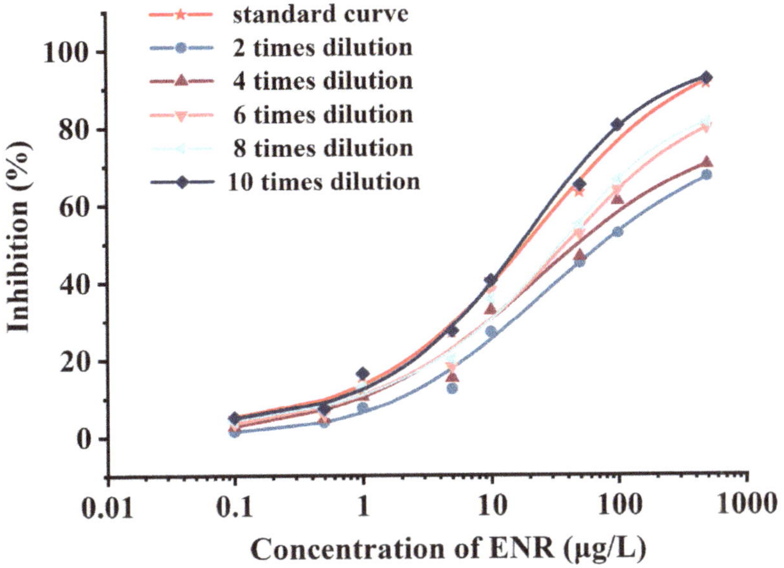

Figure 4. Matrix effect elimination curve of *Rana catesbeianus*.

Table 3. Recoveries of ENR in *Rana catesbeianus* by ic-ELISA and HPLC (n = 3).

Sample	Spiked Concn (μg/kg)	ic-ELISA			HPLC		
		Detected (Mean ± SD [a]) (μg/kg)	Recovery (%)	CV (%) [b]	Detected (Mean ± SD) (μg/kg)	Recovery (%)	CV (%)
Rana catesbeianus	0	ND [c]	–	–	ND	–	–
	50	46.94 ± 1.81	93.87	3.86	42.94 ± 3.01	85.88	7.01
	100	90.39 ± 2.08	90.39	2.30	102.70 ± 3.27	102.70	3.18
	200	181.78 ± 9.29	90.89	5.11	184.35 ± 2.81	92.18	1.52

[a] SD, standard deviation. [b] CV, coefficient of variation. [c] ND, not detected.

Moreover, since multiple FQs may be simultaneously present in the samples, we measured the concentrations of nine FQs (ENR, CIP, OFL, NOR, PEF, MAR, FLE, LOM, DAN) in commercial *Rana catesbeianus* samples by the ic-ELISA and HPLC to calculate the feasibility and practicality of the existing ic-ELISA. According to Table 4, the concentrations of ENR, CIP, OFL, NOR, PEF, MAR, FLE, LOM, and DAN in *Rana catesbeianus* evaluated by HPLC were 105.83, 118.30, 106.76, 106.18, 71.36, 96.93, 103.76, 94.37, and 94.11 μg/kg, respectively, and the CVs were 7.73%, 2.12%, 1.75%, 0.81%, 9.66%, 1.88%, 0.78%, 9.69%, and 2.87%, respectively. The overall contents of the aforementioned FQs were 897.60 μg/kg, and the CVs were all less than 10%, which was essentially consistent with the total concentration of the nine FQs measured by the ic-ELISA method (782.91 μg/kg) proposed in this paper. The CV of the above ic-ELISA was 8.53%, which proved that the ic-ELISA and HPLC had high accuracy. We found that the accuracy of both the ic-ELISA and HPLC results in determining the concentration of FQs in mixed samples was impacted by the competition amongst the FQs. The total concentration determined by HPLC was higher than the actual spiking concentration, and the total concentration determined by the ic-ELISA was lower than the actual spiking concentration, though still within the controllable range. These discoveries revealed that the ic-ELISA could be utilized to exactly and numerically identify the total amount of ENR, CIP, OFL, NOR, PM, PEF, MAR, FLE, LOM, and DAN in *Rana catesbeianus* samples.

Table 4. Determination of 9 FQs in *Rana catesbeianus* utilizing HPLC and ic-ELISA (n = 3).

Sample	FQs	HPLC			ic-ELISA	
		Detected (Mean ± SD [a]) (μg/kg)	CV [b] (%)	Total (μg/kg)	Detected (Mean ± SD) (μg/kg)	CV (%)
Rana catesbeianus	PEF	71.36 ± 6.89	9.66	897.60	782.91 ± 66.78	8.53
	DAN	94.11 ± 2.70	2.87			
	MAR	96.93 ± 1.82	1.88			
	FLE	103.76 ± 0.81	0.78			
	OFL	106.76 ± 1.87	1.75			
	NOR	106.18 ± 0.86	0.81			
	CIP	118.30 ± 2.51	2.12			
	LOM	94.37 ± 9.14	9.69			
	ENR	105.83 ± 8.18	7.73			

[a] SD, standard deviation. [b] CV, coefficient of variation.

4. Conclusions

In summary, the ic-ELISA was established employing broad-spectrum-specific antibodies with comparable recognition ability for 13 FQs, including ENR, CIP, SAR, OFL, NOR, PM, PEF, ENX, MAR, FLE, LOM, DAN, and DIF. Concurrently, the approach had a cross-reactivity of less than 0.2% when detecting the other eight antibiotic veterinary drugs and was moderately sensitive to nalidixic acid. The optimized ic-ELISA demonstrated the LOD of 0.59 μg/L for ENR in PBS. The half-maximal inhibitory concentration (IC_{50}) was 19.23 μg/L and had significant cross-reactivity values (>85.38%) for the other 12 FQs (CIP, SAR, OFL, NOR, PM, PEF, ENX, MAR, FLE, LOM, DAN, DIF). It was verified that the matrix effect was negligible with a 10-fold dilution of the optimized *Rana catesbeianus* extract samples. The outcomes of the proposed ic-ELISA were compared to those of HPLC, which showed a strong agreement between them (R^2 = 0.9936). The total content of the nine FQs in *Rana catesbeianus* samples that were commercially accessible was effectively determined by utilizing the ic-ELISA. In conclusion, the created ic-ELISA could potentially be employed in a rapid, precise, and affordable manner to screen for the drug residues of the FQs in *Rana catesbeianus*.

Supplementary Materials: The following supporting information can be downloaded at https://www.mdpi.com/article/10.3390/foods12132530/s1. Experimental methods, Figure S1: Standard inhibition curve of ic-ELISA for detecting 12 FQs (CIP, SAR, OFL, NOR, PM, PEF, ENX, MAR, FLE, LOM, DAN, DIF) in PBS; Table S1: Four-parameter equation of ic-ELISA for detecting 13 FQs (ENR, CIP, SAR, OFL, NOR, PM, PEF, ENX, MAR, FLE, LOM, DAN, DIF) in PBS; Table S2: Detection limits (IC_{10}) and half-maximal inhibitory concentration (IC_{50}) of ic-ELISA for detecting 13 FQs.

Author Contributions: Conceptualization, writing—original draft preparation, supervision, validation, B.Z.; conceptualization, methodology, writing—original draft preparation, data curation, Y.L. and D.C.; validation, B.G., Z.C., and J.C.; project administration, writing—review and editing, funding acquisition, X.Y. and X.S. All authors have read and agreed to the published version of the manuscript.

Funding: This work was supported by the Key R&D Program of Zhejiang Province (Grant Nos. 2022C02049 and 2021C02061) and the Zhejiang Provincial Natural Science Foundation of China (Grant No. LQ23C200006).

Institutional Review Board Statement: Not applicable.

Informed Consent Statement: Not applicable.

Data Availability Statement: The datasets generated for this study are available on request to the corresponding author.

Conflicts of Interest: The authors declare that they have no known competing financial interest or personal relationships that could appear to influence the work reported in this paper.

References

1. Vergalli, J.; Atzori, A.; Pajovic, J.; Dumont, E.; Malloci, G.; Masi, M.; Vargiu, A.V.; Winterhalter, M.; Réfrégiers, M.; Ruggerone, P.; et al. The Challenge of Intracellular Antibiotic Accumulation, a Function of Fluoroquinolone Influx versus Bacterial Efflux. *Commun. Biol.* **2020**, *3*, 198. [CrossRef] [PubMed]
2. Pan, Y.; Yang, H.; Wen, K.; Ke, Y.; Shen, J.; Wang, Z. Current Advances in Immunoassays for Quinolones in Food and Environmental Samples. *TrAC Trends Anal. Chem.* **2022**, *157*, 116726. [CrossRef]
3. Zhu, Y.; Li, L.; Wang, Z.; Chen, Y.; Zhao, Z.; Zhu, L.; Wu, X.; Wan, Y.; He, F.; Shen, J. Development of an Immunochromatography Strip for the Rapid Detection of 12 Fluoroquinolones in Chicken Muscle and Liver. *J. Agric. Food Chem.* **2008**, *56*, 5469–5474. [CrossRef]
4. Laxminarayan, R.; Matsoso, P.; Pant, S.; Brower, C.; Røttingen, J.-A.; Klugman, K.; Davies, S. Access to Effective Antimicrobials: A Worldwide Challenge. *Lancet* **2016**, *387*, 168–175. [CrossRef]
5. Masi, M.; Réfregiers, M.; Pos, K.M.; Pagès, J.-M. Mechanisms of Envelope Permeability and Antibiotic Influx and Efflux in Gram-Negative Bacteria. *Nat. Microbiol.* **2017**, *2*, 17001. [CrossRef]
6. Fernández, F.; Hegnerová, K.; Piliarik, M.; Sanchez-Baeza, F.; Homola, J.; Marco, M.-P. A Label-Free and Portable Multichannel Surface Plasmon Resonance Immunosensor for on Site Analysis of Antibiotics in Milk Samples. *Biosens. Bioelectron.* **2010**, *26*, 1231–1238. [CrossRef]
7. Hooper, D.C. Mechanisms of Action of Antimicrobials: Focus on Fluoroquinolones. *Clin. Infect. Dis.* **2001**, *32*, S9–S15. [CrossRef]
8. Endtz, H.P.; Ruijs, G.J.; van Klingeren, B.; Jansen, W.H.; van der Reyden, T.; Mouton, R.P. Quinolone Resistance in Campylobacter Isolated from Man and Poultry Following the Introduction of Fluoroquinolones in Veterinary Medicine. *J. Antimicrob. Chemother.* **1991**, *27*, 199–208. [CrossRef]
9. Lomaestro, B.M. Fluoroquinolone-Induced Renal Failure. *Drug-Saf.* **2000**, *22*, 479–485. [CrossRef]
10. Andreu, V.; Blasco, C.; Picó, Y. Analytical Strategies to Determine Quinolone Residues in Food and the Environment. *TrAC Trends Anal. Chem.* **2007**, *26*, 534–556. [CrossRef]
11. Appelbaum, P.C.; Hunter, P.A. The Fluoroquinolone Antibacterials: Past, Present and Future Perspectives. *Int. J. Antimicrob. Agents* **2000**, *16*, 5–15. [CrossRef] [PubMed]
12. Pasteris, S.E.; Guidoli, M.G.; Otero, M.C.; Bühler, M.I.; Nader-Macías, M.E. In Vitro Inhibition of Citrobacter Freundii, a Red-Leg Syndrome Associated Pathogen in Raniculture, by Indigenous Lactococcus Lactis CRL 1584. *Vet. Microbiol.* **2011**, *151*, 336–344. [CrossRef] [PubMed]
13. Pasteris, S.E.; Bühler, M.I.; Nader-Macías, M.E. Microbiological and Histological Studies of Farmed-Bullfrog (*Rana catesbeiana*) Tissues Displaying Red-Leg Syndrome. *Aquaculture* **2006**, *251*, 11–18. [CrossRef]
14. Grogan, L.F.; Robert, J.; Berger, L.; Skerratt, L.F.; Scheele, B.C.; Castley, J.G.; Newell, D.A.; McCallum, H.I. Review of the Amphibian Immune Response to Chytridiomycosis, and Future Directions. *Front. Immunol.* **2018**, *9*, 2536. [CrossRef] [PubMed]
15. Hu, W.; Dong, B.; Kong, S.; Mao, Y.; Zheng, R. Pathogen Resistance and Gene Frequency Stability of Major Histocompatibility Complex Class IIB Alleles in the Giant Spiny Frog Quasipaa Spinosa. *Aquaculture* **2017**, *468*, 410–416. [CrossRef]
16. Huang, F.; An, Z.; Moran, M.J.; Liu, F. Recognition of Typical Antibiotic Residues in Environmental Media Related to Groundwater in China (2009−2019). *J. Hazard. Mater.* **2020**, *399*, 122813. [CrossRef]
17. Gonzalez Ronquillo, M.; Angeles Hernandez, J.C. Antibiotic and Synthetic Growth Promoters in Animal Diets: Review of Impact and Analytical Methods. *Food Control.* **2017**, *72*, 255–267. [CrossRef]
18. Muñoz-Olivas, R. Screening Analysis: An Overview of Methods Applied to Environmental, Clinical and Food Analyses. *TrAC Trends Anal. Chem.* **2004**, *23*, 203–216. [CrossRef]
19. Trullols, E.; Ruisánchez, I.; Rius, F.X. Validation of Qualitative Analytical Methods. *TrAC Trends Anal. Chem.* **2004**, *23*, 137–145. [CrossRef]
20. Mitani, K.; Kataoka, H. Determination of Fluoroquinolones in Environmental Waters by In-Tube Solid-Phase Microextraction Coupled with Liquid Chromatography–Tandem Mass Spectrometry. *Anal. Chim. Acta* **2006**, *562*, 16–22. [CrossRef]
21. Yu, H.; Mu, H.; Hu, Y.-M. Determination of Fluoroquinolones, Sulfonamides, and Tetracyclines Multiresidues Simultaneously in Porcine Tissue by MSPD and HPLC–DAD. *J. Pharm. Anal.* **2012**, *2*, 76–81. [CrossRef] [PubMed]
22. Schneider, M.J.; Donoghue, D.J. Multiresidue Determination of Fluoroquinolone Antibiotics in Eggs Using Liquid Chromatography–Fluorescence–Mass Spectrometryn. *Anal. Chim. Acta* **2003**, *483*, 39–49. [CrossRef]

23. Delatour, T.; Savoy, M.-C.; Tarres, A.; Bessaire, T.; Mottier, P.; Desmarchelier, A. Low False Response Rates in Screening a Hundred Veterinary Drug Residues in Foodstuffs by LC-MS/MS with Analyte-Specific Correction of the Matrix Effect. *Food Control.* **2018**, *94*, 353–360. [CrossRef]
24. Jiang, W.; Wang, Z.; Beier, R.C.; Jiang, H.; Wu, Y.; Shen, J. Simultaneous Determination of 13 Fluoroquinolone and 22 Sulfonamide Residues in Milk by a Dual-Colorimetric Enzyme-Linked Immunosorbent Assay. *Anal. Chem.* **2013**, *85*, 1995–1999. [CrossRef] [PubMed]
25. Zhang, B.; Zhou, R.; Zhang, H.; Cai, D.; Lin, X.; Lang, Y.; Qiu, Y.; Shentu, X.; Ye, Z.; Yu, X. A Smartphone Colorimetric Sensor Based on Pt@Au Nanozyme for Visual and Quantitative Detection of Omethoate. *Foods* **2022**, *11*, 2900. [CrossRef]
26. Lang, Y.; Zhang, B.; Cai, D.; Tu, W.; Zhang, J.; Shentu, X.; Ye, Z.; Yu, X. Determination Methods of the Risk Factors in Food Based on Nanozymes: A Review. *Biosensors* **2023**, *13*, 69. [CrossRef] [PubMed]
27. Van Coillie, E.; De Block, J.; Reybroeck, W. Development of an Indirect Competitive ELISA for Flumequine Residues in Raw Milk Using Chicken Egg Yolk Antibodies. *J. Agric. Food Chem.* **2004**, *52*, 4975–4978. [CrossRef]
28. Fernández, F.; Pinacho, D.G.; Gratacós-Cubarsí, M.; García-Regueiro, J.-A.; Castellari, M.; Sánchez-Baeza, F.; Marco, M.-P. Immunochemical Determination of Fluoroquinolone Antibiotics in Cattle Hair: A Strategy to Ensure Food Safety. *Food Chem.* **2014**, *157*, 221–228. [CrossRef]
29. Du, B.; Wen, F.; Guo, X.; Zheng, N.; Zhang, Y.; Li, S.; Zhao, S.; Liu, H.; Meng, L.; Xu, Q.; et al. Evaluation of an ELISA-Based Visualization Microarray Chip Technique for the Detection of Veterinary Antibiotics in Milk. *Food Control* **2019**, *106*, 106713. [CrossRef]
30. Wang, Z.; Zhu, Y.; Ding, S.; He, F.; Beier, R.C.; Li, J.; Jiang, H.; Feng, C.; Wan, Y.; Zhang, S.; et al. Development of a Monoclonal Antibody-Based Broad-Specificity ELISA for Fluoroquinolone Antibiotics in Foods and Molecular Modeling Studies of Cross-Reactive Compounds. *Anal. Chem.* **2007**, *79*, 4471–4483. [CrossRef]
31. Hu, K.; Huang, X.; Jiang, Y.; Fang, W.; Yang, X. Monoclonal Antibody Based Enzyme-Linked Immunosorbent Assay for the Specific Detection of Ciprofloxacin and Enrofloxacin Residues in Fishery Products. *Aquaculture* **2010**, *310*, 8–12. [CrossRef]
32. Duan, J.; Yuan, Z. Development of an Indirect Competitive ELISA for Ciprofloxacin Residues in Food Animal Edible Tissues. *J. Agric. Food Chem.* **2001**, *49*, 1087–1089. [CrossRef] [PubMed]
33. Hu, S.; Huang, Z.; Wang, C.; Peng, J.; Lai, W.-H. Using Hapten Cross-Reactivity to Screen Heterologous Competitive Antigens for Improving the Sensitivity of ELISA. *Food Chem.* **2020**, *303*, 125379. [CrossRef]
34. Tufa, R.A.; Pinacho, D.G.; Pascual, N.; Granados, M.; Companyó, R.; Marco, M.P. Development and Validation of an Enzyme Linked Immunosorbent Assay for Fluoroquinolones in Animal Feeds. *Food Control* **2015**, *57*, 195–201. [CrossRef]
35. Sheng, W.; Zhang, B.; Zhao, Q.; Wang, S.; Zhang, Y. Preparation of a Broad-Spectrum Heterocyclic Aromatic Amines (HAAs) Antibody and Its Application in Detection of Eight HAAs in Heat Processed Meat. *J. Agric. Food Chem.* **2020**, *68*, 15501–15508. [CrossRef]
36. The Ministry of China Agriculture Announcement No.1025−8−2008. Available online: https://www.moa.gov.cn/nybgb/2008/dwuq/201806/t20180609_6151576.htm (accessed on 21 April 2023).
37. The Chinese National Standard GB 31656.3−2021. Available online: http://www.aqsc.agri.cn/zlbz/bzgg/202201/t20220126_396511.htm (accessed on 23 April 2023).
38. Quantitative Determination of Gatifloxacin, Levofloxacin, Lomefloxacin and Pefloxacin Fluoroquinolonic Antibiotics in Pharmaceutical Preparations by High-Performance Liquid Chromatography. *J. Pharm. Biomed. Anal.* **2006**, *40*, 179–184. [CrossRef] [PubMed]
39. Mukunzi, D.; Isanga, J.; Suryoprabowo, S.; Liu, L.; Kuang, H. Rapid and Sensitive Immunoassays for the Detection of Lomefloxacin and Related Drug Residues in Bovine Milk Samples. *Food Agric. Immunol.* **2017**, *28*, 599–611. [CrossRef]

foods

MDPI

Article

Rapid Determination of Oxygenated and Parent Polycyclic Aromatic Hydrocarbons in Milk Using Supercritical Fluid Chromatography-Mass Spectrometry

Limin Zhang, Wei Li and Shimin Wu *

Department of Food Science and Technology, School of Agriculture and Biology, Shanghai Jiao Tong University, 800 Dongchuan Road, Shanghai 200240, China
* Correspondence: wushimin@sjtu.edu.cn; Tel./Fax: +86-21-34205717

Abstract: Liquid milks are consumed worldwide in large amounts, especially by adolescents and infants. Thus, their health quality linked with polycyclic aromatic hydrocarbon (PAH) contamination has attracted great concern. This study developed a rapid and sensitive supercritical fluid chromatography (SFC)-MS method to determine two typical oxygenated PAHs (OPAHs) and EU 15+1PAHs except for benzo[k]fluoranthene (BkF) in three types of liquid milks: 10 ultra heat treated (UHT) milks, 8 pasteurized milks, and 4 extended-shelf-life pasteurized milks. The instrumental analysis was 15 min with a recovery of 67.66–118.46%, a precision of 1.45–14.68%, detection limits of 0.04–0.24 μg/kg, and quantification limits of 0.13–0.78 μg/kg. We found 9-fluorenone, anthraquinone, 15 EU priority PAHs, and benzo[a]pyrene toxic equivalent quantity (BaPeq) in the 22 milk samples, which were 0.32–1.56 μg/kg, 0.40–1.74 μg/kg, 0.57–8.48 μg/kg, and 0.01–17.42 μg/kg, respectively. The UHT milks and whole fat milks showed higher PAH concentrations than other investigated samples, where the maximum levels of BaP and PAH4 were 0.77 and 3.61 μg/kg, respectively. PAH4 dominantly contributed to the PAH8 concentration and was detected in 73% and 32% of samples at more than 1.0 and 2.0 μg/kg, respectively. The results suggest that raw milks should be strictly monitored and extensively investigated for PAH4 and BaP concentrations for future risk assessment, limitations, and dietary guidance.

Keywords: supercritical carbon dioxide; pasteurization; fat content; toxic equivalency factor; dibenzopyrenes; carcinogen

1. Introduction

Polycyclic aromatic hydrocarbons (PAHs) and their oxygenated derivatives (OPAHs) have been shown to be carcinogenic, teratogenic, and mutagenic. They are ubiquitous and widely present in soil, water, air, sediment, dust, and food. Currently, 24 well-known parent PAHs (the US EPA 16 PAHs and EU 15+1PAHs) are of academic concern worldwide [1]. OPAHs are considered more harmful than their parent PAHs despite their undisclosed toxic mechanism.

Milk is favored by consumers worldwide due to its indispensable nutritional value. In China, nearly 400 million people eat dairy products every day. In 2020, China produced 34.4 million tons of milk, and imported dairy products were translated into approximately 19 million tons of fresh milk. According to the Chinese Nutrition Society's recommendation of a daily 300 g milk intake, there is still a gap of nearly two-thirds of the total nutritional demand for dairy products in China. In fact, milk consumers in China have been rapidly growing. Moreover, milk is the main inseparable ingredient of many foods and beverages, including highly sensitive infant formula. Therefore, high concern must be assigned to persistent organic pollutants such as PAHs in milk and milk products. Foods of animal origin are considered the main source of dietary intake of PAHs, including milk and dairy products [1].

PAHs in milks mainly come from two routes: pollution in the process of pasture feeding [2] and generation or introduction in the process of production, processing, and transportation [3]. At present, the safety limit of benzo[a]pyrene (BaP) and PAH4 in liquid milk has not been regulated worldwide but considering that the BaP limit in drinking water in China and many developed countries is 0.01 μg/L, the content of carcinogenic PAHs such as BaP in fresh milk and sterilized liquid milk should be highly vigilant. EU has extremely strict limits on PAHs for infant formula, among which the limits of BaP and PAH4 (Benz[a]anthracene, chrysene, benzo[b]fluoranthene, BaP) are not allowed to exceed 1.0 μg/kg. Our previous study indicated that BaP in 31 commercial infant formula samples was below 1.0 μg/kg, but the PAH4 content in seven samples was 1.08–2.41 μg/kg, which exceeded the limit value of the EU [4]. It is necessary to strengthen the monitoring of BaP, PAH4, PAH8 (PAH4, benzo[k]fluoranthene, indeno-[1,2,3-cd]pyrene, dibenz[a,h]anthracene, benzo[ghi]-perylene), EU 15+1PAH (PAH8, benzo[c]fluorene, cyclopenta[cd]pyrene, 5-methylchrysene, benzo[k]fluoranthene, benzo[j]fluoranthene, dibenz o[a,l]pyrene, dibenzo[a,e]pyrene, dibenzo[a,i]pyrene, and dibenzo[a,h]pyrene) and OPAHs in raw milk and liquid milk.

The improvement of analytical methods for trace contaminants in food is endless, and the analysis of PAHs is no exception. Current quantitative analysis methods to detect PAH in milks and milk products are mostly based on chromatographic separation. The most commonly used methods are GC–MS [4–9] and HPLC-FLD [10–12]. SFC combines the advantages of both GC and UPLC and is compatible with various types of detectors. The combination of SFC and MS greatly improves the sensitivity and efficiency of analysis. SFC-MS uses supercritical CO_2 as the mobile phase, which effectively reduces toxic and expensive organic solvents. Supercritical CO_2 and n-hexane have similar polarities, which is suitable for the analysis of weakly polar substances such as PAHs [13]. Supercritical CO_2 is significantly less viscous than liquid, and the diffusivity of analyte is much higher than that in liquid [14]. Thus, SFC analysis can be performed at a high flow rate without increasing the pressure, which significantly shortens the analysis time. Increasing attention has been paid to SFC due to the solvent reduction characteristic, environmentally friendly nature, and green chemistry requirements [15,16].

Compared with traditional HPLC and GC, SFC exhibits faster and better separation ability. For example, the quantification of EU 15+1PAHs in coffee and dark beer can be completed within only 14 min using SFC-MS with detection limits below 1 μg/kg [16]. Another example reported that SFC-MS enabled rapid online monitoring of EPA PAH16 in particulate matter samples using an online supercritical fluid extraction-MS detection system [17].

The purpose of this study was to develop a rapid and sensitive analytical method to determine EU 15+1PAHs and OPAHs in milk by using SFC-QqQ-MS. The method was optimized and validated to investigate the PAH concentrations of 22 real milk samples, including 10 UHT milks, 8 pasteurized milks, and 4 extended-shelf-life pasteurized milks. The influencing factors of PAHs in different types of milks were also discussed.

2. Materials and Methods

2.1. Reagents and Samples

All organic solvents in this study were of HPLC grade. Acetonitrile, acetone, n-hexane, and methanol were supplied by CNW Technologies (Dusseldorf, Germany). Deionized water was purified using a Milli-Q water purification system (Millipore, MA, USA). The mixed EU 15+1PAH standard cyclohexane solution (>99.8%) was ordered from ANPEL Laboratory Technologies (Shanghai, China). It consists of benzo[c]fluorene (BcF), benz[a]anthracene (BaA), cyclopenta[cd]pyrene (CP), chrysene (Chr), 5-methylchrysene (5-MChr), benzo[b]fluoranthene (BbF), benzo[k]fluoranthene (BkF), benzo[j]fluoranthene (BjF), benzo[a]pyrene (BaP), indeno-[1,2,3-cd]pyrene (IP), dibenz[a,h]anthracene (DahA), benzo[ghi]-perylene (BghiP), dibenzo[a,l]pyrene (DalP), dibenzo[a,e]pyrene (DaeP), dibenz o[a,i]pyrene (DaiP), and dibenzo[a,h]pyrene (DahP). Standard cyclohexane solutions (>99.8%) of two OPAHs, i.e., 9-fluorenone (9-FO) and anthraquinone (ATQ), were purchased from

Dr. Ehrenstorfer GmbH (Augsburg, Germany). All standard solutions were diluted to 1 mg/mL using n-hexane and subsequently stored at −20 °C in darkness.

The QuEChERS extraction tube, EMR-Lipid tube, and polish tube were obtained from Agilent Technologies (Santa Clara, CA, USA). The extraction tube included 6 g magnesium of sulfate and 1.5 g of sodium acetate. The EMR-Lipid tube included 1 g of adsorbents (compositions unknown), and the polish tube included 1.6 g of magnesium sulfate and 0.4 g of sodium chloride.

Twenty-two commercial milk samples were collected from local supermarkets in Shanghai, China. The milk samples were commonly consumed, including two from Germany and one from Australia. They consisted of 10 UHT milks (M1–10), 8 pasteurized milks (M11–18), and 4 extended-shelf-life pasteurized milks (M19–22) with different fat contents. The UHT milks were stored at room temperature, and the other two types of milk were stored at 4 °C until analysis.

2.2. Sample Pretreatment

The procedures of PAH extraction and cleanup were conducted according to our previous studies [4,5]. In brief, milk samples were extracted by a mixture of acetonitrile/acetone (2:1, *v/v*) and subsequently successively purified by a QuEChERS extraction salt pack, EMR-Lipid tube, and polish tube. Finally, the extract was concentrated using a nitrogen stream to approximately 20 mg and reconstituted using n-hexane to 200 µL.

2.3. SFC-QqQ-MS Analysis

Separation: A Waters Acquity UPC2 instrument (Milford, MA, USA) was used to separate the n-hexane extract. The SFC separation column was ACQUITY HSS C18 SB (3.0 × 100 mm, 1.8 µm) supplied by Waters (MA, USA), and its temperature was held at 40 °C. The injection volume was 5 µL. Solvents A (CO_2) and B (methanol) were selected as mobile phases, and the gradient elution was as follows: 1% B (0–0.5 min), 10% B (8 min), 20% B (10.5 min), 35% B (11 min), 50% B (11.1–12.5 min), and 1% B (13–15 min). The flow rates were set to the gradient mode as follows: 1.5 mL/min (0–11 min), 1.0 mL/min (11.1–13 min), and 1.5 mL/min (13.5–15 min). The back-pressure was 1600 psi.

Identification: A Xevo TQ-XS (Waters, MA, USA) was linked to UPC2 to identify MS. Methanol with 0.1% formic acid was used as the make-up solvent for MS at a flow rate of 0.2 mL/min. The APCI ion source was used to detect MS in the positive mode. The probe temperature was set at 450 °C, and the desolvation gas flow rate was 1000 L/h. The mass detector was operated in the selected-ion monitoring (SIM) mode with a corona pin voltage value of 2.0 kV. The SIM conditions are shown in Table 1.

Table 1. SIM parameters of target analytes using UPCC-QqQ-MS.

PAHs	Retention Time (min)	Quantification Ion (m/z)	Dwell Time (ms)	Scan Start Time (min)	Corn Voltage (V)
9-FO	1.55	181	100	0.5	20
ATQ	1.73	209	100	0.5	15
BcF	2.02	217	100	1.0	50
BaA	2.77	227	100	2.0	40
Chr	2.94	227	100	2.5	10
5-MChr	3.17	243	100	2.5	10
CP	3.31	229	100	2.5	20
BbF	4.26	253	100	4.0	40
BjF	4.40	253	100	4.0	60
BaP	5.21	253	100	4.5	20
DahA	5.96	279	200	5.5	35
IP	7.31	279	200	7.0	35
BghiP	8.52	277	200	7.5	30
DalP	8.86	303	200	8.0	25
DaeP	10.38	303	200	9.5	40
DaiP	10.65	303	100	9.5	40
DahP	10.82	303	100	9.5	50

2.4. Method Validation

Six UHT milk samples with low concentrations of PAHs were spiked in two levels of 5.0 and 10.0 μg/kg for recovery and precision tests. The calculation of recovery has deducted the concentrations of PAHs in the pseudo blank samples. PAH standard solutions at levels of 1, 2, 5, 10, 20, 40, 60, and 100 μg/kg were tested for linearity. The limits of determination (LODs) were determined using S/N = 3, and the limits of quantification (LOQs) were calculated using S/N = 10. The LODs and LOQs were estimated with 10 blank milk samples and spiked samples (0.5 μg/kg), and PAH calibration curves at 0.2, 0.4, 0.6, 0.8, and 1.0 μg/kg were used.

2.5. Statistical Analysis

The analysis of the sample was conducted in triplicate ($n = 3$). The statistical calculations were performed by SPSS 24.0 software.

3. Results and Discussion

3.1. Optimization of the Conditions of SFC Separation and MS Ionization

By using an HSS C18 SB column, the separation and elution of two OPAHs and EU 15+1PAHs could be completed within 11 min. Since the separation of benzofluoranthenes (BbF, BkF, and BjF) and dibenzopyrenes (DaeP, DaiP, and DahP) was not favorable, further optimization was conducted to obtain the ideal separation parameters.

Usually, a mixture of CO_2 and organic cosolvent is used as the mobile phase to improve the solubility of analytes and avoid precipitation during the injection. During the elution process, organic cosolvents are one of the key factors that affect the peak shape and retention time. Therefore, selecting an appropriate cosolvent is necessary for satisfactory separation [18]. In this study, the composition of the mobile phase was optimized with three different organic cosolvents: methanol, acetonitrile, and formic acid. To compare the separation ability of mobile phases with different polarities and pH values, methanol, methanol/acetonitrile (1:1, *v/v*), and acetonitrile (0.5% formic acid) were selected as mobile phase B. The retention time varied with the change in polarity and pH. When methanol with a strong polarity was used, the resolution of PAHs was good except for BbF, BkF, and BjF. When methanol/acetonitrile (1:1, *v/v*) with a weaker polarity was used, the solubility of PAHs increased with decreasing polarity. The total elution time was shortened by approximately 1 min, but the resolutions of the two OPAHs and two groups of isomers (BaA and Chr, DaeP and DaiP and DahP) were poorer (Figure 1). When acetonitrile (0.5% formic acid) was used, the separation of the two groups of isomers deteriorated, which indicates that the reduction in pH was not conducive to the separation of PAHs. Thus, methanol was selected as mobile phase B.

After the mobile phase was determined, the flow rate and column temperature were chosen at 1.5 mL/min and 40 °C, respectively. During the condition adjustment, we observed that the resolution between the isomers slightly decreased after the initial flow rate decreased from 1.5 mL/min to 1.2 mL/min. The reason may be that the decrease in the flow rate widened the peaks and resulted in a decrease in resolution. We also found that the separation of dibenzopyrenes at 40 °C was better than that at 50 °C because high-molecular-weight PAHs were less soluble in supercritical CO_2 at higher temperatures [19].

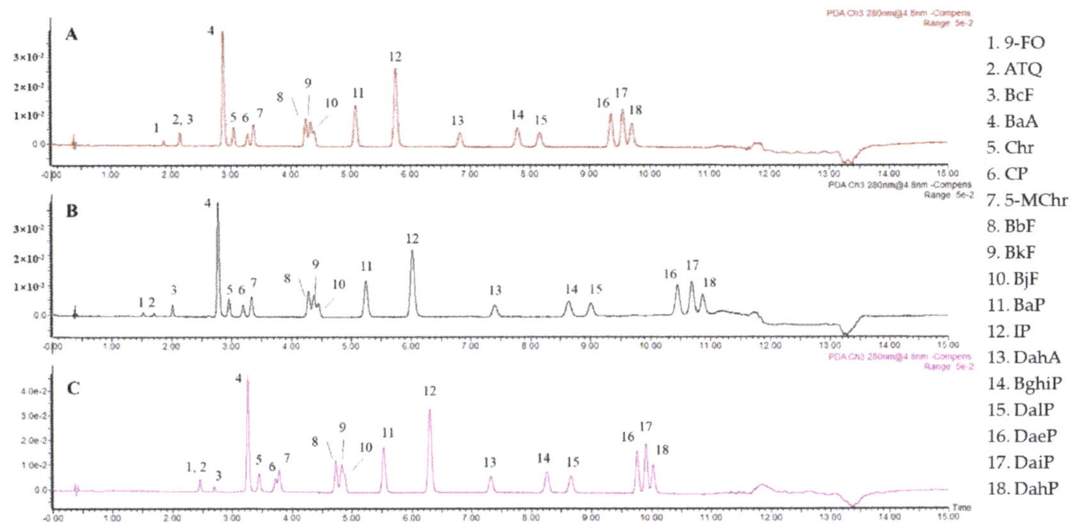

Figure 1. SFC chromatography of 18 PAHs in different mobile phases (1000 μg/kg). (**A**) Methanol/acetonitrile (1:1, *v/v*), (**B**) methanol, and (**C**) acetonitrile (0.5% formic acid), and 1–18 peaks represent 9-FO, ATQ, BcF, BaA, Chr, CP, 5-MChr, BbF, BkF, BjF, BaP, IP, DahA, BghiP, DalP, DaeP, DaiP, and DahP, respectively.

Finally, the MS response was optimized. With increasing corona pin voltage (2.0–2.5 kv), the MS responses of ATQ, BcF, and BaP were stronger, but the responses of other analytes decreased. The responses of BcF improved if the corona pin voltage continued to increase. However, the responses of the other 14 PAHs (except for ATQ and BaP) were reduced. Thus, we chose 2 kV as the corona pin voltage, which also ensured strong intensities of ATQ and BaP. Methanol with different proportions of formic acid was used as the make-up solvent to improve ionization by adjusting the pH. When the composition of formic acid in methanol increased from 0.1% to 0.2%, the response of PAHs decreased, which indicates that the reduction of pH was not conducive to ionization. Similarly, when the flow rate increased from 0.2 mL/min to 0.4 mL/min, the response of each PAH decreased. Therefore, methanol containing 0.1% formic acid was selected as the make-up solvent, and the flow rate was set at 0.2 mL/min.

After the above-mentioned optimization, an acceptable chromatographic separation of the 17 PAHs was achieved (Figure 2). Except for three benzofluoranthenes (BbF, BkF, and BjF), the separation of other neighboring PAHs is satisfactory with resolution values greater than 1.0, which indicates that most of the PAH peaks can be well separated. In comparison, the resolution values of BbF/BkF and BkF/BjF were 0.82 and 0.73, respectively. The quantitative ions of these three isomers are identical (m/z = 253), so their unsatisfactory resolution will inevitably affect the quantitative analysis. Nevertheless, the influence of the poor separation of BkF on its quantitation can be ignored due to its trace lack of detection in milk samples [4] and weak response in this method. The quantitation of BbF and BjF can be well determined since their resolution is above 1.0. If necessary, special columns and mobile phases can be explored for the BkF well separation from its two isomers (BbF and BjF) when using SFC-QqQ-MS for a synchronous quantitative analysis of EU 15+1PAHs.

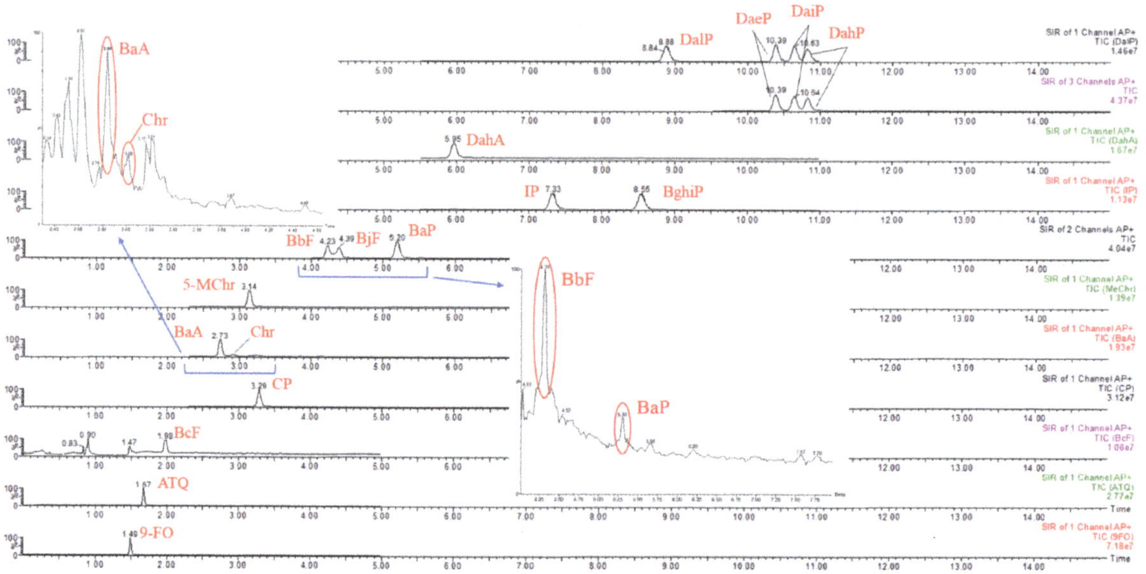

Figure 2. Optimized SFC-QqQ-MS chromatograms of 17 PAHs standards (200 µg/kg) and PAH4 in real milk samples.

3.2. Method Validation

The results are shown in Table 2. All analytes had satisfactory linearity with the R^2 value > 0.99 at the concentration range of 1.0−100.0 µg/kg. At the 5-µg/kg spike level, the recoveries of 17 target compounds were 67.66–118.46%. This recovery conformed to the requirements of EU regulations (EU No 836/2011) on trace analyte quantification. The calculated LOD and LOQ values were 0.05–0.28 µg/kg and 0.13–0.73 µg/kg, respectively. The LOQs of three dibenzopyrenes (DaeP, DaiP, and DahP) significantly improved compared with our GC–MS method [4]. This result is mainly attributed to the better dissolution ability of supercritical CO_2 for PAHs with a high molecular weight and high boiling point. Thus, the chromatographic response of the four dibenzopyrenes was enhanced. This is of great significance for the quantification of PAHs that contain six or more benzene rings in foods, water, beverages, and drinks with ppb to trace PAH levels. The only disadvantage of this method compared to GC–MS is that BkF cannot be determined. This is due to the poor chromatographic separation of BkF with a weak response covered by adjacent peaks in the SFC chromatogram. For milk, it is difficult to achieve accurate quantification of BkF in real samples considering its trace occurrence based on our previous investigation [4]. Therefore, the following application analysis on quantification of PAHs in milk samples targeted only the two OPAHs and EU 15+1PAHs except for BkF.

Table 2. Validation results of the established UPCC-QqQ-MS method.

PAHs	R^2	Low Spike Level 5 µg/kg (%)			Low Spike Level 10 µg/kg (%)			LOD (µg/kg)	LOQ (µg/kg)
		Recovery	RSD_r [a]	RSD_R [b]	Recovery	RSD_r [a]	RSD_R [b]		
9-FO	0.9974	67.66	8.94	12.93	72.46	6.50	6.19	0.04	0.13
ATQ	0.9948	87.77	11.44	13.25	76.80	4.19	8.33	0.11	0.37
BcF	0.9914	118.00	10.23	13.60	106.00	3.23	8.84	0.16	0.53
BaA	0.9935	100.46	3.39	10.89	108.11	12.05	12.70	0.06	0.20
Chr	0.9967	81.83	0.65	6.03	78.40	1.45	5.34	0.18	0.60
5-MChr	0.9969	103.54	1.78	4.72	99.66	4.41	8.65	0.10	0.32
CP	0.9948	110.00	0.32	4.52	93.09	4.59	5.31	0.12	0.40
BbF	0.9963	118.46	8.73	14.59	86.29	4.38	5.86	0.22	0.73
BjF	0.9964	112.23	3.72	8.41	94.97	7.00	11.69	0.24	0.78
BaP	0.9941	114.74	2.67	5.86	112.23	4.47	4.81	0.10	0.33
DahA	0.9919	108.91	5.98	8.42	103.66	6.20	10.87	0.12	0.38
IP	0.9963	109.83	4.88	8.24	110.51	6.05	14.68	0.12	0.41
BghiP	0.9967	109.03	7.14	9.05	98.51	4.25	8.55	0.13	0.42
DalP	0.9965	70.06	4.36	5.90	73.71	3.84	5.19	0.13	0.43
DaeP	0.9932	88.46	4.85	6.83	86.29	4.65	6.79	0.10	0.34
DaiP	0.9940	93.60	6.95	9.56	87.77	5.16	7.98	0.08	0.25
DahP	0.9933	90.29	3.56	6.14	107.66	4.09	6.36	0.15	0.51

[a] The relative standard deviation of repeatability, $n = 6$. [b] The relative standard deviation of intermediate precision, $n = 6$.

3.3. Application of the Developed Method for Real Sample Analysis

Twenty-two commercial milk samples were successfully analyzed using the established SFC-QqQ-MS method. Table 3 shows the PAH4 and PAH8 concentrations and the BaP toxic equivalent quantity (BaPeq) of the 22 samples [2]. Because there were no reliable toxic equivalency factor (TEF) data, the calculation of BaPeq did not contain BcF, 9-FO, or ATQ. The results indicate that the total concentrations of 17 PAHs in the 22 tested samples were 1.63–11.01 µg/kg. The average PAH4 concentrations in the UHT milks, pasteurized milks, and extended-shelf-life pasteurized milks were 2.27 µg/kg, 1.08 µg/kg, and 1.28 µg/kg, respectively. Seven samples contained more than 2.0 µg/kg PAH4, which was approximately 32% of the investigated products. As a PAH4, BaP was present at concentrations of 0.41–0.77 µg/kg in 9 UHT samples, which consisted of a 90% BaP rate in all tested UHT products. In contrast, BaP was not detected in pasteurized milks or extended-shelf-life pasteurized milks. The BaP concentrations in all samples were lower than 1.0 µg/kg, while the PAH4 concentrations in 73% of the samples were higher than 1.0 µg/kg.

This result suggests that raw milks should be strictly monitored for PAH4 and BaP concentrations if the infant formula considers the legal limitations of BaP and PAH4 by the EU (<1.0 µg/kg). Theoretically, according to the conversion rate of fresh milk into milk powder, fresh milk containing both BaP and PAH4 should not exceed 0.14 µg/kg [4]. In this view, all 22 samples were not suitable for infant formula production, since their PAH4 concentrations were much more than 0.14 µg/kg. Three other PAHs (Chr, BaA, and BbF) showed detection rates of 86%, 59%, and 50%, respectively.

DahA was detected in two samples, so its health risk should be highly considered due to its TEF at 5.0. Four dibenzopyrenes (DalP, DaeP, DaiP, and DahP) were detected in seven samples. The detection rates of DalP, DaeP, DaiP, and DahP were 27%, 14%, 14%, and 5%, respectively. Although these four analytes had generally low concentrations, their TEFs were relatively high. DalP belongs to class-2A carcinogens, while DaeP, DaiP, and DahP belong to class-3 and -2B carcinogens, respectively. DaeP had a TEF of 1.0, but all TEFs of DalP, DaiP, and DahP were 10.0, which increased BaPeq of the samples where dibenzopyrenes were detected. For the two samples with BaPeq above 10.0, the contributions of dibenzopyrenes were 71% and 92%. In fact, dibenzopyrenes were the main contributors of BaPeq in each sample, which is consistent with the results of environmental samples [20]. In particular, these four analytes were simultaneously detected in one milk sample, and three of them were simultaneously detected in the other two milk samples.

Table 3. OPAH and PAH concentrations of 22 commercial milk samples by UPCC- QqQ-MS.

Samples	Concentration (μg/kg)						
	OPAHs		Parent PAHs				
	9-FO	ATQ	BaP	PAH4	PAH8	EU PAHs [a]	BaPeq [b]
M1	0.91	1.62	0.41	2.01	3.51	8.48	17.42
M2	0.59	1.74	0.50	1.44	2.05	5.28	12.18
M3	1.15	1.10	0.47	2.80	3.30	6.69	6.44
M4	0.78	1.13	0.49	2.37	2.37	4.42	1.66
M5	1.56	0.87	0.46	3.61	3.61	4.64	5.40
M6	0.88	1.52	0.45	3.28	3.28	5.41	4.30
M7	1.15	1.60	0.54	1.66	1.66	3.92	7.90
M8	0.74	1.29	Nd [c]	1.98	1.98	3.03	3.91
M9	0.82	1.03	0.66	1.21	1.21	2.42	0.72
M10	0.68	0.74	0.77	2.38	2.96	4.77	0.94
M11	0.96	0.56	nd [c]	1.43	1.43	1.80	0.44
M12	0.43	1.27	nd [c]	2.02	2.38	4.69	6.06
M13	1.09	0.56	nd [c]	0.98	0.98	1.89	0.53
M14	0.39	0.40	nd [c]	0.57	0.57	0.57	0.57
M15	0.50	0.51	nd [c]	0.83	0.83	0.83	0.01
M16	0.44	0.63	nd [c]	0.84	0.84	0.84	0.01
M17	0.46	0.60	nd [c]	1.02	1.02	2.87	0.49
M18	0.45	0.56	nd [c]	0.92	0.92	0.92	0.01
M19	0.77	0.66	nd [c]	1.86	1.86	2.21	0.46
M20	0.32	0.56	nd [c]	1.14	1.14	1.92	0.37
M21	0.87	0.58	nd [c]	0.85	0.85	1.31	0.01
M22	0.86	0.49	nd [c]	1.26	1.26	2.84	0.62

[a] EU 15+1 PAHs except for BkF. [b] BaPeq was calculated using the 15 parent PAHs. [c] nd: not detected.

The detection rate of the two OPAHs was 100%, which is consistent with our previous results [4]. The parent PAH of 9-FO is fluorene, and the parent PAH of ATQ is anthracene, both of which belong to the US EPA 16PAHs. Our previous study investigated 89 milk samples from 9 different countries, which indicates a much higher average concentration of anthracene than that of other PAHs [5]. The high content of anthracene may cause a high concentration of ATQ.

3.4. Influencing Factors of PAH Content in Different Types of Milks

Pasteurized milks are commonly considered better in flavor and nutrition, but they are limited by shorter shelf life (seven days) and lower storage temperature than UHT milks. Currently, the dominant liquid milk products in the Chinese market are UHT milks. In general, UHT milks enjoy six to twelve months of shelf life. Alternatively, a new product named extended shelf life milks appeared in recent years [21]. They combined pasteurization treatment with new processing technology to extend their shelf life to 14–20 days and can be considered a special type of pasteurized milk that also must be stored at 4 °C.

Figure 3A shows the average concentrations of 17 PAHs in different types of milks. The top concentrations of the three PAHs were ATQ, 9-FO, and BcF in UHT milk; Chr, ATQ, and 9-FO in pasteurized milk; and Chr, 9-FO, and ATQ in extended-shelf-life pasteurized milk. Two OPAHs were detected in all tested samples, and no significant difference was observed in the concentrations of different types of milks. In pasteurized and extended-shelf-life pasteurized milks, the Chr concentration was higher than that in UHT milks; the BaA, CP, and BbF contents were significantly lower than those in UHT milk ($p < 0.05$); and the BaP, BghiP, DaeP, DaiP, and DahP concentrations were all below the LOQ. In previous studies, pasteurized milk also had significantly higher Chr concentrations and significantly lower BaA and IP concentrations than UHT milk [5,12]. In addition, Naccari et al. [3] found that unheated raw milk had significantly lower PAH concentrations than pasteurized or UHT-treated milk. The reason is that PAHs in raw milk only came from the environment, while PAHs in pasteurized milk were the sum of PAHs from the environment and the heating process. Rawash et al. [22] also found that raw milk had lower PAH concentrations than pasteurized milk and UHT milk in the detection of PAH pollution in Egyptian milk. Overall, these results confirm that the factory sterilization processing of milk may form PAHs and OPAHs. How to minimize the influence of processing on PAHs and OPAHs in milks requires further study.

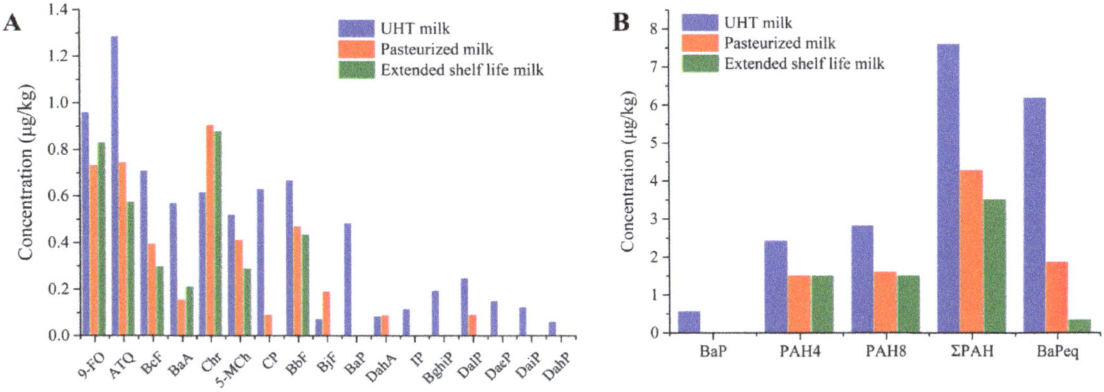

Figure 3. PAH concentrations in milks with different sterilization processes. (**A**) individual PAH analytes. (**B**) Concentrations of BaP, PAH4, PAH8, 17 PAHs, and BaPeq.

Figure 3B shows the difference in average PAH concentrations in three different types of milks. UHT milks had a higher total PAH concentration than pasteurized milk and extended-shelf-life pasteurized milk. In this study, the detection rates of heavy PAHs in pasteurized milks were lower than those in UHT milks. Therefore, the total PAH concentration and BaPeq of UHT milks were significantly higher. In addition to the influence of sterilization temperature and time, different storage times may affect PAHs and OPAHs in the milk samples. A previous study found that the concentrations of PAHs in soybean oil and rapeseed oil significantly increased with storage time, and the concentrations increased faster at room temperature than at 4 °C [23]. Pasteurized milk and extended-shelf-life pasteurized milks had a shelf life of less than 20 days and were stored at 4 °C, while UHT milks had a shelf life of at least six months and were stored at room temperature. During storage, oxidation, migration, and transformation of PAHs occur in milks, which changed the PAH and OPAH concentrations. Overall, the extended-shelf-life pasteurized milks had lower total PAH concentrations than pasteurized milks. Based on the total PAH concentration and storage time data of 14 whole milks, we calculated the Pearson correlation coefficient between storage time and total PAH concentrations and obtained 0.89 (Figure 4). This result suggests that the concentration of PAHs in milks increased with storage time.

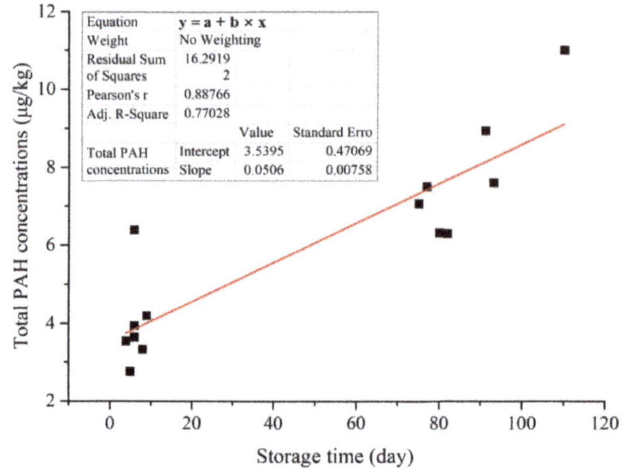

Figure 4. Correlation analysis of storage time and total PAH concentration in milks. Red line: the linear regression model.

Figure 5 compares the concentrations of PAH4, total PAH, and BaPeq in milk samples with different fat contents. In the pasteurized milks and UHT milks, the concentrations of PAHs in whole milks were higher than those in the corresponding skim and semi-skim samples. Various studies have proven that the concentration of PAHs in milks is positively correlated with fat content [4,24,25]. This study calculated the BaPeq value between whole milks and semi-skim/skim milks, which indicates significant differences between whole milks and semi-skim/skim milks. Thus, more attention should be given to the potential PAH health risks of whole milks.

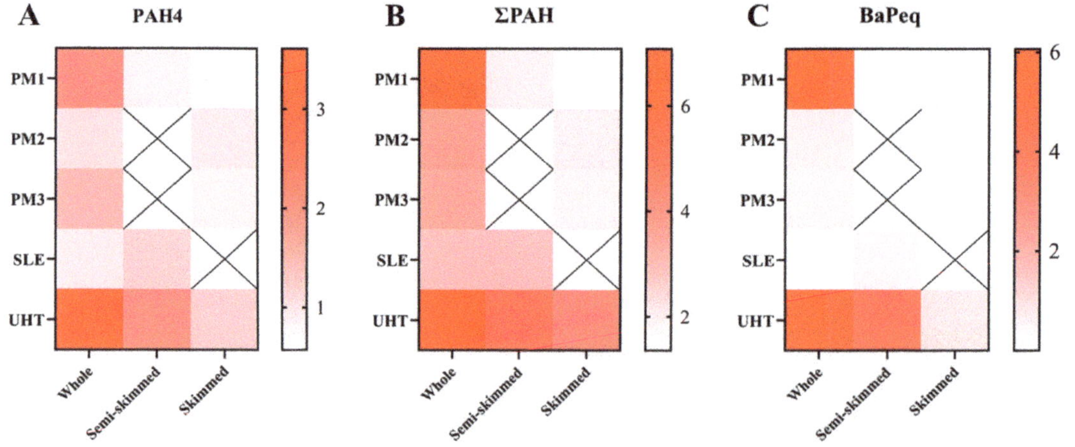

Figure 5. Comparison of PAHs among milk samples with different fat contents. PM: pasteurized milk, ESL: extend-shelf-life pasteurized milks. (**A**) concentration of PAH4 (BaP, BaA, Chr, BbF), (**B**) Concentration of 17 PAHs, (**C**) BaPeq of 17 PAHs.

4. Conclusions

This study established an SFC-QqQ-MS method with the advantage of rapid, sensitive, and solvent-saving nature to determine PAH and OPAH in milks. The EU 15+1PAHs except for BkF and two OPAHs (9-FO and ATQ) were well separated and determined in

22 different commercial milks. Compared with GC–MS, the instrument analysis time was shortened by 1/2, and the LOQs of four dibenzopyrenes (DalP, DaeP, DaiP, and DahP) were significantly lower. Higher PAH concentrations were detected in the UHT milks compared to pasteurized milks and whole milks compared to semi-skim/skim milks. This result suggests that more attention should be given to the potential PAH health risks of UHT milks and whole milks. The noticeable BaPeq in the 22 milk samples indicates that the maximum liquid milk intake should be defined for populations of different ages. Further studies are necessary to explain the lower PAH level in extended-shelf-life pasteurized milks, elucidate the occurrence of dibenzopyrenes in milks, and minimize the influence of processing on PAHs and OPAHs in milks, including sterilization, skimming, and storage conditions.

Author Contributions: L.Z.: Data curation, Formal analysis, Investigation, Methodology, Validation, Software, Writing-original draft. W.L.: Data curation, Formal analysis, Writing-original draft. S.W.: Conceptualization, Funding acquisition, Project administration, Supervision, Writing-review & editing. All authors have read and agreed to the published version of the manuscript.

Funding: This work was supported by the National Natural Science Foundation of China (No. 32061160476).

Data Availability Statement: No data was used for the research described in the article.

Conflicts of Interest: The authors declare no conflict of interest.

Abbreviations

ATQ	anthraquinone
BaA	benz[a]anthracene
BaP	benzo[a]pyrene
BbF	benzo[b]fluoranthene
BcF	benzo[c]fluorene
BghiP	benzo[ghi]perylene
BjF	benzo[j]fluoranthene
BkF	benzo[k]fluoranthene
Chr	chrysene
CP	cyclopenta[cd]pyrene
DaeP	dibenzo[a,e]pyrene
DahA	dibenz[a,h]anthracene
DahP	dibenzo[a,h]pyrene
DaiP	dibenzo[a,i]pyrene
DalP	dibenzo[a,l]pyrene
9-FO	9-fluorenone
IP	indeno [1,2,3-cd]pyrene
5-MChr	5-methylchrysene
QuEChERS	quick, easy, cheap, effective, rugged, and safe

References

1. Amirdivani, S.; Khorshidian, N.; Dana, M.G.; Mohammadi, R.; Mortazavian, A.M.; de Souza, S.L.Q.; Rocha, H.B.; Raices, R. Polycyclic aromatic hydrocarbons in milk and dairy products. *Int. J. Dairy Technol.* **2019**, *72*, 120–131. [CrossRef]
2. Usman, A.T.; Abugu, H.O.; Okoye, C.O.B. Environmental impact and human health risk assessment of polycyclic aromatic hydrocarbons (PAHs) in raw milk from free-ranging cattles in northwest nigeria. *J. Environ. Health Sci.* **2021**, *19*, 1523–1534. [CrossRef]
3. Naccari, C.; Cristani, M.; Giofre, F.; Ferrante, M.; Siracusa, L.; Trombetta, D. PAHs concentration in heat-treated milk samples. *Food Res. Int.* **2011**, *44*, 716–724. [CrossRef]
4. Yan, K.; Wu, S.M.; Gong, G.Y.; Xin, L.; Ge, Y.X. Simultaneous Determination of Typical Chlorinated, Oxygenated, and European Union Priority Polycyclic Aromatic Hydrocarbons in Milk Samples and Milk Powders. *J. Agric. Food Chem.* **2021**, *69*, 3923–3931. [CrossRef]
5. Sun, Y.Q.; Yan, K.; Wu, S.M.; Gong, G.Y. Occurrence, spatial distribution and impact factors of 16 polycyclic aromatic hydrocarbons in milks from nine countries. *Food Control* **2020**, *113*, 107197. [CrossRef]

6. Sun, Y.Q.; Wu, S.M. Analysis of PAHs in oily systems using modified QuEChERS with EMR-Lipid clean-up followed by GC-QqQ-MS. *Food Control* **2019**, *109*, 106950. [CrossRef]
7. Colón, L.P.; Rascón, A.J.; Ballesteros, E. Trace-Level Determination of Polycyclic Aromatic Hydrocarbons in Dairy Products Available in Spanish Supermarkets by Semi-Automated Solid-Phase Extraction and Gas Chromatography–Mass Spectrometry Detection. *Foods* **2022**, *11*, 713. [CrossRef]
8. Sonego, E.; Bhattarai, B.; Duedahl-Olesen, L. Detection of Nitrated, Oxygenated and Hydrogenated Polycyclic Aromatic Compounds in Smoked Fish and Meat Products. *Foods* **2022**, *11*, 2446. [CrossRef] [PubMed]
9. Racovita, R.C.; Secuianu, C.; Israel-Roming, F. Quantification and risk assessment of carcinogenic polycyclic aromatic hydrocarbons in retail smoked fish and smoked cheeses. *Food Control* **2021**, *121*, 107586. [CrossRef]
10. Barp, L.; Moret, S.; Purcaro, G. Monitoring and Occurrence of Heavy PAHs in Pomace Oil Supply Chain Using a Double-Step Solid-Phase Purification and HPLC-FLD Determination. *Foods* **2022**, *11*, 2737. [CrossRef] [PubMed]
11. Santonicola, S.; Albrizio, S.; Murru, N.; Ferrante, M.C.; Mercogliano, R. Study on the occurrence of polycyclic aromatic hydrocarbons in milk and meat/fish based baby food available in Italy. *Chemosphere* **2017**, *184*, 467–472. [CrossRef] [PubMed]
12. Girelli, A.M.; Sperati, D.; Tarola, A.M. Determination of polycyclic aromatic hydrocarbons in Italian milk by HPLC with fluorescence detection. *Food Addit. Contam. A* **2014**, *31*, 703–710. [CrossRef] [PubMed]
13. Lesellier, E.; West, C. The many faces of packed column supercritical fluid chromatography—A critical review. *J. Chromatogr. A* **2015**, *1382*, 2–46. [CrossRef] [PubMed]
14. Velde, B.V.; Guillarme, D.; Kohler, I. Supercritical fluid chromatography—Mass spectrometry in metabolomics: Past, present, and future perspectives. *J. Chromatogr. B* **2020**, *1161*, 122444. [CrossRef] [PubMed]
15. Kaplitz, A.S.; Berger, T.A.; Berger, B.K.; Schug, K.A. A review of fraction collection technology for supercritical fluid chromatography. *TrAC-Trend. Anal. Chem.* **2022**, *151*, 116588. [CrossRef]
16. Yoshioka, T.; Nagatomi, Y.; Harayama, K.; Bamba, T. Development of an analytical method for polycyclic aromatic hydrocarbons in coffee beverages and dark beer using novel high-sensitivity technique of supercritical fluid chromatography/mass spectrometry. *J. Biosci. Bioeng.* **2018**, *126*, 126–130. [CrossRef]
17. Tang, S.D.; Liu, S.; Frank, B.P. Feasibility of supercritical fluid extraction-supercritical fluid chromatography mass spectrometry for the determination of polycyclic aromatic hydrocarbons in particulate matter samples. *J. Sep. Sci.* **2021**, *44*, 3717–3726. [CrossRef]
18. Huang, Y.; Tang, G.; Zhang, T.; Fillet, M.; Crommen, J.; Jiang, Z. Supercritical fluid chromatography in traditional Chinese medicine analysis. *J. Pharm. Biomed. Anal.* **2018**, *147*, 65–80. [CrossRef]
19. Amziane, A.; Monteau, F.; el djalil Lalaouna, A.; Alamir, B.; Le Bizec, B.; Dervilly, G. Optimization and validation of a fast supercritical fluid chromatography tandem mass spectrometry method for the quantitative determination of a large set of PFASs in food matrices and human milk. *J. Chromatogr. B* **2022**, *1210*, 123455. [CrossRef] [PubMed]
20. Jiang, Y.F.; Shi, L.P.; Mu, Z.F.; Sun, H.; Zhan, H.Y.; Wu, Y.Q. Occurrence of and risk posed by highly carcinogenic dibenzo[def,p]chrysene isomers in urban soil. *Toxicol. Environ. Chem.* **2017**, *99*, 824–836. [CrossRef]
21. Rankin, S.A.; Lopez-Hernandez, A.; Westover, A.R. Super-Pasteurized Milk (Extended Shelf-Life Milk). In *Encyclopedia of Dairy Sciences*, 3rd ed.; Elsevier: Amsterdam, The Netherlands, 2022; pp. 470–476. [CrossRef]
22. Rawash, E.S.A.; Mohamed, G.G.; Souaya, E.R.; Khalil, L.H.; El-Chaghaby, G.A.; El-Gammal, M.H. Distribution and Health Hazards of Polycyclic Aromatic Hydrocarbons in Egyptian Milk and Dairy-Based Products. *Beverages* **2018**, *4*, 63. [CrossRef]
23. Zhao, X.; Gong, G.Y.; Wu, S.M. Effect of storage time and temperature on parent and oxygenated polycyclic aromatic hydrocarbons in crude and refined vegetable oils. *Food Chem.* **2018**, *239*, 781–788. [CrossRef] [PubMed]
24. Battisti, C.; Girelli, A.M.; Tarola, A.M. Polycyclic aromatic hydrocarbons (PAHs) in yogurt samples. *Food Addit. Contam. B* **2015**, *8*, 50–55. [CrossRef] [PubMed]
25. Londono, V.A.G.; Garcia, L.P.; Scussel, V.M.; Resnik, S. Polycyclic aromatic hydrocarbons in milk powders marketed in Argentina and Brazil. *Food Addit. Contam. A* **2013**, *30*, 1573–1580. [CrossRef] [PubMed]

foods

MDPI

Article

The Effects of Volatile Organic Compounds (VOCs) on the Formation of Heterocyclic Amines (HAs) in Meat Patties, under Different Smoking Temperatures and Durations

Xing Shen [1,2], Yang Chen [2], Jacob Ojobi Omedi [3], Emel Oz [4], Fatih Oz [4], Chunwang Xiao [1,*], Yijun Zhou [1], Jie Chen [2] and Maomao Zeng [2,*]

1 College of Life and Environmental Sciences, Minzu University of China, Beijing 100081, China
2 State Key Laboratory of Food Science and Technology, Jiangnan University, Wuxi 214122, China
3 School of Food Science and Technology, Jiangnan University, Wuxi 214122, China
4 Department of Food Engineering, Faculty of Agriculture, Ataturk University, Erzurum 25240, Turkey
* Correspondence: cwxiao@muc.edu.cn (C.X.); mmzeng@jiangnan.edu.cn (M.Z.)

Abstract: In this study, UPLC-MS/MS was used to study the effects of smoking duration and temperature on the formation of heterocyclic amines (HAs) in smoke-processed meat patties. Four kinds of free HAs—including F-7,8-DiMeIQx; F-MeAαC; F-Harman and F-Norharman—and six kinds of protein-bound HAs—including B-AαC; B-7,8-DiMeIQx; B-Glu-p-1; B-MeAαC; B-Harman and B-Norharman—were detected and quantified. Among the free HAs, we observed a 23-fold content increase ($p < 0.05$), from 0–4 h (at 0 h and 4 h they were 4.24 ng·g^{-1} and 98.33 ng·g^{-1}, respectively), and the content of the free HAs decreased to 78.80 ng·g^{-1}, at 5 h. At the same time, the free HAs content increased from 53.52 ng·g^{-1}, at 50 °C, to 127.16 ng·g^{-1}, at 60 °C, and then decreased continuously. The content of the free HAs was the highest at 60 °C. For the protein-bound HAs, their content was found to generally decrease with the increase in smoking duration and temperature. However, at 5 h, the content of protein-bound HAs slightly increased to 984.2 ng·g^{-1}. Meanwhile, at 90 °C, it increased to 1643.53 ng·g^{-1}. Additionally, a total of 16 volatile organic compounds (VOCs) were found in all of the meat samples, of which 10 VOCs (one acid, three aldehydes and seven phenols) were significantly related to the formation of free HAs. These findings showed that all the different types of HAs were produced under low-temperature processing, which provided scientific insights into the potential generation of HAs during meat smoking processes and could be used as a reference to minimize the risks of cancer related to the consumption of smoked meat products.

Keywords: heterocyclic amines; smoking durations; smoking temperatures; volatile organic compounds; UPLC-MS/MS; meat patties

1. Introduction

Food smoking is one of the oldest food preservation techniques and belongs to a typical thermal processing food treatment method. During the smoking of meat products, the process flow generally involves curing, air drying, ripening and smoking [1]. With increasing consumer awareness, there has been increased research interest within the food industry on the quality and improvement of smoked meat [2,3]. To improve the quality of smoked meat, researchers have explored the significance of the smoking parameters—e.g., the chemical compositions of the smoke, and the smoking temperature and duration—in generating smoked foods, which are attributed to the diverse compounds present in smoked products due to the incomplete combustion of wood; the interactions between the smoked chemical components and the meat components; the effects of the heat generated by the wood combustion on the meat components (e.g., lipid, protein and carbohydrates); and the complex effects of smoking on the unique sensory and aroma qualities of meat products [4,5]. The published literature shows the different smoking conditions and smoked constituents that have been commonly explored to enhance the quality of smoked meat. For instance, phenolic compounds in smoked food have been found to be partly responsible for the flavor of meat and conferred antioxidant properties to smoked meat during storage, while carbonyl compounds have been associated with the attractive brown color formed on smoked meat's surfaces [2]. However, the processing of smoked foods has been associated with the formation of and/or the production of hazardous substances such as polycyclic aromatic hydrocarbons and acrylamide [6,7]. With consumer safety and quality awareness in mind, little is known about the presence of heterocyclic amines (HAs) in smoked foods [8].

HAs are heterocyclic aromatic hydrocarbons with known carcinogenic and mutagenic properties. Presently, 30 HAs have been isolated and identified [9]. The International Agency for Research on Cancer (IARC) defined MeIQ, MeIQx, PhIP, Glu-P-1, AαC and MeAαC as class 2B carcinogens, and IQ as a class 2A carcinogen [10]. A high intake of HAs was reported to significantly increase the risks of cancer, particularly the incidence of colorectal cancer in a human case-control study [11]. A variety of HAs have been detected in several heat-processed meat products, such as sausages [12], grilled fish [13], smoked and roasted poultry [8], etc., indicating that humans are at risk of continuous exposure to HAs through their dietary intake. Ordinary thermal processing, such as hot air drying or baking, can cause water to migrate on food surfaces, enhance the migration of HAs' precursors and increase water mobility during the drying effect in the smoking process, resulting in the accumulation and formation of HAs on food surfaces. However, the effects of smoking on the formation of HAs differs from other thermal processing processes. The constituents of the smoke during the smoking process have an additional influence on the carbonyl compounds of HAs' intermediates and can lead to the introduction of phenolic compounds (HA inhibitors) [14]. Various phenolic compounds were detected in smoke fueled by jarrah, karri, marri, oats and pine, with their content varying greatly, based on the type of fuel used [15]. Recent studies on the formation mechanism have revealed that reactive carbonyls, such as phenylacetaldehyde, acrolein and crotonaldehyde, were intermediate products of HAs' formation. An increase or decrease in these intermediates [16] can change the levels of HAs, which may significantly affect the Maillard reaction pathway, resulting in the uncertainty of HAs' formation. Thus, studying the formation and accumulation of HAs in smoked meat might offer several practical significances.

Considering that exploring the influence of different smoking processing methods on the formation of HAs in smoked meat could provide a basis for food quality control in the smoking process stage—which could be adaptable in the industrialized production of smoked food—the objectives of this study were to analyze the formation of HAs during different meat smoking conditions, including different temperatures (50, 60, 70, 80 and 90 °C) and durations (1, 2, 3, 4 and 5 h), and to study the VOCs in meat patties under different processing conditions. The relationship between each VOC and HAs' formation was ascertained.

2. Materials and Methods

2.1. Chemicals and Materials

HAs' standards—including 2-amino-9H-pyrido[2,3-b]indole (AαC); 2-amino-3-methyl-9H-pyrido[2,3-b]indole (MeAαC); 1-methyl-9H-pyrido[3,4-b]indole (Harman); 9H-pyrido indole (Norharman); 2-amino-3-methylimidazo[4,5-f]quinoline (IQ); 2-amino-3,4-dimethylimidazo[4,5-f]quinoline (MeIQ); 2-amino-1-methylimidazo[4,5-b]quinoline (IQ[4,5-b]); 2-amino-1-methyl-6-phenylimidazo[4,5-b]pyridine (PhIP); 2-amino-1,6-dimethylimidazo[4,5-b]pyridine (DMIP); 2-amino-1,5,6-trimethylimidazo[4,5-b]pyridine (1,5,6-TMIP); 2-amino-3-methyl-3H-imidazo[4,5-f]quinoxaline (IQx); 2-amino-3,8-dimethylimidazo[4,5-f]quinoxaline (MeIQx); 2-amino-3,4,8-trimethylimidazo[4,5-f] quinoxaline (4,8-DiMeIQx); 2-amino-3,7,8-trimethyl-3H-imidazo[4,5-f]quinoxaline (7,8-Di-MeIQx); 2-amino-3,4,7,8-tetramethyl-3H-imidazo[4,5-f]quinoxaline (4,7,8-TriMeIQx); 2-amino-5-phenylpyridine (Phe-P-1); and 2-amino-6-methyldipyrido[1,2-α:3′,2′-d]imidazole

(Glu-P-1)—were purchased from Santa Cruz Biotechnology, Inc. (Santa Cruz, CA, USA). All standards had a purity level >99.9%. All 17 HAs were diluted with chromatographic grade methanol to a standard solution (125 mg·L^{-1}). UPLC-MS/MS-grade methanol, acetonitrile and formic acids were obtained from Thermo Fisher Scientific (Waltham, MA, USA). Analytical-grade diatomaceous earth was purchased from Sinopharm Chemical Reagent Co., Ltd. (Shanghai, China). Oasis MCX cartridges (3 cc, 60 mg^{-1}) were purchased from Waters (Shanghai, China). Pork and salt were purchased from a local supermarket (Wuxi, China).

2.2. Preparation of Meat Patties

Pork (tenderloin) was thoroughly ground using a meat mincer (BJRJ-12T, Jiaxing, China). The meat tendons were removed, followed by adding 1.5% NaCl, rolling the mixture for 1 h in a vacuum curing machine (Sunhow KA-6189A, Shenzhen, China) and pickling under vacuum refrigeration at 4 °C, for 18–20 h. Lastly, the minced meat (80 g) was pressed into 10 cm diameter meat patties, using a meat patty forming machine (Hualing ZF-100, Maanshan, China).

2.3. Smoking of Meat Patties

The meat patties were dried for 0.5 h in the Self-Cooking Center of RATIONAL SCC 61 (Landsberg, Germany), whose oven was pre-heated to 70 °C. The measurement was conducted in triplicate. The meat patties were then naturally cooled at room temperature. After cooling, they were separated into the following two treatment groups: Group 1, in which the meat patties were smoked for 0, 1, 2, 3, 4 and 5 h, at 70 °C; and Group 2, in which the meat patties were smoked for 3 h at ambient temperatures (AMB), 50, 60, 70, 80 and 90 °C. As each sample took a long time to be smoked, it took 2–3 days to complete the total processing of each group. Therefore, the meat of the two groups was purchased separately. In all treatments, the smoke was produced by VarioSmoker 61-201 (Landsberg, Germany). VarioSmoker was used together with Self-Cooking Center. In the experiment, the weight of woodchips used for each fumigation was 120 g, and the VarioSmoker was put in the Self-Cooking Center, as shown in Figure S1. A total of 12 kinds (Figure 1) and 36 smoked meat patties were produced, with each treatment conducted in triplicate. After smoking, the meat patties were cooled at room temperature and weighed. Lastly, the smoked meat patties were crushed in a food processor, freeze-dried and stored at −20 °C for further analysis.

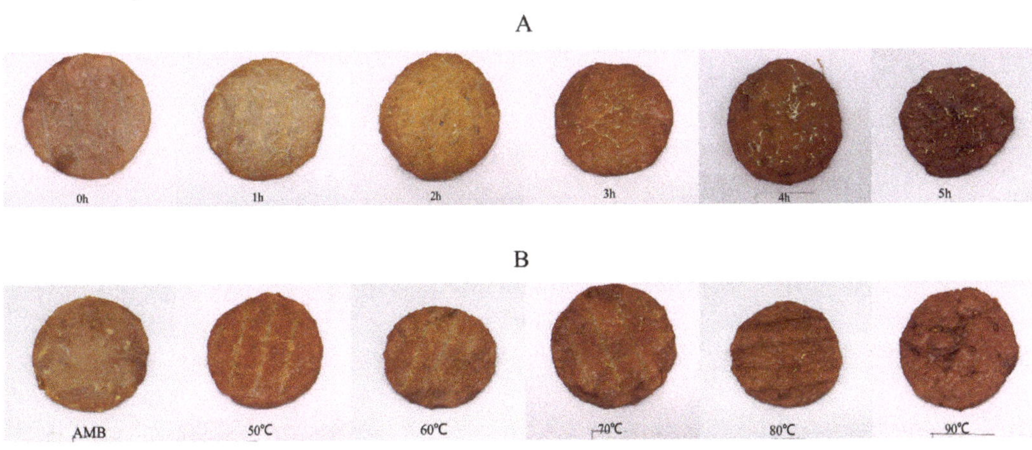

Figure 1. Smoked meat patties at different smoking durations (**A**) and temperatures (**B**). Group 1, in which the meat patties were smoked for 0, 1, 2, 3, 4 and 5 h, at 70 °C; and Group 2, in which the meat patties were smoked for 3 h, at ambient temperatures (AMB), 50, 60, 70, 80 and 90 °C.

2.4. Analysis of Conventional Physicochemical Properties of Smoked Pork Patties

The surface color of the smoked meat patties was determined according to the method of Zhang et al. [17], but we made a slight change. An automatic colorimeter (Kangguang WB-2000IXA, Beijing, China) was used to measure the color of freeze-dried smoked meat patties. The instrument was calibrated using a standard whiteboard (X = 83.24, Y = 87.98, Z = 91.61). A standard D65 light source, with a 10° field of view, a Φ18 mm reflection measurement aperture and a Φ10 mm projection measurement aperture was also used. After calibration, the meat patty was tiled on the round measuring port for measurement. Then, the lightness (L^*), redness (a^*) and yellowness (b^*) of the sample were recorded. Each sample was analyzed in triplicate.

The water activity (Aw, AQUALAB 4TEV, Jinan, China) and cooking loss were detected according to the method described by Zhang et al. [17].

The changes in meat pH (METTLER TOLEDO FE20, Shanghai, China) were determined using a previously described assay [18].

The texture properties of the pork patties were measured by TA. XT plus texture analyzer (Stable Micro Systems Ltd., Godalming, UK) was used according to the previously described methods, with some modifications [19]. A flat-bottomed cylindrical probe, P/36R (36 mm diameter), was used to perform texture profile analysis (TPA). The test conditions used were as follows: 3 mm·s^{-1} pre-test speed, 2 mm·s^{-1} test speed, 3 mm·s^{-1} post-test speed, 50% strain, 5 s dwell time interval and a trigger force of 5 g.

2.5. Determination of the VOCs in Smoked Pork Patties

The VOCs in the pork patties were identified by GC-MS (GCMS-QP2020NX, Shimadzu, Kyoto, Japan), according to the methods described by Rizzo et al. [20], with some modifications. Briefly, 2 g pork patty powder was added to 20 mL glass vials, with Teflon silicon septa lids. An internal standard of 20 μL of 2-methyl-3-heptanone (118.2 μg·L^{-1} in methanol) was added to each vial before SPME fiber exposure. The SPME fibers used were 1.0 cm in length with DVB/CAR/PDMS coating (Supelco, Bellefonte, PA, USA). Before exposure to the sample bottle, the SPME fiber was pre-heated at 270 °C, for 30 min, in the GC inlet. The sample bottles were equilibrated to 50 °C, for 30 min, during exposure. The samples were then manually injected at a depth of 5.0 cm into a split-less GC inlet, at 250 °C, onto an SH-WAX column (30 m × 0.25 mm, 0.25 μm; Shimadzu), with a constant flow rate of 0.8 mL·min^{-1} of helium. The condition of GC-MS was set as described by Rizzo et al. [20], with some modifications. The GC oven was initially set to 40 °C and held for 3 min. The temperature ramp began at a rate of 6 °C·min^{-1} to 100 °C, then accelerated to 10 °C·min^{-1} to 230 °C, which was held for 5 min. The purge time was set for 1 min. The MS transfer line was held at 250 °C, with the ion source held at 200 °C.

2.6. Sample Preparation for HAs

Free HAs in smoked meat were extracted and concentrated according to the method described by Zeng et al. [21], with some modifications. Protein-bound HAs were extracted according to the method described by Chen et al. [22]. Briefly, 3 g of smoked meat powder and 30 mL sodium hydroxide (1 mol·L^{-1}) solution was added and homogenized (12,000 rpm, 1 min) (RayKol AH-30, Xiamen, China). Then, 13 g of diatomite was added to the homogenate and mixed, followed by adding and mixing 50 mL ethyl acetate and ultrasonic (40 KH_Z, 250 W, SCIENTZ SB-5200, Ningbo, China) treatment, for 30 min. The extraction procedure was repeated twice, followed by centrifugation (10,000× *g*, 10 min) to obtain a supernatant, which was concentrated with nitrogen to 10 mL. Furthermore, the filtrate was collected and added to a 48 mL thick-walled, pressure-resistant bottle, followed by the addition of 40 mL of 6 mol·L^{-1} HCl to the bottle, bubbling nitrogen through for 1 min and heating at 110 °C, for 24 h. Next, they were centrifuged and the supernatant collected was diluted with ultrapure water to 100 mL, from which 10 mL was used for solid phase extraction (SPE).

2.7. SPE Procedure

For free HAs, Oasis MCX cartridges were activated in advance with 6 mL water, 6 mL methanol and 6 mL ethyl acetate, by automatic SPE (RayKol Fotector-04HT, Xiamen, China). For protein-bound HAs, Oasis MCX cartridges were activated in advance with 6 mL methanol, 6 mL water and 6 mL 0.1 mol·L^{-1} HCl. Ten milliliters (10 mL) of extract were added to the SPE column, the effluent was discarded and they were rinsed with 6 mL 0.1 M HCl and 6 mL methanol. The retained HAs were eluted with 6 mL methanol-ammonia mixture (19:1, v/v), at a flow rate of 0.6 mL·min^{-1} [22]. The eluted mixtures were evaporated to dryness under nitrogen at 50 °C, dissolved in 250 μL methanol and filtered through a 0.22 μm syringe filter for UPLC-MS/MS analysis. All tests were performed in triplicate.

2.8. UPLC-MS/MS Analysis

The HAs extracted from smoked meat were identified and quantified using a UPLC system, equipped with a triple quadrupole mass spectrometer (Waters, Milford, MA, USA). Acquity BEH C18 column (2.1 × 100 mm, particle size 1.6 μm, Waters) was used. The mobile phases used consisted of A: 100% acetonitrile and B: ultrapure water + 0.1% formic acid. The elution gradient at a 45 °C column temperature was 2% A (0–2 min), 20% A (12 min), 100% A (14 min), 2% (17 min) and 2% A (20 min) [23]. The flow rate was 0.3 mL·min^{-1} and the injection volume was 2 μL. The mass spectrometry detection was conducted in multiple-reaction monitoring (MRM) scan mode, under positive electrospray ionization (ESI), as described by Li et al. [24], with some modifications. The operating conditions of mass spectrometry were as follows: 100 °C ion source temperature; 400 °C dissolved gas temperature; 700 L·h^{-1} flow rate; 3.5 kV capillary voltage; 50 L·h^{-1} cone flow (nitrogen); 0.15 mL·min^{-1} flow rate of collision gas (argon); and 2-2000 Da scanning range. According to the peak shape and intensity of different treatments, the channel reaction, dwell time, cone voltage, collision energy and delay time of 17 HAs in the MRM mode of this study were optimized (Table S1).

2.9. Statistical Analysis

The UPLC-MS/MS results were analyzed using the Masslynx v4.1 software (Waters). Data were reported as means ± standard deviation and analyzed using the one-way analysis of variance to determine significant differences ($p < 0.05$) among the treatments. Statistical analysis was performed using the general linear model program of the Statistic software v9.0 (Analytical Software, Tallahassee, FL, USA). HAs' data were imported into SIMCA 14.1 (Umetrics, Umea, Sweden) for principal component analysis (PCA). Spearman rank-order correlations were assessed using the Origin Pro 2021b software (Northampton, NC, USA). The Origin Pro 2021b and Anaconda Jupyter Notebook v6.4.8 (Austin, TX, USA) software were used for drawing.

3. Results and Discussion

3.1. Conventional Physicochemical Properties of Smoked Meat Patties

The physicochemical properties of the smoked meat patties are presented in Table 1. The results showed that increasing the smoking duration and temperature significantly decreased the L^* and increased the a^* value of the smoked meat. This was attributed to the melanoids that formed on the surface of the meat through the Maillard reaction pathway [17]. These observations were consistent with those reported in similar research studies [25]. The pH value of the pork patties that were smoked at different durations was significantly lower than the control sample (0 h), which was possibly due to the natural decrease in the pH value of the meat products during thermal processing and an increase in the meat's exposure to acidic smoke metabolites due to the increased smoking time [26]. Moreover, the increased duration and temperature during the smoking led to significantly increased water processing loss and decreased water content in the meat patties. At the same time, the hardness, springiness and chewiness increased ($p < 0.05$) in all the pork meat patties, which were associated with the changes in the water, protein and lipids' constituents. The heating of the meat was associated with the thermal denaturation of the proteins and a simultaneous change in the binding of the water and oxidation of the lipids, which were reflected by the changes in the gel properties [27]. These assertions were confirmed in an earlier study, where the heat treatment resulted in protein coagulation in the interstitial spaces, collagen and lipids, forming a gel; the denaturation and coagulation of the myofibrillar and sarcoplasmic proteins; and the shrinking of the intermuscular connective tissue [28]. Furthermore, similar observations were confirmed in a recent study, where an electronic nose, electronic tongue and HS-SPME/GC-MS were used to evaluate cooked chicken legs, which showed that the water content, water activity, pH value and L^* value decreased from 71.26% to 65.23%, from 0.987 to 0.979, from 6.66 to 5.36 and from 61.84 to 52.34, respectively ($p < 0.05$) [29].

Table 1. The conventional physical and chemical indicators of smoked pork patties, with different smoking durations and temperatures.

		Hardness (g)	Springiness (%)	Chewiness (g)	Resilience (%)	L^*	a^*	b^*	a_w	Water Processing Loss (g·100 g^{-1})	Water Content (g·100 g^{-1})	pH
Duration	0 h	863.21 ± 104.96 d	0.63 ± 0.02 b	357.23 ± 73.54 c	0.23 ± 0.01 b	69 ± 0.93 a	6.7 ± 0.14 c	21.25 ± 0.56 c	0.97 ± 0 a	29.84 ± 0.35 f	66.13 ± 0.1 a	5.82 ± 0.01 a
	1 h	2425.06 ± 274.87 c	0.74 ± 0.05 a	1098.44 ± 127.31 bc	0.25 ± 0.01 ab	65.17 ± 0.38 b	8.31 ± 1.15 c	29.28 ± 1.16 ab	0.97 ± 0 b	32.97 ± 0.46 e	63.75 ± 0.45 b	5.66 ± 0.03 b
	2 h	3362.42 ± 469.36 bc	0.71 ± 0.01 ab	1474.3 ± 218.88 b	0.22 ± 0.01 b	55.74 ± 0.25 c	11.39 ± 0.91 b	28.23 ± 0.67 ab	0.96 ± 0 b	39.63 ± 0.45 d	59.26 ± 0.32 c	5.49 ± 0.01 c
	3 h	4607.45 ± 533.33 b	0.72 ± 0.04 ab	2022.51 ± 186.17 b	0.22 ± 0.01 b	50.42 ± 1.15 d	13.11 ± 1.2 ab	27.09 ± 0.86 b	0.96 ± 0 c	43.2 ± 0.73 c	53.34 ± 0.6 d	5.38 ± 0.01 d
	4 h	8719.97 ± 830.69 a	0.78 ± 0.01 a	4866.18 ± 645.94 a	0.27 ± 0.01 a	44.8 ± 1.15 e	15.26 ± 0.71 a	30.02 ± 1.24 a	0.95 ± 0 d	47.84 ± 0.64 b	50.58 ± 0.53 e	5.28 ± 0.03 e
	5 h	8689.31 ± 331.12 a	0.79 ± 0.06 a	4555.57 ± 509.58 a	0.24 ± 0.02 ab	31.04 ± 2.27 f	14.35 ± 0.22 a	23.55 ± 0.94 c	0.92 ± 0 e	57.31 ± 0.6 a	44.84 ± 0.19 f	5.23 ± 0.02 e
Temperature	AMB	1679.25 ± 106.28 c	0.7 ± 0.02 b	741.84 ± 100.64 b	0.26 ± 0.02 a	57.3 ± 0.34 a	14.15 ± 0.7 c	35.55 ± 1.17 a	0.97 ± 0 a	30.17 ± 0.24 e	62.96 ± 0.23 a	5.42 ± 0.02 a
	50 °C	2588.34 ± 127.49 bc	0.7 ± 0.05 b	1172.55 ± 60.61 b	0.24 ± 0.02 a	54.41 ± 0.72 b	16.21 ± 0.46 bc	35.81 ± 1.97 a	0.96 ± 0 b	34.97 ± 0.75 d	60.45 ± 0.51 b	5.12 ± 0.01 c
	60 °C	2453.55 ± 183.92 bc	0.75 ± 0.03 ab	1215.88 ± 68.46 b	0.26 ± 0 a	51.63 ± 0.48 c	16.17 ± 0.92 bc	33.92 ± 0.89 ab	0.96 ± 0 c	37.24 ± 0.46 c	56.84 ± 0.16 c	5.03 ± 0.01 d
	70 °C	4163.07 ± 486.14 b	0.69 ± 0.02 b	1760.74 ± 205.96 b	0.23 ± 0.02 a	48.06 ± 1.87 d	15.75 ± 0.38 c	32.33 ± 0.76 b	0.95 ± 0 d	42.29 ± 0.62 b	53.16 ± 0.22 d	5.38 ± 0.01 b
	80 °C	8969.06 ± 1306.85 a	0.8 ± 0.02 a	5803.19 ± 798.57 a	0.26 ± 0.01 a	33.87 ± 0.93 e	20.93 ± 2.04 a	28.13 ± 0.82 c	0.94 ± 0 e	49.19 ± 1.33 a	47.69 ± 0.27 e	5.43 ± 0.02 a
	90 °C	10288.56 ± 791.83 a	0.81 ± 0.04 a	5602.76 ± 815.01 a	0.27 ± 0 a	27.77 ± 0.89 f	18.82 ± 0.54 ab	25.76 ± 0.95 c	0.94 ± 0 e	51.15 ± 0.5 a	46.46 ± 0.57 f	5.43 ± 0.02 a

Group 1: the meat patties were smoked for 0, 1, 2, 3, 4 and 5 h at 70 °C. Group 2: the meat patties were smoked for 3 h, at ambient temperatures (AMB), 50, 60, 70, 80 and 90 °C. Data are presented as means ± standard deviations (n = 3). Different letters in each treatment group represented significant differences at $p < 0.05$.

3.2. Effect of Smoking Duration and Temperature on the Formation of HAs

The effects of the smoking duration and temperature on the formation of HAs are presented in Figure 2. Four kinds of free HAs—namely F-7,8-DiMeIQx (0.15–2.89 ng·g^{-1}); F-MeAαC (0–1 ng·g^{-1}); F-Harman (0.2–3.6 ng·g^{-1}) and F-Norharman (1.15–123.99 ng·g^{-1})—were detected in the smoked meat patties (Figure 2A,B). The results showed that the content of free HAs initially increased, then decreased with the increase in the smoking duration, reaching a maximum value of 98.33 ng·g^{-1}, at 4 h, which was 23 times the total content, compared with the non-smoked meat (0 h) (4.24 ng·g^{-1}). During the smoking and heating process, water was gradually lost and acted as a reaction and transmission medium for several precursor molecules on the surface of the meat patties. On the surface, the precursors were exposed to higher temperatures, resulting in the formation of HAs [30]. Compared with AMB, the total amount of free HAs decreased with an increase in the temperature, from 127.16 ng·g^{-1} to 47.82 ng·g^{-1}. Among the four kinds of HAs, the content of F-Norharman was the highest, while the content of the other HAs changed little with the increase in the temperature. The amount of total free HAs decreased gradually, primarily due to a decrease in F-Norharman. A similar trend was observed with the formation of HAs in trout fillets that were cooked by sous-vide, under different cooking temperature conditions [31]. We hypothesized that the content of HAs could be explained by the rate of the thermal decomposition of some HAs being faster than their formation with an increase in the smoking temperature and time [32], which were concordant with the findings of Xi et al. [25], who reported that the rate of HAs' formation significantly decreased with an increased temperature and time at some nodes. Furthermore, six kinds of protein-bound HAs—including B-AαC (2.53–20.83 ng·g^{-1}); B-7,8-DiMeIQx (1.33–5.63 ng·g^{-1}); B-Glu-p-1 (14.55–66.65 ng·g^{-1}); B-MeAαC (54.37–191.3 ng·g^{-1}); B-Harman (139.57–571.53 ng·g^{-1}) and B-Norharman (333–1140 ng·g^{-1})—were detected in the smoked meat patties (Figure 2C,D). With an increase in the smoking duration, the content of B-Norharman initially increased slowly, and then decreased (600–800 ng·g^{-1}), while the content of B-Harman decreased gradually (Figure 2C). At the same time, the content of B-Harman and B-Norharman first decreased, then increased with an increase in the smoking temperature (Figure 2D). The contents of the other four kinds of HAs were low, with no obvious change observed with the changes in the smoking duration and temperature. These observations were attributed to the effects of the temperature on the increased rate of the sample's dehydration. In addition, the Maillard reaction and a series of complex reactions could have affected the formation of HAs. Harman and Norharman are the main HAs that are widely found in high contents in animal- [33] and plant-based [34] food products. This is due to Harman and Norharman being endogenous components—which are formed in natural mild conditions [35], and which could be detected in raw meat [36]—and their content may increase continuously with an increase in smoking temperature or time [25], until their degradation temperatures are reached [37]. In this study, protein-bound HAs were mainly formed in two ways: the bonding of free HAs with protein, or the involvement of amino acid residues on protein side chains in reactions that generated protein-bound HAs. In addition to HAs forming directly as a result of smoking, the changes in meat components, such as tryptophan and glutamate [38], and the effects of the smoke [39] on the protein and oil during smoking are related to free and protein-bound HAs. Generally, it is considered that HAs are produced at high temperatures; however, some HAs have been found to accumulate significantly at low temperatures. For instance, in an earlier study, the total content of HAs—based on AαC, Phe-P-1, DMIP, Harman and Norharman—was found to increase up to 1.41 μg·g^{-1} in raw sausage that was dried at 50 °C, for 10 min [40]. Furthermore, Norharman was produced in temperatures as low as 40 °C [41]. These findings imply that smoked meat products that are produced at relatively low temperatures might contain HAs. For quality control, it is, therefore, very important to control the production of HAs to help minimize the risk of cancer.

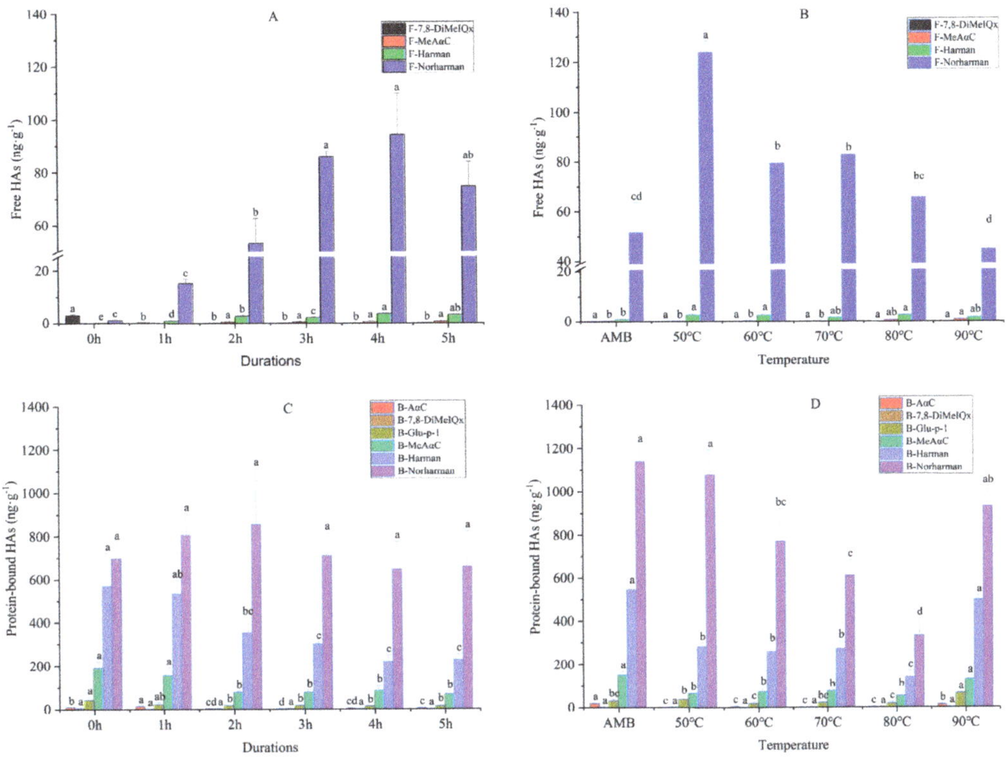

Figure 2. Changes in free HAs (**A**,**B**) and protein-bound HAs (**C**,**D**) in smoked meat patties, treated at different smoking durations and temperatures ($ng \cdot g^{-1}$). Data are presented as means ± standard deviations (n = 3). AMB represents ambient temperatures. Different letters in each treatment group represent significant differences at $p < 0.05$.

3.3. The VOCs of the Smoked Pork Patties

There were 52 kinds of VOCs in the control sample of pork patties (0 h). An increase in the smoking duration from 1 h to 5 h was associated with the 63, 74, 72, 73 and 60 kinds of VOCs contained in the smoked pork patties, respectively (Figure 3). Zhenba bacon had the lowest VOC content when fresh. The type and content of VOCs increased with an increase in the treatment duration, resulting in a large number of phenolic substances [4]. There were 36 kinds of common VOCs and three kinds of VOCs, which included 10 ketones, 9 phenols and 5 aldehydes, which accounted for the main VOC. At the same time, compared with the 66 kinds of VOCs under AMB, an increase in smoking temperature from 50 °C to 90 °C resulted in 62, 69, 72, 76 and 67 kinds of VOCs, respectively, in the smoked pork patties. Among them, 44 kinds of VOCs were common at all temperatures. Similar to the duration gradient, ketones (11 kinds), phenols (11 kinds) and aldehydes (five kinds) were the main VOCs. Phenolic substances were mainly produced by a gradual decomposition of lignin and were unique volatile flavor substances in smoking [42]. It was mainly reflected in the generation of VOCs during the heating and decomposition of smoking substances, as well as phenolic substances formed after smoking [2]. Harman and Norharman are generally considered the products of protein pyrolysis, while aldehydes are reported to be the intermediates of β-carbolines HAs. Therefore, the presence of aldehydes in smoking can be regarded as the precursors that promote the formation of nonpolar HAs [43]. The VOCs in heated meat rise via major reaction mechanisms such as the Maillard reaction and lipid oxidation. A cluster analysis showed that the repeatability between the smoked pork patties was good, regardless of the duration or temperature of smoking. The six smoking durations were divided into three categories (Figure 3A), as follows: (i) 0 h; (ii) 1, 2 and 3 h; and (iii) 4 and 5 h. Similarly, the different smoking temperatures (Figure 3C) were divided into the following three categories: (i) AMB and 50 °C; (ii) 70, 80 and 90 °C; and (iii) 60 °C. Generally, the difference between the control (0 h) and other treatments was large, mainly because the control was not smoked and the VOCs were mainly the odor substance of the meat. The differences between the meat patties that were smoked for 1, 2 and 3 h and those that were smoked for 4 and 5 h may have arisen because the ingredients of the VOCs needed time to penetrate into the meat patties, and they then reacted with the ingredients of the meat patties. At the same time, the VOCs with a higher content in the control patties than in the smoked meat patties were mainly aldehydes and alcohols. With an increase in the smoking duration, the content of ketones, terpenes and phenols accumulated and increased in the meat patties. However, when the temperature was above 70 °C, the difference in the VOCs in the smoked meat was not obvious.

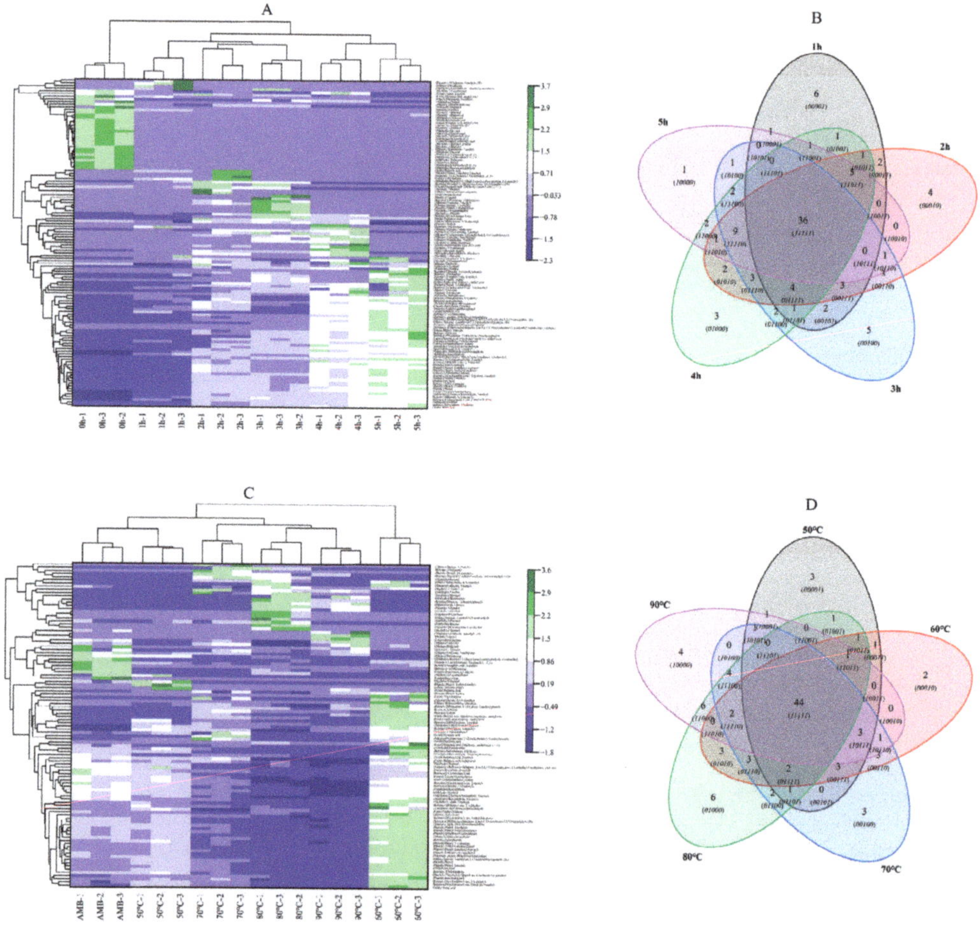

Figure 3. Cluster analysis diagram of VOCs in smoked pork patties at different durations (**A**) and temperatures (**C**). Wayne diagram of VOCs in smoked pork patties at different durations (**B**) and temperatures (**D**). AMB represents ambient temperatures.

3.4. Correlation between VOCs and HAs in Meat Patties at Different Durations and Temperatures

The Wayne mapping showed that there were 16 VOCs, including one acid, one alcohol, four aldehydes, two alkanes, one ketone and seven phenols, under different durations and temperatures (Figure 4A)—specifically comprising acetic acid; 1-Octen-3-ol; 2-Furancarboxaldehyde,5-methyl-; Benzaldehyde; Furfural Nonanal; Dodecane; Tetradecane; 5-Hepten-2-one,6-methyl-; 2-Methoxy-5-methylphenol; p-Cresol; Phenol; Phenol,2-methoxy-; Phenol,2-methoxy-4-propyl-; Phenol,4-ethyl-2-methoxy-; and trans-Isoeugenol. The proportion of phenol was very high. Different methods were used to distinguish the VOCs in the dry-cured hams, under different processing methods. The content of phenols, aromatic hydrocarbons and acids found in the smoked–cured, dry-cured hams was high and demonstrated the characteristics of smoking aroma [44]. A total of 27 VOCs with odor activity were identified as the key aroma compounds of smoked, cooked pork loin. The highest flavor dilution factor was 2-methoxyphenol; 2-methoxy-4-(propyl-2-enyl) phenol; and 2-methoxy-4-E)-(propyl-1-en-1-yl) phenol [45]. Among them, seven kinds of phenols were positively correlated with the free HAs (rs = 0.61–0.71, $p \leq 0.001$), including five methoxyphenols. In principle, this confirmed that a darker color was associated with a higher content of HAs, which might be related to some phenolic compounds in smoking that promoted the formation of HAs. For instance, smoking-associated carbonyl compounds might provide intermediate compounds that promote the formation of HAs [11]. Mixed phenolic compounds have different effects on single HAs, some of which can effectively reduce their level, while others can effectively improve them [46]. It was also reported that the inhibitory activity of phenols on the formation of HAs was, at least partially, related to the content of phenols [47]. Among them, the obvious changes were as follows: (i) the content of 2-Methoxy-5-methylphenol, which increased from 9.13 $\mu g \cdot g^{-1}$, at 0 h, to 2935.41 $\mu g \cdot g^{-1}$, at 5 h; and (ii) Phenol,2-methoxy-, which increased from 11.72 $\mu g \cdot g^{-1}$, at 0 h, to 2789.87 $\mu g \cdot g^{-1}$, at 5 h. The content increased by 322 and 238 times, respectively, and gradually increased with the increase in the smoking duration (Figure S2A). With the gradual increase in the temperature, 2-Methoxy-5-methylphenol increased from 2367.75 $\mu g \cdot g^{-1}$, at AMB, to 4305.16 $\mu g \cdot g^{-1}$, at 60 °C, and then rapidly decreased to 642.58 $\mu g \cdot g^{-1}$, at 90 °C. Phenol,2-methoxy- followed the same trend, increasing from 2377.83 $\mu g \cdot g^{-1}$, at AMB, to 3393.7 $\mu g \cdot g^{-1}$, at 60 °C, and then dropping rapidly to only 803.27 $\mu g \cdot g^{-1}$, at 90 °C (Figure S2B). All of these changes were consistent with the results observed after the cluster analysis (Figure 3C). At 60 °C, the content of phenols was the highest and the effect was the most significant. The possible reason was that the pine woodchips enhanced the dealkylation and de-methoxylation of methoxyphenol with a rise in the smoking temperature [48]. At the same time, the acetic acid; 2-Furancarboxaldehyde,5-methyl-; Benzaldehyde; and Furfural were also positively correlated with the free HAs, with correlation coefficients (rs) of 0.77, 0.61, 0.59 and 0.57, respectively ($p \leq 0.001$). Furthermore, 2-Methoxy-5-methylphenol; p-Cresol; Phenol; Phenol,2-methoxy-; Phenol,2-methoxy-4-propyl-; Phenol,4-ethyl-2-methoxy-; and trans-Isoeugenol showed a strong positive correlation with the free HAs, with correlation coefficients of 0.63, 0.68, 0.71, 0.61, 0.66, 0.64 and 0.57, respectively ($p \leq 0.001$) (Figure 4B). However, different concentrations of phenols had quite different or opposite effects on the formation of HAs [49]. The possible reason was the different content of phenol. In roast beef patties, the addition of 0.05% and 0.10% chlorogenic acid significantly promoted the formation of PhIP ($p < 0.05$). Among them, tetradecane deserved special attention as it had a positive correlation with F-Harman (rs = 0.62, $p \leq 0.001$) and a strong negative correlation with B-Glu-P-1 and B-Harman (rs = −0.66 and −0.57, $p \leq 0.001$). In general, the smoking process significantly promoted the formation of HAs in meat products, due to the involvement of the smoking ingredients in producing HAs in meat products [50].

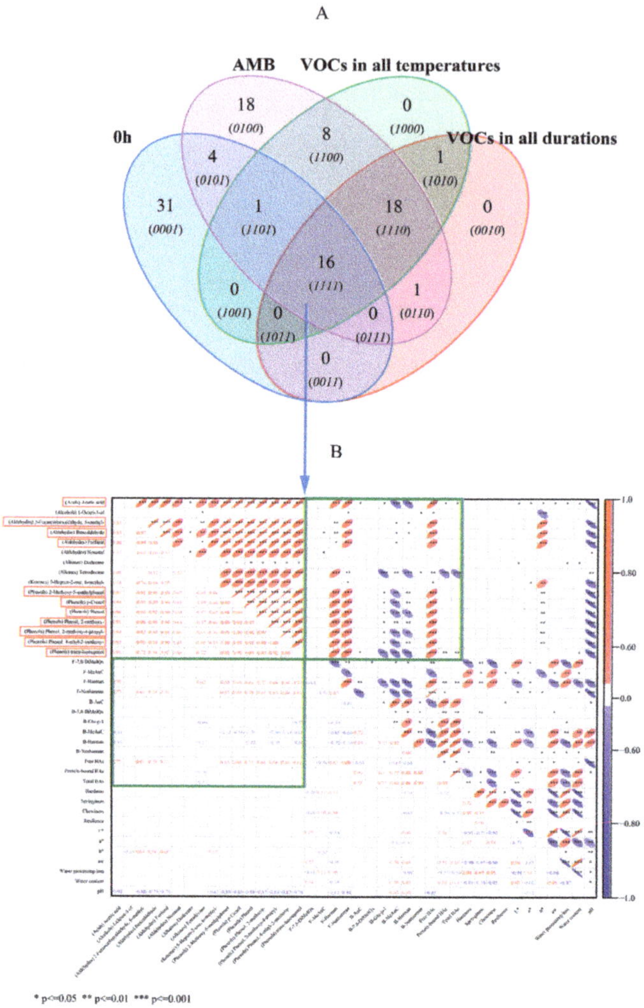

Figure 4. Wayne diagram of VOCs in non-smoked meat patties, at all durations and all temperatures of smoked meat patties (**A**). Correlation analysis of VOCs in all processed meat patties, HAs and other physical and chemical indexes (**B**). AMB represents ambient temperatures.

3.5. PCA Analysis of HAs, VOCs and Other Indexes in Smoked Meat Patties at Different Smoking Durations and Temperatures

Two principal components were extracted, representing 65% total variance, with component one (PC1) and component two (PC2) contributing 42.1 and 22.9% of the variance, respectively. As shown in the score scatter plot (Figure 5A), clear clusters and separations among the samples were seen. For instance, as the duration increased, the samples were scattered from left to right along the X axis, whereas, with the increase in temperature, the samples were dispersed from the top to the bottom along the Y axis. Unsmoked meat (0 h, control) was positioned at the far left of the first quadrant, while those after 4 and 5 h of smoking were located in the third quadrant. AMB was at the top, while those heated at 50 and 60 °C were located in the second quadrant, and those heated at 90 °C were located in the lower left corner. The results showed that the smoking duration and the temperature significantly influenced the content of HAs in the smoked meat patties. Furthermore, the loading scatter plot (Figure 5B) showed that, except for B-Glu-p-1, the other five types of protein-bound HAs were located in the first quadrant—similar to the positions of 0 h, 1 h and AMB in the score scatter plot—implying that the content of the protein-bound HAs was higher in the unsmoked meat patties when the smoking was short and at low smoking temperatures. The positions of F-Harman and F-Norharman on the loading scatter plot were similar to those at 4 h and 5 h on the score scatter plot. An increase in the smoking duration significantly increased the content of free HAs, compared with a short smoking duration. It was particularly interesting that VOCs positively correlated with free HAs—they were concentrated in the second quadrant and were in a similar position at 60 °C on the score scatter plot, indicating that the content of VOCs was high at 60 °C. Yang et al. [12] found that the total amount of free HAs in 60 °C smoked sausage was 5.49 ng·g^{-1} ($p < 0.05$), in which the Harman content was 0.85 ng·g^{-1} and the Norharman content was 5.81 ng·g^{-1}. The TPA indexes were all concentrated in the third quadrant, with the same quadrant of 80 °C samples representing a higher processing temperature or 4 h and 5 h of a longer processing duration in the loading scatter plot. An increase in the smoking duration and temperature made the texture of the smoked meat patties hard and significantly deteriorated their chewiness, elasticity and resilience.

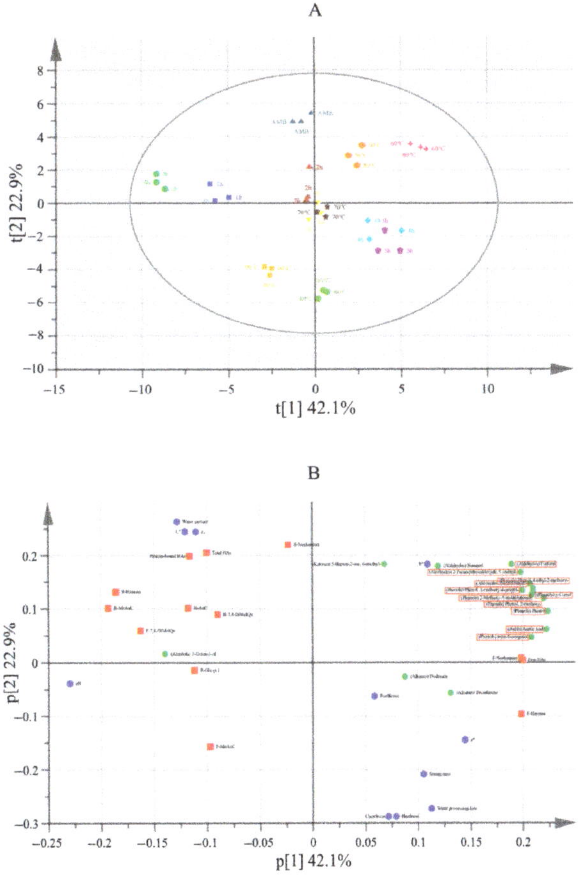

Figure 5. Scores (**A**) and loadings (**B**) scatter plot of PCA analysis of smoked meat, treated under different smoking durations and temperatures. AMB represents ambient temperatures.

4. Conclusions

A low temperature for meat processing significantly increased the generation of various kinds of HAs. During the smoking of meat, some components may induce the production of HAs in the smoked meat. The 16 VOCs that existed under different conditions deserved special attention. In this study, the concentration of various HAs varied with the changes in the smoking durations and temperatures. In particular, VOCs with a high correlation with free HAs were concentrated in the second quadrant, corresponding to the position of 60 °C in the scores scatter plot of PCA. This also coincided with the situation that 60 °C was the most different from the other smoking temperature conditions in the cluster analysis. Our results showed that the smoking process significantly affected the formation of HAs in meat products. Therefore, the smoking of meat should be carefully controlled to minimize cancer risks.

Supplementary Materials: The following supporting information can be downloaded at, https://www.mdpi.com/article/10.3390/foods11223687/s1: Figure S1—schematic diagram of smoke experiment device; Figure S2—VOC content ($ng \cdot g^{-1}$) of smoked meat patties at different durations (A) and temperatures (B); And Table S1—mass spectrometric parameters of HAs in MRM mode of UPLC-MS/MS.

Author Contributions: Conceptualization, X.S.; funding acquisition, M.Z.; methodology, X.S., J.C. and M.Z.; project administration, M.Z.; software, Y.C., J.O.O., C.X., Y.Z., J.C. and M.Z.; supervision, E.O., F.O., C.X. and M.Z.; writing—original draft, X.S.; writing—review & editing, Y.C. All authors have read and agreed to the published version of the manuscript.

Funding: This work was funded by the National Natural Science Foundation of China (Grant No. 31871905) and the National First-Class Discipline Program of Food Science and Technology (Grant No. JUFSTR20180201).

Institutional Review Board Statement: Not applicable.

Informed Consent Statement: Not applicable.

Data Availability Statement: The results obtained for all experiments performed are shown in the manuscript and Supplementary Materials, the raw data will be provided upon request.

Acknowledgments: We thank Xinlei Hu for his great help in the experimental operation.

Conflicts of Interest: The authors declare no competing financial interest.

References

1. Zeng, W.; Wen, W.; Deng, Y.; Tian, Y.; Sun, H.; Sun, Q. Chinese ethnic meat products: Continuity and development. *Meat Sci.* **2016**, *120*, 37–46. [CrossRef] [PubMed]
2. Zhou, B.; Luo, J.; Quan, W.; Lou, A.; Shen, Q. Antioxidant Activity and Sensory Quality of Bacon. *Foods* **2022**, *11*, 236. [CrossRef] [PubMed]
3. Du, H.; Chen, Q.; Liu, Q.; Wang, Y.; Kong, B. Evaluation of flavor characteristics of bacon smoked with different woodchips by HS-SPME-GC-MS combined with an electronic tongue and electronic nose. *Meat Sci.* **2021**, *182*, 108626. [CrossRef] [PubMed]
4. Xi, L.J.; Zhang, J.; Wu, R.X.; Wang, T.; Ding, W. Characterization of the Volatile Compounds of Zhenba Bacon at Different Process Stages Using GC-MS and GC-IMS. *Foods* **2021**, *10*, 2869. [CrossRef] [PubMed]
5. Poljanec, I.; Radovcic, N.M.; Petricevic, S.; Karolyi, D.; Listes, E.; Medic, H. Proteolysis and protein oxidation throughout the smoked dry-cured ham process. *Food Chem.* **2021**, *362*, 130207. [CrossRef] [PubMed]
6. Hur, S.J.; Yoon, Y.; Jo, C.; Jeong, J.Y.; Lee, K.T. Effect of Dietary Red Meat on Colorectal Cancer Risk—A Review. *Compr. Rev. Food Sci. Food Saf.* **2019**, *18*, 1812–1824. [CrossRef]
7. Chen, L.; Liu, R.; Wu, M.; Yu, H.; Ge, Q.; Zhang, W. Nitrosamines and Polycyclic Aromatic Hydrocarbons in Smoke-Cured Bacon (Larou) of Artisanal and Industrial Origin. *Foods* **2021**, *10*, 2830. [CrossRef]
8. Zhang, L.; Du, H.; Zhang, P.; Kong, B.; Liu, Q. Heterocyclic aromatic amine concentrations and quality characteristics of traditional smoked and roasted poultry products on the northern Chinese market. *Food Chem. Toxicol.* **2020**, *135*, 110931. [CrossRef]
9. Zamora, R.; Hidalgo, F.J. Formation of heterocyclic aromatic amines with the structure of aminoimidazoazarenes in food products. *Food Chem.* **2020**, *313*, 126128. [CrossRef]
10. IARC. *Monographs on the Evaluation of Carcinogenic Risk to Humans*; World Health Organization: Geneva, Switzerland; International Agency for Research on Cancer (IARC): Lyon, Frence, 1993; pp. 163–242.

11. Gibis, M. Heterocyclic Aromatic Amines in Cooked Meat Products: Causes, Formation, Occurrence, and Risk Assessment. *Compr. Rev. Food Sci. Food Saf.* **2016**, *15*, 269–302. [CrossRef]
12. Yang, D.; He, Z.; Gao, D.; Qin, F.; Deng, S.; Wang, P.; Xu, X.; Chen, J.; Zeng, M. Effects of smoking or baking procedures during sausage processing on the formation of heterocyclic amines measured using UPLC-MS/MS. *Food Chem.* **2019**, *276*, 195–201. [CrossRef] [PubMed]
13. Oz, E. The impact of fat content and charcoal types on quality and the development of carcinogenic polycyclic aromatic hydrocarbons and heterocyclic aromatic amines formation of barbecued fish. *Int. J. Food Sci. Technol.* **2021**, *56*, 954–964. [CrossRef]
14. Varlet, V.; Knockaert, C.; Prost, C.; Serot, T. Comparison of odor-active volatile compounds of fresh and smoked salmon. *J. Agric. Food Chem.* **2006**, *54*, 3391–3401. [CrossRef] [PubMed]
15. Kelly, D.; Zerihun, A.; Singh, D.P.; von Eckstaedt, C.V.; Gibberd, M.; Grice, K.; Downey, M. Exposure of grapes to smoke of vegetation with varying lignin composition and accretion of lignin derived putative smoke taint compounds in wine. *Food Chem.* **2012**, *135*, 787–798. [CrossRef]
16. Hidalgo, F.J.; Zamora, R. Carbonyl Chemistry and the Formation of Heterocyclic Aromatic Amines with the Structure of Aminoimidazoazaarene. *J. Agric. Food Chem.* **2021**, *70*, 79–86. [CrossRef]
17. Zhang, L.; Chen, Q.; Liu, Q.; Xia, X.; Wang, Y.; Kong, B. Effect of different types of smoking materials on the flavor, heterocyclic aromatic amines, and sensory property of smoked chicken drumsticks. *Food Chem.* **2022**, *367*, 130680. [CrossRef]
18. Chen, Y.; Xi, J.; Chen, Z.N.; Fu, Y. Effect of Different Roasting Conditions and Coreopsis Extract on the Heterocyclic Amine Formation in Roast Lamb Products. *J. Food Prot.* **2022**, *85*, 1107–1113. [CrossRef]
19. Dolores Romero de Ávila, M.; Isabel Cambero, M.; Ordóñez, J.A.; de la Hoz, L.; Herrero, A.M. Rheological behaviour of commercial cooked meat products evaluated by tensile test and texture profile analysis (TPA). *Meat Sci.* **2014**, *98*, 310–315. [CrossRef]
20. Rizzo, P.V.; Del Toro-Gipson, R.S.; Cadwallader, D.C.; Drake, M.A. Identification of aroma-active compounds in Cheddar cheese imparted by wood smoke. *J. Dairy Sci.* **2022**, *105*, 5622–5640. [CrossRef]
21. Zeng, M.M.; He, Z.Y.; Zheng, Z.P.; Qin, F.; Tao, G.J.; Zhang, S.; Gao, Y.H.; Chen, J. Effect of Six Chinese Spices on Heterocyclic Amine Profiles in Roast Beef Patties by Ultra Performance Liquid Chromatography-Tandem Mass Spectrometry and Principal Component Analysis. *J. Agric. Food Chem.* **2014**, *62*, 9908–9915. [CrossRef]
22. Chen, J.; He, Z.; Qin, F.; Chen, J.; Zeng, M. Formation of Free and Protein-Bound Heterocyclic Amines in Roast Beef Patties Assessed by UPLC-MS/MS. *J. Agric. Food Chem.* **2017**, *65*, 4493–4499. [CrossRef] [PubMed]
23. Deng, P.; Xue, C.Y.; He, Z.Y.; Wang, Z.J.; Qin, F.; Oz, E.; Chen, J.; Sheikha, A.F.E.; Proestos, C.; Oz, F.; et al. Synergistic Inhibitory Effects of Selected Amino Acids on the Formation of 2-Amino-1-methyl-6-phenylimidazo 4,5-b pyridine (PhIP) in both Benzaldehyde- and Phenylacetaldehyde-Creatinine Model Systems. *J. Agric. Food Chem.* **2022**, *70*, 10858–10871. [CrossRef] [PubMed]
24. Li, Y.; He, J.L.; Quan, W.; He, Z.Y.; Qin, F.; Tao, G.J.; Wang, Z.J.; Zeng, M.M.; Chen, J. Effects of polyphosphates and sodium chloride on heterocyclic amines in roasted beef patties as revealed by UPLC-MS/MS. *Food Chem.* **2020**, *326*, 127016. [CrossRef] [PubMed]
25. Xi, J.; Chen, Y. Analysis of the relationship between heterocyclic amines and the oxidation and thermal decomposition of protein using the dry heated soy protein isolate system. *Lwt-Food Sci. Technol.* **2021**, *148*, 111738. [CrossRef]
26. Sokamte Tegang, A.; Mbougueng, P.D.; Sachindra, N.M.; Douanla Nodem, N.F.; Tatsadjieu Ngoune, L. Characterization of volatile compounds of liquid smoke flavourings from some tropical hardwoods. *Sci. Afr.* **2020**, *8*, e00443. [CrossRef]
27. Tokifuji, A.; Matsushima, Y.; Hachisuka, K.; Yoshioka, K. Texture, sensory and swallowing characteristics of high-pressure-heat-treated pork meat gel as a dysphagia diet. *Meat Sci.* **2013**, *93*, 843–848. [CrossRef]
28. Tornberg, E. Effects of heat on meat proteins—Implications on structure and quality of meat products. *Meat Sci.* **2005**, *70*, 493–508. [CrossRef] [PubMed]
29. Zhang, L.; Hu, Y.; Wang, Y.; Kong, B.; Chen, Q. Evaluation of the flavour properties of cooked chicken drumsticks as affected by sugar smoking times using an electronic nose, electronic tongue, and HS-SPME/GC-MS. *Lwt Food Sci. Technol.* **2021**, *140*, 110764. [CrossRef]
30. Overvik, E.; Kleman, M.; Berg, I.; Gustafsson, J.A. Influence of creatine, amino acids and water on the formation of the mutagenic heterocyclic amines found in cooked meat. *Carcinogenesis* **1989**, *10*, 2293–2301. [CrossRef]
31. Oz, F.; Seyyar, E. Formation of Heterocyclic Aromatic Amines and Migration Level of Bisphenol-A in Sous-Vide-Cooked Trout Fillets at Different Cooking Temperatures and Cooking Levels. *J. Agric. Food Chem.* **2016**, *64*, 3070–3082. [CrossRef]
32. Chiu, C.P.; Chen, B.H. Stability of heterocyclic amines during heating. *Food Chem.* **2000**, *68*, 267–272. [CrossRef]
33. Pan, H.; Wang, Z.; Guo, H.; Ni, N.; Zhang, D. Heterocyclic aromatic amines in meat products consumed in China. *Food Sci. Biotechnol.* **2014**, *23*, 2089–2095. [CrossRef]
34. Xi, J.; Chen, Y. Effects of tetrahydro-curcumin on the formation of β-carboline heterocyclic amines in dry-heated soy protein isolate in the presence of glucose. *LWT* **2021**, *152*, 112362. [CrossRef]
35. Wrońska, A.K.; Boguś, M.I. Harman and norharman, metabolites of the entomopathogenic fungus *Conidiobolus coronatus* (Entomophthorales), affect the serotonin levels and phagocytic activity of hemocytes, insect immunocompetent cells, in *Galleria mellonella* (Lepidoptera). *Cell Biosci.* **2019**, *9*, 29. [CrossRef] [PubMed]

36. Wu, X.; Zhang, Z.; He, Z.; Wang, Z.; Qin, F.; Zeng, M.; Chen, J. Effect of Freeze-Thaw Cycles on the Oxidation of Protein and Fat and Its Relationship with the Formation of Heterocyclic Aromatic Amines and Advanced Glycation End Products in Raw Meat. *Molecules* **2021**, *26*, 1264. [CrossRef] [PubMed]
37. Liu, W.; Yang, Z.; Shi, L.; Cui, Z.; Li, Y. Degradation of β-Carbolines Harman and Norharman in Edible Oils during Heating. *Molecules* **2021**, *26*, 7018. [CrossRef] [PubMed]
38. Hou, C.; Wang, Z.; Wu, L.; Chai, J.; Song, X.; Wang, W.; Zhang, D. Effects of breeds on the formation of heterocyclic aromatic amines in smoked lamb. *Int. J. Food Sci. Technol.* **2017**, *52*, 2661–2669. [CrossRef]
39. Du, H.; Liu, Q.; Chen, Q.; Xia, X.; Xu, M.; Kong, B. Effect of woodchip types on heterocyclic aromatic amine formation and quality characteristics of smoked bacon. *Food Biosci.* **2022**, *47*, 101709. [CrossRef]
40. Diaodiao, Y.; Jing, J.; Zhigang, Z.; Zhiyong, H.; Fang, Q.; Jie, C.; Maomao, Z. Accumulation of heterocyclic amines across low-temperature sausage processing stages as revealed by UPLC-MS/MS. *Food Res. Int.* **2020**, *137*, 109668.
41. Ziegenhagen, R.; Boczek, P.; Viell, B. Formation of the comutagenic beta-carboline norharman in a simple tryptophan-containing model system at low temperature (40 degrees C–80 degrees C). In *Tryptophan, Serotonin and Melatonin: Basic Aspects and Applications;* Huether, G., Kochen, W., Simat, T.J., Steinhart, H., Eds.; Advances in Experimental Medicine and Biology; Springer: Boston, MA, USA, 1999; Volume 467, pp. 693–696.
42. Oberholster, A.; Wen, Y.; Dominguez Suarez, S.; Erdmann, J.; Cauduro Girardello, R.; Rumbaugh, A.; Neupane, B.; Brenneman, C.; Cantu, A.; Heymann, H. Investigation of Different Winemaking Protocols to Mitigate Smoke Taint Character in Wine. *Molecules* **2022**, *27*, 1732. [CrossRef]
43. Herraiz, T.; Papavergou, E. Identification and Occurrence of Tryptamine- and Tryptophan-Derived Tetrahydro-β-carbolines in Commercial Sausages. *J. Agric. Food Chem.* **2004**, *52*, 2652–2658. [CrossRef] [PubMed]
44. Dominguez, R.; Purrinos, L.; Perez-Santaescolastica, C.; Pateiro, M.; Barba, F.J.; Tomasevic, I.; Bastianello Campagnol, P.C.; Lorenzo, J.M. Characterization of Volatile Compounds of Dry-Cured Meat Products Using HS-SPME-GC/MS Technique. *Food Anal. Methods* **2019**, *12*, 1263–1284. [CrossRef]
45. Kosowska, M.; Majcher, M.A.; Jelen, H.H.; Fortuna, T. Key Aroma Compounds in Smoked Cooked Loin. *J. Agric. Food Chem.* **2018**, *66*, 3683–3690. [CrossRef] [PubMed]
46. Khan, I.A.; Luo, J.; Shi, H.; Zou, Y.; Khan, A.; Zhu, Z.; Xu, W.; Wang, D.; Huang, M. Mitigation of heterocyclic amines by phenolic compounds in allspice and perilla frutescens seed extract: The correlation between antioxidant capacities and mitigating activities. *Food Chem.* **2022**, *368*, 130845. [CrossRef] [PubMed]
47. Cheng, K.-W.; Chen, F.; Wang, M. Inhibitory activities of dietary phenolic compounds on heterocyclic amine formation in both chemical model system and beef patties. *Mol. Nutr. Food Res.* **2007**, *51*, 969–976. [CrossRef] [PubMed]
48. Kim, Y.-M.; Jae, J.; Myung, S.; Sung, B.H.; Dong, J.-I.; Park, Y.-K. Investigation into the lignin decomposition mechanism by analysis of the pyrolysis product of Pinus radiata. *Bioresour. Technol.* **2016**, *219*, 371–377. [CrossRef]
49. Zeng, M.; Li, Y.; He, Z.; Qin, F.; Chen, J. Effect of phenolic compounds from spices consumed in China on heterocyclic amine profiles in roast beef patties by UPLC-MS/MS and multivariate analysis. *Meat Sci.* **2016**, *116*, 50–57. [CrossRef] [PubMed]
50. Gibis, M.; Kruwinnus, M.; Weiss, J. Impact of different pan-frying conditions on the formation of heterocyclic aromatic amines and sensory quality in fried bacon. *Food Chem.* **2015**, *168*, 383–389. [CrossRef]

foods

MDPI

Article

The Influence of Deep Eutectic Solvents Extract from Ginger on the Formation of Heterocyclic Amines and Advanced Glycation End Products in Roast Beef Patties

Yang Xu [1], Ye Jiao [2], Jie Luo [1], Zhiyong He [3], Maomao Zeng [3], Qingwu Shen [1], Jie Chen [3,4] and Wei Quan [1,3,*]

1 College of Food Science and Technology, Hunan Agricultural University, Changsha 410128, China
2 School of Food science and Bioengineering, Changsha University of Science & Technology, Changsha 410114, China
3 State Key Laboratory of Food Science and Technology, Jiangnan University, Wuxi 214122, China
4 International Joint Laboratory on Food Safety, Jiangnan University, Wuxi 214122, China
* Correspondence: reus_quan@hunau.edu.cn; Tel./Fax: +86-510-8591-9065

Abstract: Heterocyclic amines (HAs) and advanced glycation end products (AGEs) are important harmful products formed simultaneously during the thermal processing of food. In order to develop a green, efficient method that can be used to control the production of two harmful products simultaneously in food processing. In the present study, deep eutectic solvents (DESs) were used to extract ginger, and this method produced significantly higher levels of total phenolic and flavonoid content as well as an antioxidant activity than ginger extracted using conventional solvents. Herein, we further investigated the inhibitory effects of DES extracts from ginger on the generation of HAs and AGEs in roast beef patties. All the nine DES extracts reduced the formation of HAs and AGEs, and the application of choline chloride–lactic-acid-based DES extract caused a signification reduction of 44.33%, 29.38%, 50.95%, 78.61%, 21.94%, and 17.52% of the PhIP, MeIQx, MeIQ, 4,8-DiMeIQx, Harmane, and Norhamane content, and those for Nε-(carboxymethyl)lysine (CML) and Nε-(carboxyethyl)lysine (CEL) were 49.08% and 58.50%, respectively. Furthermore, the proximate and texture profile changes of beef patties as well as the precursors (creatine, creatinine, and glucose) of HAs and AGEs were evaluated to determine the mechanism of ginger DES extracts on the formation of HAs and AGEs and the physical/chemical changes of ginger DES extracts on beef patties. This study develops a new method for reducing the amount of HAs and AGEs in meat, which will help food manufacturers produce healthier meat products.

Keywords: ginger; deep eutectic solvents; heterocyclic amines; advanced glycation end products; beef patties

1. Introduction

With the changing lifestyle around the world, there is an increasing trend in the consumption of high-temperature, processed foods [1]. High-temperature processing can impart unique flavors, colors, and textures to foods as well as increase nutrient digestibility or reduce antinutritional elements [2]. In addition to these beneficial effects, heat treatment can also lead to adverse outcomes, such as the production and accumulation of some harmful substances, including heterocyclic amines (HAs) and advanced glycation end products (AGEs), which have become a research hotspot in the field of food safety in recent years [3,4].

HAs are a class of mutagenic and carcinogenic compounds with various structures formed through Maillard reaction when creatinine, carbohydrates, and amino acids are heated at high temperatures, in which the free radical pathway and the carbonyl pathway are considered to be two main pathways for the formation of HAs [5,6]. Based on the results of epidemiological and animal studies, the IARC have classified most HAs as Class 2A and 2B human carcinogens and recommended limiting the intake of these compounds [7]. AGEs are harmful chemicals formed in nonenzymatic reactions, which refer to the polymerization, condensation, and other reactions between the free amino groups of proteins and the active carbonyls of reducing sugars or intermediate carbonyl compounds of the Maillard reaction [8]. Researchers showed that endogenous AGEs from the daily diet can accumulate in vivo and causing harmful effects in humans, such as increased oxidative stress, inflammation, and the risk of some chronic diseases [8,9].

Given the potential risks and the mechanisms by which HAs and AGEs are formed [10], several effective inhibition strategies such as adjusting cooking conditions have been studied in various food systems to reduce their formation [11]. Since free radicals are believed to be involved in the formation of HAs and AGEs through the Maillard reaction, chemical intervention methods involving the use of various food sources of antioxidants, including spices and herbal extracts, are considered to be promising and effective to inhibit the formation of HAs and AGEs [12–14]. Among them, ginger has received increased attention due to its widespread use as a spice and ingredient. It not only can impart special flavor to food but also contains a variety of functional compounds with strong free radical scavenging properties such as polyphenols, flavonoids, and terpenoids [15]. Our previous study found that ginger dose-dependently inhibited HAs and AGEs simultaneously, in which 1.5% of ginger exhibited a strong inhibition capacity against HAs and AGEs (inhibition rates: 27.42% and 35.64%) via their quenching activities of free radical during the Maillard reaction [16]. However, a large amount of spice could adversely affect the flavor and taste of the heat-processed food. Therefore, to improve the inhibition efficiency of natural products against HAs and AGEs in heat-treated foods, researchers have focused on the preparation of natural product extracts with higher antioxidant activities [17,18].

Currently, in food ingredients industries, some conventional organic solvents with strong extraction and dissolving ability are still widely used to extract natural antioxidant compounds. However, since most organic solvents are toxic, volatile, and cause pollution to the environment, their application in the food industry is therefore limited [19]. Deep eutectic solvents (DESs), a novel class of green solvents, are made of elements found in nature that serve as hydrogen bond acceptors (HBAs) or hydrogen bond donors (HBDs) [19,20]. Since DES has the benefits of easy preparation, high purity, no waste generation, and compliance with green chemistry principles, it has already found significant application potential in the extraction of natural ingredients for food fields such as anthocyanins, flavonoids, and isoflavones [20,21].

In the present study, various DESs have been designed and prepared with the ultrasonic bath method, in which choline chloride, betaine, and L-carnitine have been selected as HBA and glycerol, lactic acid, and xylitol as HBD. The aim of the present study to explore how different DES affected the antioxidant capacity of ginger extract and examine how ginger DES extract affected the production of HAs and AGEs in roast beef patties, which provides a theoretical basis for the development of a green, efficient method that can be used to simultaneously control the production of two harmful products in food processing.

2. Materials and Methods

2.1. Reagents and Chemicals

The ginger powder and raw beef were purchased from local market (Changsha, China), deep eutectic solvents, including choline chloride, L-carnitine, betaine, glycerol, lactic acid, and xylitol, were purchased by Sigma Chemical Co. (St. Louis, MO, USA). Folin–Ciocalteu's phenol reagent, rutin, gallic acid, 6-hydroxy-2,5,7,8-tetra-methylchroman-2-carboxylic acid (Trolox), 1,1-diphenyl-2-picrylhydrazyl (DPPH), 2,2-azinobis-(3-ethylbenzthiazoline-6-sulfonate) (ABTS), and 2,4,6-tris-(2-pyridyl)-s-triazine (TPTZ) were purchased from Sigma-Aldrich (St. Louis, MO, USA). AGEs standards, namely CML, CEL, Nε-(1-carboxymethyl)-L-lysine-d4 (CML-d4), and Nε-(1-carboxyethyl)-L-lysine-d4 (CEL-d4), and HA standards, namely PhIP, IQ, MeIQ, MeIQx, 4,8-DiMeIQx, 7,8-DiMeIQx, IQx, Harmane, and Norharmane, were supplied by Santa Cruz Biotechnology, Inc. (Santa Cruz, CA, USA).

2.2. Preparation of DES

The previously described heating and stirring procedure was used to prepare DES [15,22]. As donors of hydrogen bonds, the solutions of choline chloride, L-carnitine, and betaine were mixed with acceptors of hydrogen bonds (glycerol, lactic acid, and xylitol) in specific molar ratios and then stirred at 150 rpm and heated at 75 °C to obtain transparent solutions. Table 1 shows comprehensive details on the compositions and molar ratios of the produced DES1-9.

Table 1. The composition of nine kinds of deep eutectic solvents.

Name of DES	Component 1 (HBA)	Component 2 (HBD)	Molar Ratio
DES1	Choline chloride	Glycerol	1:2
DES2	Choline chloride	Lactic acid	1:2
DES3	Choline chloride	Xylitol	2:1
DES4	Betaine	Glycerol	1:2
DES5	Betaine	Lactic acid	1:2
DES6	Betaine	Xylitol	2:1
DES7	L-carnitine	Glycerol	1:2
DES8	L-carnitine	Lactic acid	1:2
DES9	L-carnitine	Xylitol	2:1

2.3. Ultrasonication-Assisted Extraction of Ginger with DES

The ginger powder was combined with a concentration of 75% DES in a 30:1 solvent to solid ratio, and then, each mixture was ultrasonicated for 20 min at 35 °C, 40 kHz, and 600 W by a KQ-300E ultrasonic [15]. After being centrifuged at 1000× *g* for 10 min in a centrifuge, the extract was collected for further testing. Meanwhile, some conventional extraction solvents including water, ethanol, and 80% ethanol solutions (*v*/*v*) were used for comparison.

2.4. Determination of Total Phenolic (TPC) and Flavonoid Content (TFC)

The Folin–Ciocalteu technique was used to calculate the TPC [23]. First, 1 mL of Folin–Ciocalteu reagent (0.2 M) and 2 mL of Na_2CO_3 solution (75 g/L) were combined with 0.05 mL of each extract. After the combination was reacted at 25 °C in the dark for 2 h, the absorbance of the reacted solutions was measured at 765 nm by a SpectraMax 190 microplate reader. The TPC was calculated and represented as mg GAE/g dw based on the standard curve created with gallic acid (GA).

A colorimetric experiment was conducted to determine the TFC [24]. Then, 2.0 mL samples were mixed with 0.3 mL $NaNO_2$ (0.05 g/mL) and incubation for 5 min. Next, 500 μL of $AlCl_3$ solution (10%) and 2 mL of NaOH solution (1 mol/L) was added and reacted for 6 min and 10 min at room temperature sequentially. Using a microplate reader, the solution's absorbance was determined at 510 nm. Rutin was used as the reference substance, and the TFC was calculated based on a calibration curve reacted with rutin (RE) and expressed as mg RE/g dw.

2.5. Antioxidant Activity Assays

According to Quan et al. [25], after mixing 20 μL extract solutions with 380 μL of ABTS working solution, the mixture was reacted at room temperature for 10 min. A UV-5300PC spectrophotometer was used to measure the absorbance of the reacting solution at 734 nm.

According to our prior report, the extracts' ability to scavenge DPPH free radicals was assessed [25]. Briefly, 500 μL samples were mixed with 3 mL of DPPH working solution. A 517 nm absorbance measurement was conducted after the mixtures were incubated at 30 °C for 30 min in the dark.

Based on a previous description by Qie et al., we calculated the ferric ion-reducing antioxidant power (FRAP) [26]. Extract solutions (10 μL) and 190 μL FRAP working solution was reacted for 30 min at 35 °C, and the absorbance of reacted solutions was measured at 593 nm.

Based on Gillespie et al.'s instructions, oxygen radical absorbance capacity (ORAC) assays were conducted [27], and extract solutions were mixed with fluorescein solution (0.2 μM) and 150 mM AAPH (25 μL) and reacted at 37 °C for 10 min. Fluorescence of reacted solutions were read at 485 nm excitation and 530 nm emission.

Based on the standard curve created with Trolox as a standard calibration method, the antioxidant activity of different ginger extracts including ABTS and DPPH free radical scavenging ability, FRAP, and ORAC were calculated and represented as the Trolox equivalent antioxidant capacity.

2.6. Meat Preparation and Cooking

Ginger extract from DES and conventional solvents were added to ground beef (with addition levels 0 or 1.0%), and beef patties (40 ± 0.1 g) were prepared using a Petri dish (Φ 6 cm × 1.5 cm) [14,16]. Then, in a SCC61E oven (RATIONAL, Munich, Germany), all patties were roasted at 220 °C for 10 min on each side and stored at −80 °C before further analysis. Beef patties with addition of ginger extract from DES (DES1-9) and conventional solvents (water, ethanol, and 80% ethanol solutions) were named as DES1-9, water, EtOH, and 80EtOH, respectively, while beef patties without addition of ginger extract were defined as control.

2.7. Composition, Cooking Loss, and Texture Profile Analysis

According to AOAC procedures [28], the proximate composition of raw and cooked samples, including protein and ash, was analyzed. With the use of a digital pH meter, the pH values of the samples were calculated. Moreover, raw and cooked beef patties were weighed to calculate the cooking loss.

According to our earlier studies, the texture profile of roasted beef patties (1 × 1 × 1 cm) was measured using a TA-XT plus texture analyzer (Godalming, UK) equipped with a P/50 cylinder probe [29]. The parameters for texture profile analysis (TPA) were set as follow: pre-test (4 mm/s), test (3 mm/s), and post-test (4 mm/s) and trigger force (5.0 g with 50% strain for 5 s).

2.8. Determination of HAs

We applied our previously reported technique for the extraction of HAs [12,16,29]. In brief, 2 g beef powder, 30 mL NaOH solution (2 M), and 15 mL ethyl acetate were mixed and ultrasonic extracted at 50 kHz, 40 °C, for 40 min. The sample was centrifuged at 3000× *g* for 10 min, and the supernatant was collected for solid-phase extraction [12,16,29].

HAs were analysis by a Waters UPLC-Q-TOF-MS (Milford, USA) with an Acquity UPLC BEH C18 column (50 × 2.1 mm i.d., 1.7 m). A gradient elution method were used to separate the target HAs with a flow rate of 0.3 mL/min, and acetonitrile and ammonium acetate solution (5 mM, pH 6.8) set as mobile phase A and B, respectively. The gradient program was set as: 0 min, 90% B; 10 min, 85% B; 12 min, 0% B; 16 min, 90% B. For the MS analysis, the instrument was set as: capillary voltage (3.0 kV), source temperature (120 °C), desolvation temperature (350 °C), cone gas flow rate (65 L/h), desolvation gas flow rate (650 L/h), and collision gas flow rate (0.15 mL/min). Table 2 shows a summary of the TQD parameters for HAs that were optimized using HA standards.

Table 2. Description of the MRM parameters for the analysis of HAs and AGEs in beef patties.

	Precursor Ion (m/z)	Product Ion (m/z)	Cone Voltage (V)	Collision Voltage (eV)	Dwell Time (s)
HAs					
DMIP	163	148	30	25	0.10
Phe-p-1	171	127	30	35	0.10
1,5,6-TMIP	177	162	30	25	0.15
Harmane	183	115	30	30	0.15
Norharmane	169	115	30	35	0.10
AαC	183	140	30	35	0.15
MeAαC	198	181	30	30	0.15
Glu-p-1	199	145	30	35	0.10
IQ	199	130	30	35	0.15
IQ [4,5-b]	199	115	30	35	0.10
IQx	200	185	30	35	0.10
MeIQ	213	198	30	30	0.10
MeIQx	214	131	30	35	0.10
PhIP	225	210	30	35	0.15
4,8-DiMeIQx	228	212	30	30	0.15
7,8-DiMeIQx	228	213	30	35	0.15
4,7,8-DiMeIQx	242	227	30	30	0.10
AGEs					
CML	205	84	20	18	0.15
d_4-CML	209	88	20	18	0.15
CEL	219	84	22	20	0.15
d_4-CEL	223	88	22	20	0.15

2.9. Determination of AGEs

With a few minor modifications, we used the method we previously described to extract the bound AGEs [12,16]. To remove the lipid, 3 mL n-hexane was added to 30 mg sample and centrifuged for 15 min at 4000× *g*. The defatted samples were next reduced with 3 mL sodium borate buffer and 2 mL $NaBH_4$ at 4 °C for 8 h. The mixture was heated at 110 °C for 24 h to facilitate hydrolysis after 5 mL HCl was added. After filtering, solid-phase extraction was performed on 2 mL of redissolved hydrolysates.

The parameters for the UPLC-Q-TOF-MS analysis of AGEs were referred to our previous studies [12,16]. In brief, 5μL of samples were separated in a HSS T3 column (150 × 2.1 mm, 3.5 μm) at 40 °C with gradient elution program: 0–3 min, 1% A; 3–7 min, 3%–100% A; 7–9 min, 100% A; 9–10 min, 100%–1% A, in which acetonitrile and 0.1% formic acid were selected as mobile phase A and B, respectively, and the flow rate was set as 0.3 mL/min. For MS analysis, ESI+ mode and MRM were used for the acquisition of the mass spectrometric data (as shown in Table 2), and the parameters of the MS instrument were the same as in our previous studies [12,16].

2.10. Determination of Creatine, Creatinine, and Glucose in Roasted Beef Patties

The content of creatinine and glucose in beef patties was measured by commercial kits using the peroxidase and glucose oxidase method, respectively [17,30]. All experimental operations were performed according to the instructions of commercial kits. According to a method described by Zhang [30], the contents of creatine were measured as following program: 0.5 g samples with 100 mL 30 g/L TCA were homogenized for 1 min and filtered to remove the precipitated protein. Twenty microliters of the extract were defatted with 10 mL diethyl ether. Defatted extract, 2mL diacetyl (0.2 g/L), and 2 mL L-naphthol (25 g/L) were mixed and heated at 50 °C for 6 min, and then, the absorbance was measured at 520 nm.

2.11. Measurements of Protein and Lipid Oxidation of the Roasted Beef Patties

The beef powder was used to extract the total lipid. The samples' extracted lipids were then weighed and kept in the dark at a temperature of −80 °C for later use. As per the literature, POV and TBARS were measured [16,17], and the results were expressed as mmol/kg lipid and mg MDA equivalents/kg lipid, respectively. Protein oxidation was reflected and evaluated by a protein carbonyl content detection kit, which adopted the method derivatization with dinitrophenylhydrazine (DNPH) [31].

2.12. Statistical Analysis

All samples were tested three times independently, and data were analyzed by Statistix 9.0 software (Tallahassee, FL, USA) using an analysis of variance to determine whether there were significant differences between treatments. For statistical significance, $p < 0.05$ was considered. The relative amount was calculated as follows: HAs or AGEs levels of beef patties treated with DES/HAs or AGEs levels of control.

3. Results and Discussion

3.1. TPC, TFC, and Antioxidant Capacity of Ginger Extracts

As a widely used spice and ingredient in China, ginger is rich in polyphenols and flavonoids. In this study (Figure 1A), the TPC in the ginger extract from conventional solvent ranged from 15.3 ± 1.10 to 29.6 ± 1.69 mg GAE/g DW, in which ginger extracted from water and 80% ethanol showed the highest and lowest TPC, respectively. However, compared with the conventional solvents, the TPC of the DESs extracts of ginger was significantly higher than that of the conventional solvent extract of ginger except for glycerol-based DESs (DES1, DES4, and DES 7). Among those DESs, DES2 showed that the highest TPC reached 50.6 ± 2.98 mg GAE/g DW, followed by DES 5, DES 6, and DES8. Moreover, the TFC of ginger extracts was also determined and showed similar results to that of TPC. As shown in Figure 1B, ginger extract from water and ethanol also showed the lowest TFC, which only reached 0.71 ± 0.13 mg RE/g DW and 3.04 ± 0.34 mg RE/g DW, respectively. Although the TFC of 80EtOH was significantly higher than that of the other two conventional solvents, it was still significantly lower than that of DESs, especially for DES2, DES5, DES6, and DES8, which ranged from 8.15 ± 0.59 to 9.80 ± 0.58 mg RE/g DW. In general, in comparison with different DES, lactic-acid-based DESs including DES 2, DES5, and DES8, especially for DES 2, showed significantly higher TPC and TFC than other DES. While glycerol-based DES including DES1, DES4, and DES 7, especially for DES1, showed a TPC and TFC comparable to conventional solvents and even slightly lower than 80EtOH, which is significantly lower than that of another DES. Our results are particularly consistent with earlier research, which found that the extraction yield of gingerols (an important phenolic compound from ginger) was significantly higher in 15 different DESs than in ethanol and water and that lactic-acid-based DESs also demonstrated the highest extraction yield of gingerols [15].

The ginger extract that was prepared under the ideal extraction conditions was then subjected to four antioxidant assays (DPPH, ABTS, FRAP, and ORAC) in the current study. According to earlier research, in vitro, antioxidant activity assays are frequently used to assess the efficacy of DES extracts since they are practical and simple to use [15,20,22]. As shown in Figure 2, the antioxidant activity of ginger extract with nine kinds of DESs was evaluated and compared with three kinds of conventional solvents (water, EtOH, and 80EtOH). As Figure 2A–C shows, the free radical scavenging activities of ginger extract from water were significantly ($p < 0.05$) lower than other solvents in the three antioxidant assays, which is consistent with our abovementioned results that ginger extract from water showed the lowest TPC and TFC. On the contrary, the ABTS, DPPH, and oxygen radical scavenging activities of lactic acid and xylitol-based DESs were significantly higher than that of conventional solvents, in which DES2 and DES 6 exhibited the highest DPPH and ABTS radical scavenging activities, followed by DES5, DES8, and DES9. Similarly, DES2, DES6, and DES8 also exhibited the highest oxygen radical scavenging activities, followed by DES5, DES8, and DES9. These results show that both lactic-acid- and xylitol-based DESs have excellent antioxidant activity. Despite that, the free radical scavenging activities of glycerol-based DESs (DES1, DES4, and DES 7) were the same as that of EtOH or even significantly lower than that of 80 EtOH. The respective range of DES1, DES4, and DES7 in the three antioxidant assays were 22.3 ± 2.52 to 32.7 ± 4.43 mg Trolox/g in the DPPH assay, 21.2 ± 1.47 to 41.2 ± 1.56 mg Trolox/g in the ABTS assay, and 66.3 ± 5.88 to 143 ± 6.03 mg Trolox/g in the ORAC assay.

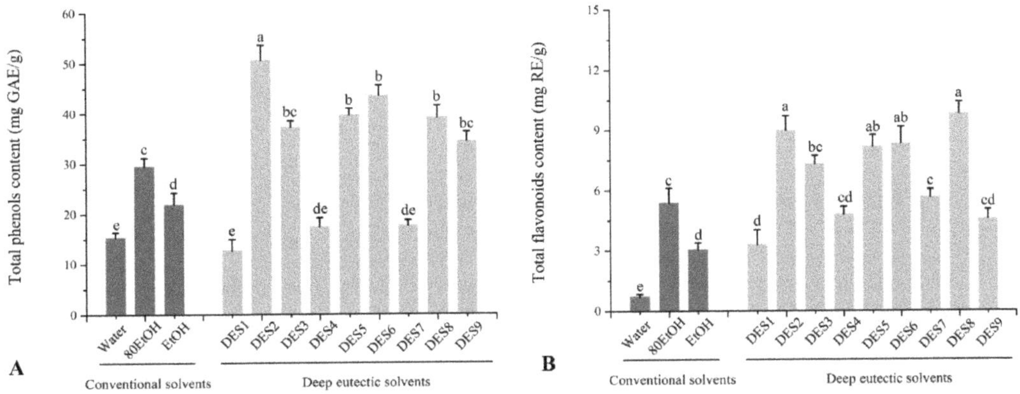

Figure 1. Effect of extraction solvent (conventional and deep eutectic solvents) on (**A**) total phenolic and (**B**) flavonoids contents of ginger extract. The corresponding to the solvent's abbreviations can be consulted in Table 1. Values are expressed as mean ± SD ($n = 3$). Different letters in each series indicate significant differences ($p < 0.05$); lowercase letters denote significant differences ($p < 0.05$) between solvents.

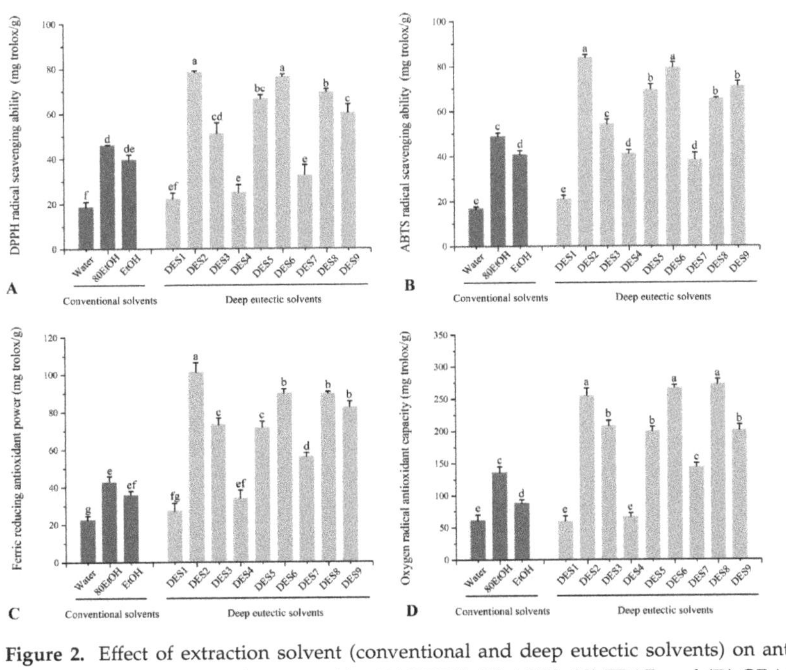

Figure 2. Effect of extraction solvent (conventional and deep eutectic solvents) on antioxidant capacity of ginger extract determined by (**A**) DPPH, (**B**) ABTS, (**C**) FRAP, and (**D**) ORAC assays. The corresponding solvents' abbreviations can be consulted in Table 1. Values are expressed as mean ± SD ($n = 3$). Different letters in each series indicate significant differences ($p < 0.05$); lowercase letters denote significant differences ($p < 0.05$) between solvents.

The FRAP values of ginger extract show a similar trend to the free radical scavenging activities observed in the ABTS, DPPH, and ORAC tests for ginger extracted from various solvents. These results also showed that both lactic-acid- and xylitol-based DESs have excellent antioxidant activity. However, the FRAP value of DES2 was significantly higher than that of other DESs. The increased TPC and TFC of the DES2 compared to other DESs may explain its higher activity (Figure 1). Similar results showed that DES extracts had better antioxidant activity than those from using conventional solvents [19–21]. Although most of DESs are not thought to have antioxidant action [32]. However, studies have reported that some DES components can boost the antioxidant activity of extracts via a synergy between DES and soluble compounds [33].

3.2. Proximate and Texture Profile Analysis after Roasting

As Table 3 shown, the pH values, cooking loss, protein, and ash contents of the raw beef were 5.58 ± 0.13, 50.9% ± 1.22%, 42.9 ± 1.74 g/100 g, and 4.46% ± 0.08%, and these values were comparable with those reported in our previous studies for raw beef meat [12,14,16]. Raw beef samples were not measured for fat content since fat was removed before cooking in order to avoid its influence on the formation of HAs and AGEs. In each sample group, the cooking loss, protein, and ash content were comparable among the groups and showed a non-significant ($p > 0.05$) change between the control and patties treated with ginger extract. However, patties treated with lactic-acid-based DESs (DES2, DES4, and DES8), which varied from 4.61 ± 0.12 to 5.01 ± 0.19, were significantly ($p < 0.05$) lower than those of control patties.

Table 3. Chemical composition, cooking loss, and pH of the beef patties with different ginger extracts.

Group	pH	Protein (g/100g)	Cooking Loss (%)	Ash (%)
Control	5.58 ± 0.13 a	42.9 ± 1.74 ab	50.9 ± 1.22 a	4.46 ± 0.08 a
Water	5.77 ± 0.09 a	43.4 ± 0.57 a	48.3 ± 2.96 a	4.31 ± 0.01 a
80EtOH	5.74 ± 0.07 a	45.1 ± 2.56 a	52.3 ± 1.05 a	4.16 ± 0.09 a
EtOH	5.62 ± 0.04 a	44.3 ± 2.53 a	49.9 ± 2.90 a	4.53 ± 0.03 a
DES1	5.69 ± 0.05 a	46.2 ± 2.98 a	50.2 ± 1.17 a	4.27 ± 0.05 a
DES2	5.01 ± 0.19 b	44.0 ± 1.94 a	52.3 ± 1.08 a	4.33 ± 0.12 a
DES3	5.73 ± 0.16 a	45.2 ± 2.38 a	49.9 ± 2.46 a	4.86 ± 0.04 a
DES4	5.81 ± 0.03 a	46.7 ± 1.08 a	52.9 ± 1.38 a	4.88 ± 0.20 a
DES5	4.61 ± 0.12 b	46.2 ± 2.19 a	48.0 ± 1.80 a	4.85 ± 0.07 a
DES6	5.88 ± 0.05 a	46.9 ± 0.67 a	52.5 ± 1.21 a	4.45 ± 0.13 a
DES7	5.33 ± 0.02 a	43.1 ± 1.04 ab	48.7 ± 2.10 a	4.55 ± 0.04 a
DES8	4.67 ± 0.09 b	47.7 ± 1.96 a	48.1 ± 2.89 a	4.79 ± 0.05 a
DES9	5.78 ± 0.17 a	45.9 ± 1.26 a	52.7 ± 1.36 a	4.86 ± 0.07 a

Mean ± SD (n = 3). Means with different letters (a, b) in the same column are significantly different ($p < 0.05$).

The results of a textural profile analysis are shown in Table 4. The beef patties with or without addition of ginger extract showed non-significant ($p > 0.05$) difference concerning hardness, springiness, cohesiveness, gumminess, cohesiveness, and chewiness, which indicates that the different ginger extracts showed no significant ($p > 0.05$) change in the texture profile of beef patties.

3.3. Effects of Ginger Extract on the Formation of HAs in Beef Patties

As shown in Figure 3, six HAs (including PhIP, MeIQx, MeIQ, 4,8-MeIQx, Harmane, and Norharmane) were determined, and the effects of ginger extract on the formation of those HAs in the beef patties were reflected as results of the relative amount compared to the control group. Except for Harmane, which increased by 20%, water extract could lower HAs generation by 5–19%. Ginger extracted from the other two conventional solvents also reduced the formation of PhIP, MeIQx, MeIQ, and 4,8-MeIQx, in which MeIQ and 4,8-MeIQx were significantly decreased by 24–39% and 37–41%, respectively. This result is consistent with the study by Xue et al., who found that ginger can reduce the formation of HAs in a dose-dependent way by 0.5%, 1.0%, and 1.5%, inhibiting the synthesis of total

HAs by 3.61%, 24.88%, and 27.42%, respectively [16]. On the contrary, Xue et al., indicated that the inhibition rates of 1.5% ginger on Norharmane and Harman were 28.56% and 27.92%, respectively [16]. However, EtOH and 80EtOH significantly increased the content of two β-carboline to 121% and 116% and 122% and 134%, respectively.

Table 4. Texture characteristics of raw beef and roast beef patties with different ginger extracts.

Group	Hardness (*N*)	Springiness (mm)	Gumminess (*N*)	Cohesiveness (*N*)	Chewiness (mJ)
Control	8298 ± 51 ab	0.39 ± 0.02 a	4954 ± 57 a	0.65 ± 0.15 ab	3728 ± 52 a
Water	8276 ± 43 b	0.49 ± 0.08 a	4832 ± 83 a	0.91 ± 0.14 a	3807 ± 77 a
80EtOH	8294 ± 59 ab	0.52 ± 0.07 a	4905 ± 30 a	0.96 ± 0.16 a	3722 ± 73 a
EtOH	8328 ± 96 a	0.32 ± 0.03 a	4859 ± 79 a	0.83 ± 0.19 a	3698 ± 49 ab
DES1	8394 ± 59 a	0.44 ± 0.07 a	4858 ± 96 a	0.73 ± 0.05 a	3672 ± 65 ab
DES2	8393 ± 103 a	0.36 ± 0.02 a	4901 ± 44 a	0.69 ± 0.11 a	3634 ± 62 ab
DES3	8478 ± 101 a	0.46 ± 0.06 a	4961 ± 48 a	0.57 ± 0.17 ab	3729 ± 65 a
DES4	8306 ± 83 ab	0.4 ± 0.07 a	4834 ± 110 a	0.74 ± 0.04 a	3799 ± 29 a
DES5	8344 ± 50 a	0.5 ± 0.02 a	4924 ± 34 a	0.78 ± 0.02 a	3710 ± 88 a
DES6	8437 ± 99 a	0.59 ± 0.10 a	4927 ± 103 a	0.75 ± 0.12 a	3842 ± 34 a
DES7	8257 ± 67 ab	0.52 ± 0.01 a	4841 ± 101 a	0.71 ± 0.01 a	3654 ± 72 ab
DES8	8455 ± 60 a	0.57 ± 0.03 a	4801 ± 98 a	0.66 ± 0.16 a	3628 ± 64 ab
DES9	8240 ± 96 ab	0.51 ± 0.06 a	4832 ± 41 a	0.60 ± 0.03 b	3733 ± 21 a

Mean ± SD (n = 3). Means with different letters (a, b) in the same column are significantly different ($p < 0.05$).

Moreover, all nine DES extracts considerably decreased the overall amount of HAs in beef patties. Of those DES extracts evaluated, the most effective in reducing the formation of HAs was DES2, which caused a signification reduction of 45% of the PhIP content, 30% of MeIQx, 51% of MeIQ, 79% of 4,8-MeIQx, 22% of Harmane, and 18% of Norharmane, and another two lactic-acid-based DESs showed similar inhibition effect on the formation of HAs. Roasting beef patties with xylitol-based DESs similarly resulted in much lower levels of the majority of HAs when compared to the control group except harmane, whose concentration rose to 109% in beef patties with DES9. For the glycerol-based DESs, the inhibition effects on HAs were significantly lower than other DESs, of which the relative amount of every HAs exhibited a similar change to that of conventional solvents. This result indicated that ginger extract exhibited inhibition ability to HAs, in which xylitol- and lactic-acid-based DESs showed better inhibition ability to HAs.

3.4. Effects of Ginger Extract on the Formation of CML and CEL in Beef Patties

According to Figure 4, the ginger extract had a significant inhibitory effect on CML and CEL. The water extract of ginger had the lowest inhibitory effect, and the content of CML and CEL was only reduced to 6.2% and 5.9%, respectively. This result is similar to the effects of ginger extract on the formation of HAs that were reported above. Another two ethanol extracts of ginger showed a better inhibition effect on the formation of CML and CEL, which decreased to 18.5–27.15% and 12.6–31.7%, respectively. Obviously, in comparison with conventional solvents, DES extracts of ginger exhibited better inhibitory capacity, which significantly decreased the formation of CML and CEL. For example, the inhibitory rate of DES2, DES5, and DES8 for bound CML in beef patties was 48.2%, 37.5%, and 41.9%, respectively, and it reached 58.5%, 52.7%, and 49.9% for bound CEL. Nonetheless, it is worth noting that glycerol-based DESs extract had a limited effect on the inhibition of CML and CEL, while the inhibitory rate of DES1, DES4, and DES7 for CML and CEL showed similar results with conventional solvent extracts.

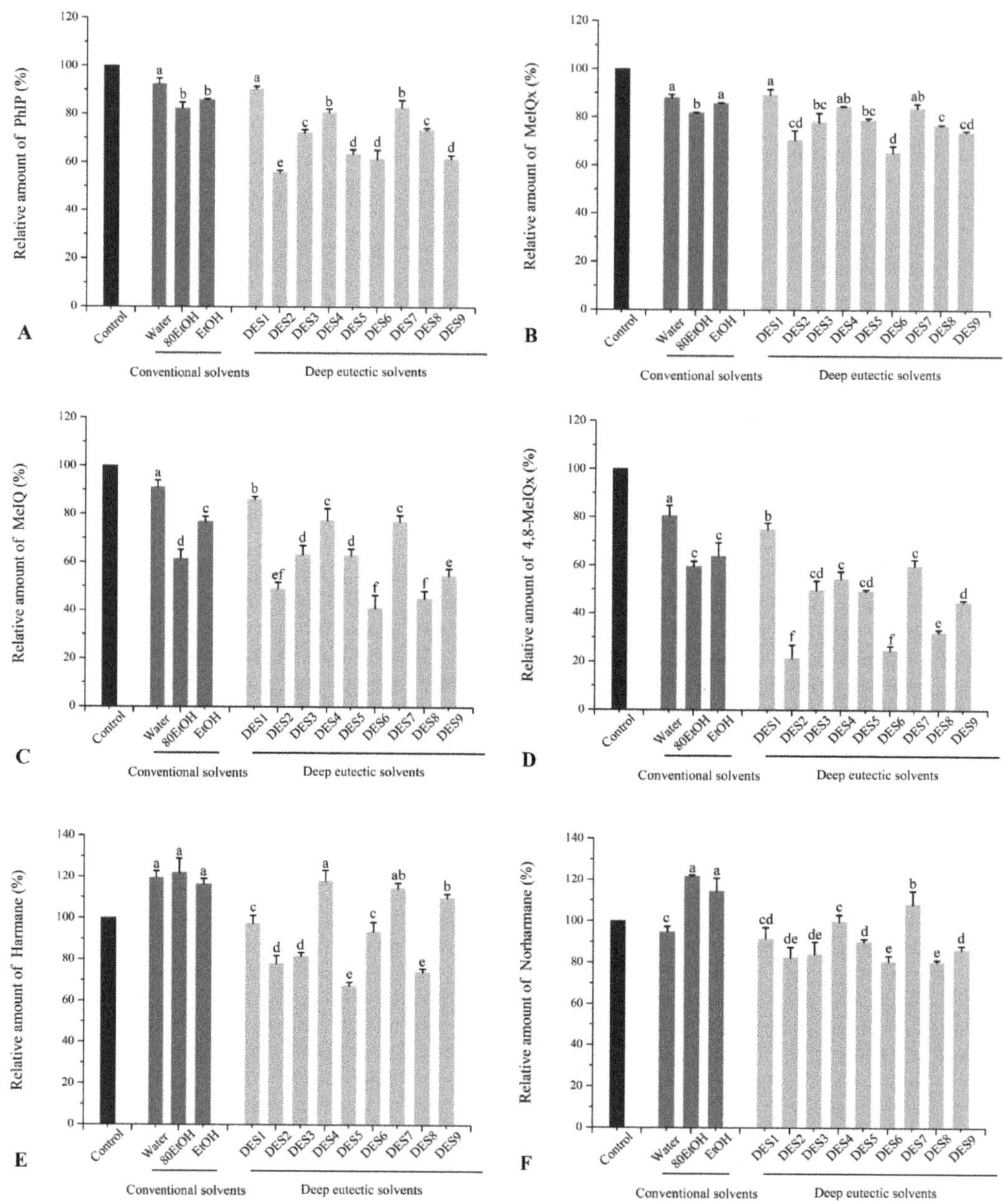

Figure 3. Relative amounts of (**A**) PhIP, (**B**) MeIQx, (**C**) MeIQ, (**D**) 4,8-DiMeIQx, (**E**) Harmane, and (**F**) Norharmane in roasted beef patties with different ginger extracts. The corresponding solvents' abbreviations can be consulted in Table 1. Values are expressed as mean ± SD ($n = 3$). Different letters in each series indicate significant differences ($p < 0.05$); lowercase letters denote significant differences ($p < 0.05$) between solvents.

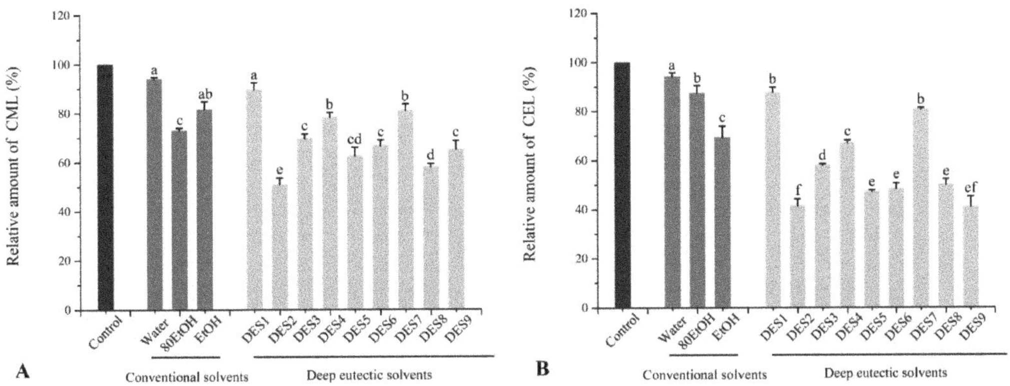

Figure 4. Relative amounts of (**A**) CML and (**B**) CEL in roasted beef patties with different ginger extracts. The corresponding to the solvent's abbreviations can be consulted in Table 1. Values are expressed as mean ± SD (n = 3). Different letters in each series indicate significant differences ($p < 0.05$); lowercase letters denote significant differences ($p < 0.05$) between solvents.

Previous studies suggested that ginger and its main ingredient curcumin might be able to inhibit HAs, CML, and CEL by removing free radicals from the system and lowering the concentration of active carbonyl intermediates [16]. Therefore, the results of the inhibitory rate of HAs and AGEs in grilled beef patties by different ginger extracts are generally consistent with the results of the antioxidant activity of different ginger extracts. Moreover, the present study was further designed to prove this speculation by detecting the changes in the content of precursor and related oxidation products in beef patties.

3.5. Change of Glucose, Creatine, and Creatinine Contents

Creatine and creatinine are both important precursors that affect the formation of HAs [6]. Meanwhile, creatine and creatinine do not affect the synthesis of β-carboline HAs and AGEs, which are mostly produced from glucose and amino acids [6]. In addition to understanding the relationship between ginger extract and the development of HAs, AGEs, creatine, creatinine, and glucose, it was necessary to look into the precursors' content changes [17,29]. In Table 5, precursor concentrations for creatine, creatinine, and glucose in roasted beef patties and beef patties treated with various ginger extracts are shown.

Table 5. Contents of precursors creatine, creatinine, and glucose in beef patties with different ginger extract.

Group	Creatine (mg/g)	Creatinine (μmol/L)	Glucose (mg/g)
Control	0.95 ± 0.04 f	1.22 ± 0.08 cd	0.11 ± 0.01 d
Water	1.21 ± 0.03 e	1.55 ± 0.19 c	0.13 ± 0.01 d
80EtOH	2.35 ± 0.07 c	1.85 ± 0.16 bc	0.21 ± 0.03 bc
EtOH	2.48 ± 0.26 c	2.45 ± 0.21 b	0.12 ± 0.01 d
DES1	1.33 ± 0.29 ef	1.19 ± 0.05 cd	0.17 ± 0.00 b
DES2	5.30 ± 0.20 a	3.26 ± 0.02 a	0.40 ± 0.08 a
DES3	2.72 ± 0.02 c	2.19 ± 0.16 b	0.26 ± 0.07 a
DES4	1.48 ± 0.28 d	2.52 ± 0.11 b	0.22 ± 0.03 b
DES5	4.80 ± 0.27 a	3.34 ± 0.21 a	0.36 ± 0.06 a
DES6	5.97 ± 0.08 a	2.54 ± 0.12 b	0.31 ± 0.09 a
DES7	1.89 ± 0.11 d	2.30 ± 0.05 b	0.27 ± 0.00 b
DES8	4.22 ± 0.04 b	3.39 ± 0.02 a	0.28 ± 0.07 a
DES9	4.44 ± 0.13 b	2.47 ± 0.03 b	0.29 ± 0.05 a

Mean ± SD (n = 3). Means with different letters (a–f) in the same column are significantly different ($p < 0.05$).

When compared to the control, those precursor levels in the water–extract-treated beef patties did not differ significantly. For EtOH and 80EtOH, the creatine level increased from 0.95 to 2.35 mg/g and 2.48 mg/g. In contrast, the batch of beef patties treated with DES extracts exhibited different phenomena, the creatine level increased from 0.95 to 1.33 mg/g and 5.97 mg/g, and both the creatinine and glucose levels ranged from 1.19 to 3.39 μmol/L and 0.17 to 0.40 mg/g, respectively. Compared to the control group and conventional solvent groups, roasted beef patties treated with DES2, DES5, DES6, and DES8 that exhibited the strongest inhibitory effect on Has and AGEs possess a significantly increasing trend of creatine, creatinine, and glucose. As a result, the inhibitory mechanism may have been caused by the DES extracts' ability to impede the reactions of the precursors of glucose, creatine, and creatinine with some amino acids and intermediates. Previous studies also pointed out that the inhibitory effect of some bioactive compounds may be due to the hindering of pyridines and pyrazines via Strecker degradation, derived from glucose and amino acid [34,35].

3.6. Inhibitory Effects of Ginger Extract on Protein and Lipid Peroxidation

Lipids and proteins are susceptible to oxidative degradation during thermal processing, leading to the formation of reactive carbonyl compounds, which could further promote the accumulation of Has and AGEs [6,31]. Based on the results of some previous studies, we hypothesized and demonstrated that ginger extract significantly inhibits the formation of HAs and AGEs from roasted beef patties by reducing the amount of free radicals produced. As previously mentioned, ginger extracts demonstrated excellent antioxidant activity. Therefore, the effect of ginger extract on lipid and protein oxidation was determined.

Protein carbonyls, which are produced during protein oxidative breakdown, show the degree of protein oxidation, while lipid oxidation is commonly measured with TBARS and POV values [31]. As shown in Figure 5, total carbonyl content, TBARS, and POV values in the control group were 5.90 ± 0.90 mmol/kg, 15.2 ± 1.13 mg MDA eq/kg lipid, and 36.7 ± 1.51 mmol/kg, respectively. All of the ginger extracts could significantly decrease the total carbonyl content, TBARS, and POV values in beef patties, in which total carbonyl content, TBARS, and POV values decreased to 3.30–5.30 mmol/kg, 11.3–14.7 mg MDA eq/kg lipid, and 22.0–32.8 mmol/kg in the conventional solvent groups, respectively. Compared with conventional solvent groups, DES1, DES3, DES4, DES7, DES8, and DES9 exhibited similar effects on protein and lipid peroxidation, but DES2, DES5, and DES6 could significantly reduce the oxidative degradation of lipid and protein in roasted beef patties, which is consistent with the results of the TPC, TFC, and in vitro antioxidant capacity of ginger extracts.

Overall, our results imply that ginger extract reduces lipid and protein peroxidation by quenching free radicals, as described above. Future experiments should investigate changes in active carbonyl intermediates.

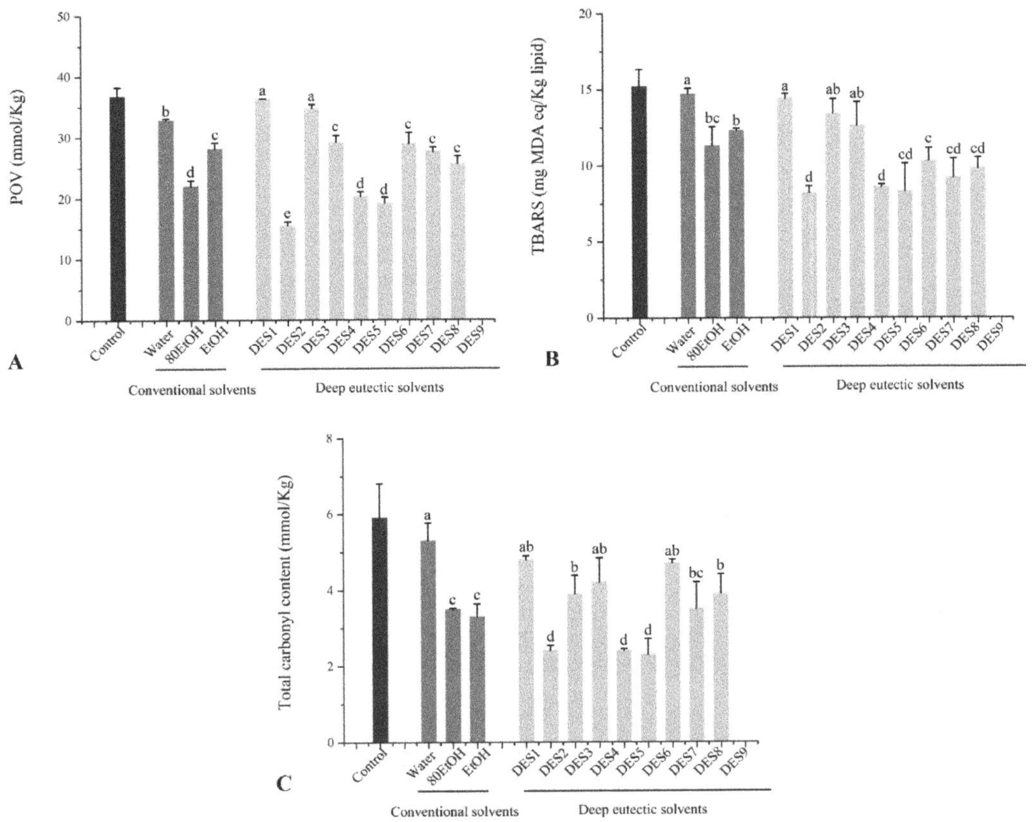

Figure 5. Effect of different ginger extract on lipid and protein oxidation of beef patties determined by (**A**) POV, (**B**) TBARS, and (**C**) total carbonyl content. The corresponding solvents' abbreviations can be consulted in Table 1. Values are expressed as mean ± SD ($n = 3$). Different letters in each series indicate significant differences ($p < 0.05$); lowercase letters denote significant differences ($p < 0.05$) between solvents.

4. Conclusions

Generally, HAs and AGEs are often produced simultaneously during the thermal processing of food. Despite this, there have been few studies exploring simultaneous inhibition methods for these two harmful compounds, which certainly cannot satisfy public demands for food safety. In the present study, ginger extracts with strong antioxidant activity were prepared with various deep eutectic and conventional solvents. Our results further indicate the use of ginger DES extracts in the preparation of beef patties as a way to lessen the generation of HAs and AGEs during roasting. As a whole, all ginger extract treatments significantly reduced the amount of HAs and AGEs to 18–79% and 41–58%, respectively. It appears that xylitol- and lactic-acid-based DES extracts reduce the amount of HAs and AGEs to the greatest extent possible. Further, we evaluated the physical/chemical changes of ginger extracts on the beef patties and clarified that inhibiting ability of ginger extracts for HAs and AGEs mainly originated from quenching of free radicals through several studies including antioxidant index, precursors of HAs and AGEs, and lipid and protein oxidation. The present study provides a theoretical framework for understanding and developing the efficient inhibition of HAs and AGEs development in thermally processed foods based on DES-prepared natural product extracts. It may be relevant to different meat systems and useful to food producers for producing beef patties with less HA and AGE formation. This has significant implications for developing healthier meat products and improving food safety.

Author Contributions: Y.X. and W.Q., investigation, writing—original draft, formal analysis, and methodology; Y.J. and J.L., validation and formal analysis; Z.H., data curation and software; M.Z., validation and software; J.C., funding acquisition, project administration, and supervision; Q.S., funding acquisition, project administration, and supervision. All authors have read and agreed to the published version of the manuscript.

Funding: This work was supported by Changsha Municipal Natural Science Foundation (No. kq2202223), the National Natural Science Foundation of China (Grant No. 3217160166), and the Innovation and Exploration Fund of State Key Laboratory of Food Science and Technology, Jiangnan University (No. SKLF-ZZA-202001).

Data Availability Statement: The data presented in this study are available on request from the corresponding author.

Conflicts of Interest: The authors declare no conflict of interest.

Abbreviations

Heterocyclic amines (HAs); advanced glycation end products (AGEs); deep eutectic solvents (DESs); Nε-(carboxymethyl)lysine (CML); Nε-(carboxyethyl)lysine (CEL); hydrogen bond acceptors (HBAs); hydrogen bond donors (HBDs); 6-hydroxy-2,5,7,8-tetra-methylchroman-2-carboxylic acid (Trolox); 1,1-diphenyl-2-picrylhydrazyl (DPPH); 2,2-azinobis-(3-ethyl-benzthiazoline-6-sulfonate) (ABTS); 2,4,6-tris-(2-pyridyl)-s-triazine (TPTZ); 2,2′-Azobis(2-amidinopropane)dihydrochloride (AAPH); dinitrophenylhydrazine (DNPH); Nε-(1-carboxymethyl)-L-lysine-d4 (CML-d4); Nε-(1-carboxyethyl)-L-lysine-d4 (CEL-d4); gallic acid (GA); Rutin (RE); ferric ion-reducing antioxidant power (FRAP); oxygen radical absorbance capacity (ORAC); texture profile analysis (TPA); thiobarbituric acid reactive substances (TBARS).

References

1. Wang, L.; Steele, E.M.; Du, M.; Pomeranz, J.L.; O'Connor, L.E.; Herrick, K.A.; Luo, H.; Zhang, X.; Mozaffarian, D.; Zhang, F.F. Trends in consumption of ultraprocessed foods among US youths aged 2–19 years, 1999–2018. *JAMA* **2021**, *326*, 519–530. [CrossRef]
2. Starowicz, M.; Zieliński, H. How Maillard reaction influences sensorial properties (color, flavor and texture) of food products? *Food Rev. Int.* **2019**, *35*, 707–725. [CrossRef]
3. Koszucka, A.; Nowak, A. Thermal processing food-related toxicants: A review. *Crit. Rev. Food Sci.* **2019**, *59*, 3579–3596. [CrossRef]
4. Quan, W.; Wu, Z.; Jiao, Y.; Liu, G.; Wang, Z.; He, Z.; Tao, G.; Qin, F.; Zeng, M.; Chen, J. Exploring the relationship between potato components and Maillard reaction derivative harmful products using multivariate statistical analysis. *Food Chem.* **2021**, *339*, 127853. [CrossRef]
5. Barzegar, F.; Kamankesh, M.; Mohammadi, A. Heterocyclic aromatic amines in cooked food: A review on formation, health risk-toxicology and their analytical techniques. *Food Chem.* **2019**, *280*, 240–254. [CrossRef]
6. Gibis, M. Heterocyclic aromatic amines in cooked meat products: Causes, formation, occurrence, and risk assessment. *Compr. Rev. Food Sci.* **2016**, *15*, 269–302. [CrossRef]
7. Turesky, R.J. Heterocyclic aromatic amines: Potential human carcinogens. In *Advances in Molecular Toxicology*; Elsevier: Amsterdam, The Netherlands, 2010; Volume 4, pp. 37–83.
8. Poulsen, M.W.; Hedegaard, R.V.; Andersen, J.M.; de Courten, B.; Bügel, S.; Nielsen, J.; Skibsted, L.H.; Dragsted, L.O. Advanced glycation endproducts in food and their effects on health. *Food Chem. Toxicol.* **2013**, *60*, 10–37. [CrossRef]
9. Quan, W.; Jiao, Y.; Xue, C.; Li, Y.; Liu, G.; He, Z.; Qin, F.; Zeng, M.; Chen, J. The effect of exogenous free N ε-(carboxymethyl) lysine on diabetic-model Goto-Kakizaki rats: Metabolomics analysis in serum and urine. *J. Agri. Food Chem.* **2021**, *69*, 783–793. [CrossRef]
10. Quan, W.; Li, Y.; Jiao, Y.; Xue, C.; Liu, G.; Wang, Z.; He, Z.; Qin, F.; Zeng, M.; Chen, J. Simultaneous generation of acrylamide, β-carboline heterocyclic amines and advanced glycation ends products in an aqueous Maillard reaction model system. *Food Chem.* **2020**, *332*, 127387. [CrossRef]
11. Rannou, C.; Laroque, D.; Renault, E.; Prost, C.; Sérot, T. Mitigation strategies of acrylamide, furans, heterocyclic amines and browning during the Maillard reaction in foods. *Food Res. Inter.* **2016**, *90*, 154–176. [CrossRef]

12. Xue, C.; Quan, W.; Li, Y.; He, Z.; Qin, F.; Wang, Z.; Chen, J.; Zeng, M. Mitigative capacity of Kaempferia galanga L. and kaempferol on heterocyclic amines and advanced glycation end products in roasted beef patties and related mechanistic analysis by density functional theory. *Food Chem.* **2022**, *385*, 132660. [CrossRef]
13. Jiao, Y.; Quan, W.; He, Z.; Gao, D.; Qin, F.; Zeng, M.; Chen, J. Effects of Catechins on N ε-(Carboxymethyl) lysine and N ε-(Carboxyethyl) lysine Formation in Green Tea and Model Systems. *J. Agri. Food Chem.* **2019**, *67*, 1254–1260. [CrossRef]
14. Xue, C.; He, Z.; Qin, F.; Chen, J.; Zeng, M. Effects of amides from pungent spices on the free and protein-bound heterocyclic amine profiles of roast beef patties by UPLC–MS/MS and multivariate statistical analysis. *Food Res. Inter.* **2020**, *135*, 109299. [CrossRef]
15. Hsieh, Y.-H.; Li, Y.; Pan, Z.; Chen, Z.; Lu, J.; Yuan, J.; Zhu, Z.; Zhang, J. Ultrasonication-assisted synthesis of alcohol-based deep eutectic solvents for extraction of active compounds from ginger. *Ultrason. Sonochem.* **2020**, *63*, 104915. [CrossRef]
16. Xue, C.; Deng, P.; Quan, W.; Li, Y.; He, Z.; Qin, F.; Wang, Z.; Chen, J.; Zeng, M. Ginger and curcumin can inhibit heterocyclic amines and advanced glycation end products in roast beef patties by quenching free radicals as revealed by electron paramagnetic resonance. *Food Control* **2022**, *138*, 109038. [CrossRef]
17. Gumus, D.; Kizil, M. Comparison of the reducing effects of blueberry and propolis extracts on heterocyclic aromatic amines formation in pan fried beef. *Meat Sci.* **2022**, *186*, 108746. [CrossRef]
18. Khan, M.R.; Busquets, R.; Azam, M. Blueberry, raspberry, and strawberry extracts reduce the formation of carcinogenic heterocyclic amines in fried camel, beef and chicken meats. *Food Control* **2021**, *123*, 107852. [CrossRef]
19. Benvenutti, L.; Zielinski, A.A.F.; Ferreira, S.R.S. Which is the best food emerging solvent: IL, DES or NADES? *Trends Food Sci. Technol.* **2019**, *90*, 133–146. [CrossRef]
20. Ali Redha, A. Review on extraction of phenolic compounds from natural sources using green deep eutectic solvents. *J. Agri. Food Chem.* **2021**, *69*, 878–912. [CrossRef]
21. Cunha, S.C.; Fernandes, J.O. Extraction techniques with deep eutectic solvents. *TrAC Trends Anal. Chem.* **2018**, *105*, 225–239. [CrossRef]
22. Długosz, O.; Ochnik, M.; Sochocka, M.; Franz, D.; Orzechowska, B.; Anna, C.-K.; Agata, D.; Banach, M. Antimicrobial and antiviral activity of selenium sulphide nanoparticles synthesised in extracts from spices in natural deep eutectic solvents (NDES). *Sustain. Mater. Technol.* **2022**, *32*, e00433. [CrossRef]
23. Sánchez-Rangel, J.C.; Benavides, J.; Heredia, J.B.; Cisneros-Zevallos, L.; Jacobo-Velázquez, D.A. The Folin–Ciocalteu assay revisited: Improvement of its specificity for total phenolic content determination. *Anal. Methods* **2013**, *5*, 5990–5999. [CrossRef]
24. Fu, Y.; Liu, W.; Soladoye, O.P. Towards innovative food processing of flavonoid compounds: Insights into stability and bioactivity. *LWT* **2021**, *150*, 111968. [CrossRef]
25. Quan, W.; Qie, X.; Chen, Y.; Zeng, M.; Qin, F.; Chen, J.; He, Z. Effect of milk addition and processing on the antioxidant capacity and phenolic bioaccessibility of coffee by using an in vitro gastrointestinal digestion model. *Food Chem.* **2020**, *308*, 125598. [CrossRef]
26. Qie, X.; Chen, Y.; Quan, W.; Wang, Z.; Zeng, M.; Qin, F.; Chen, J.; He, Z. Analysis of β-lactoglobulin–epigallocatechin gallate interactions: The antioxidant capacity and effects of polyphenols under different heating conditions in polyphenolic–protein interactions. *Food Funct.* **2020**, *11*, 3867–3878. [CrossRef]
27. Gillespie, K.M.; Chae, J.M.; Ainsworth, E.A. Rapid measurement of total antioxidant capacity in plants. *Nat. Protoc.* **2007**, *2*, 867–870. [CrossRef]
28. Horwitz, W. *Official Methods of Analysis of AOAC International. Volume I, Agricultural Chemicals, Contaminants, Drugs/Edited by William Horwitz*; AOAC International: Gaithersburg, MD, USA, 1997.
29. Li, Y.; He, J.; Quan, W.; He, Z.; Qin, F.; Tao, G.; Wang, Z.; Zeng, M.; Chen, J. Effects of polyphosphates and sodium chloride on heterocyclic amines in roasted beef patties as revealed by UPLC-MS/MS. *Food Chem.* **2020**, *326*, 127016. [CrossRef]
30. Zhang, L.; Wang, Q.; Wang, Z.; Chen, Q.; Sun, F.; Xu, M.; Kong, B. Influence of different ratios of sucrose and green tea leaves on heterocyclic aromatic amine formation and quality characteristics of smoked chicken drumsticks. *Food Control* **2022**, *133*, 108613. [CrossRef]
31. Li, Y.; Xue, C.; Quan, W.; Qin, F.; Wang, Z.; He, Z.; Zeng, M.; Chen, J. Assessment the influence of salt and polyphosphate on protein oxidation and Nε-(carboxymethyl) lysine and Nε-(carboxyethyl) lysine formation in roasted beef patties. *Meat Sci.* **2021**, *177*, 108489. [CrossRef]
32. Chen, X.-Q.; Li, Z.-H.; Liu, L.-L.; Wang, H.; Yang, S.-H.; Zhang, J.-S.; Zhang, Y. Green extraction using deep eutectic solvents and antioxidant activities of flavonoids from two fruits of Rubia species. *LWT* **2021**, *148*, 111708. [CrossRef]
33. Percevault, L.; Limanton, E.; Nicolas, P.; Paquin, L.; Lagrost, C. Electrochemical determination and antioxidant capacity modulation of polyphenols in deep eutectic solvents. *ACS Sustain. Chem. Eng.* **2020**, *9*, 776–784. [CrossRef]
34. Li, M.; Lin, S.; Wang, R.; Gao, D.; Bao, Z.; Chen, D.; Tang, Y.; Sun, N.; Zhang, S. Inhibitory effect and mechanism of various fruit extracts on the formation of heterocyclic aromatic amines and flavor changes in roast large yellow croaker (Pseudosciaena crocea). *Food Control* **2022**, *131*, 108410. [CrossRef]
35. Cao, H.; Chen, B.H.; Inbaraj, B.S.; Chen, L.; Alvarez-Rivera, G.; Cifuentes, A.; Zhang, N.; Yang, D.J.; Simal-Gandara, J.; Wang, M. Preventive potential and mechanism of dietary polyphenols on the formation of heterocyclic aromatic amines. *Food Front.* **2020**, *1*, 134–151. [CrossRef]

foods

MDPI

Article

Formation and Inhibition of Heterocyclic Amines and Polycyclic Aromatic Hydrocarbons in Ground Pork during Marinating

Yu-Wen Lai [1], Yu-Tsung Lee [2], Baskaran Stephen Inbaraj [1] and Bing-Huei Chen [1,3,*]

[1] Department of Food Science, Fu Jen Catholic University, New Taipei City 24205, Taiwan
[2] Research Center for Food and Cosmetic Safety, College of Human Ecology, Chang Gung University of Science and Technology, Taoyuan 33303, Taiwan
[3] Department of Nutrition, China Medical University, Taichung 40402, Taiwan
* Correspondence: 002622@mail.fju.edu.tw; Tel.: +886-2-29053626

Abstract: This study aims to simultaneously extract heterocyclic amines (HAs) and polycyclic aromatic hydrocarbons (PAHs) from ground pork for respective analysis by UPLC-MS/MS and GC-MS/MS, and study the effects of different flavorings and marinating time length on their formation and inhibition. Results showed that both HA and PAH contents followed a time-dependent increase during marinating, with HAs being more susceptible to formation than PAHs. The total HA contents in unmarinated pork and juice was, respectively, 61.58 and 139.26 ng/g, and rose to 2986.46 and 1792.07 ng/g after 24-h marinating, which can be attributed to the elevation of reducing sugar and creatinine contents. The total PAH contents in unmarinated pork and juice were, respectively, 34.56 and 26.84 ng/g, and increased to 55.93 and 44.16 ng/g after 24-h marinating, which can be due to the increment of PAH precursors such as benzaldehyde, 2-cyclohexene-1-one and trans,trans-2,4-decadienal. Incorporation of 0.5% (w/v) cinnamon powder or 0.5% (w/v) green tea powder was effective in inhibiting HA formation with the former showing a more pronounced effect for marinated pork, while the latter was for marinated juice. However, their addition was only effective in inhibiting PAH formation in marinated pork. Principle component analysis revealed the relationship between HA and PAH formation in ground pork and juice during marinating.

Keywords: heterocyclic amines; polycyclic aromatic hydrocarbons; marinated pork; UPLC-MS/MS; GC-MS/MS; QuEChERS; principle component analysis

1. Introduction

Both heterocyclic amines (HAs) and polycyclic aromatic hydrocarbons (PAHs) are vital toxic compounds mainly present in cooked protein-rich foods, especially meat products [1,2]. The formation mechanism of HAs and PAHs in meat/meat product during heating has been well established, with the former being generated through degradation of amino acid or protein during heating (>300 °C), or heating of amino acid, reducing sugar (6-carbon sugar) and creatine/creatinine at 100–300 °C, and the latter was produced through incomplete combustion or pyrolysis of organic materials such as fat, protein and carbohydrate at >200 °C [1,3,4]. Specifically, the Maillard reaction between amino acid and 6-carbon reducing sugar can lead to formation of α, β-dicarbonyl compound for the subsequent generation of α-amino-carbonyl compound through Strecker degradation. Next, pyridine or pyrazine can be formed through cyclization for the subsequent formation of imidazole through reaction with aldehyde and creatine/creatinine, followed by reaction with quinoline or quinoxaline to generate imidazoquinoline (IQ type HAs) and imidazoquinoxaline (IQx type HAs). While for pyrolytic HAs, both Glu-P-1 and Glu-P-2 can be produced from glutamic acid, Lys-P-1 from lysine, Phe-P-1 from phenylalanine as well as Trp-P-1, Trp-P-2, Norharman and Harman from tryptophan [5,6]. For PAHs, the benzene-containing compounds such as benzaldehyde has been shown to be a vital precursor for PAH formation through reaction with 1,3-butadiene, a degradation product from linoleic acid oxidation during heating for the subsequent Diels–Alder cycloaddition reaction [7]. According to International Agency of Research on Cancer [8], the HA IQ was classified as Group 2 A (probably carcinogenic) and PhIP, MeIQ, 8-MeIQx, AαC, MeAαC, Glu-P-1, Glu-P-2, Trp-P-1 and Trp-P-2 classified as Group 2 B (possibly carcinogenic) [9–11], while Norharman and Harman are generally considered as co-mutagen [12]. For PAHs, BaP was classified as Group 1 (carcinogenic); DBahA, CcdP and DBalP classified as Group 2 A; and NaP, CHR, BaA, MCH, BbFL, BkFL, BjF, IP, DBaiP and DBahP classified as Group 2 B [13]. As many HAs and PAHs possess potential carcinogenicity, it is urgent to learn the variety and amount of HAs and PAHs in various processed meat products.

The formation and inhibition of HAs and PAHs in meat/meat products during processing has been extensively studied. For instance, Hsu et al. [14] studied the effect of processing methods on HAs formation in duck meat and reported that both traditional oven and fan oven produced higher total HAs than the superheated steam oven, while the chicken fiber fried in lard at 180 °C/20 min was shown to generate a higher level of total HAs than in fried soybean oil [14]. In another study, Chen et al. [15] studied the effect of sugar smoking on PAH formation in meat, the level of total PAHs was found to be higher in red meat than in poultry meat, and the highly toxic BaP remained undetected in sugar-smoked meat. More recently, in a study dealing with formation of PAHs in thin slices of dried pork during roasting, soy sauce was found to be more efficient than sugar in reducing PAH formation; probably due to presence of isoflavone in the former [2]. Apparently, the formation of HAs and PAHs in meat/meat products can be affected by many factors such as meat variety, cooking time/temperature, cooking method, variety and amount of flavorings and cooking oil [4]. Thus, the inhibition of HAs and PAHs in meat products by added flavorings needs to be further investigated.

In recent years, a variety of spices/herbs and fruit extracts as well as phenolic-rich plant extracts and beverages were incorporated during meat marinating to study their inhibitory effects on formation of HAs and PAHs. By adding 0.5% of Sichuan pepper and 0.01% of sanshoamide extract, a total of HA inhibition by 70% was shown by Zeng et al. [16], while the incorporation of turmeric, curry leaf, torch ginger and lemon grass alone or in combination was reported to inhibit total HA contents ranging from 21.2 to 94.7%. In another study, Lu et al. [17] added 0.5% of garlic, onion, red chilli, paprika, ginger and black pepper powder to investigate their inhibitory effects on formation of HAs and PAHs in fried beef and chicken meat balls, and demonstrated that all the spice powders significantly reduced the formation of total HAs but not PAHs, with the ginger powder showing the most pronounced effect. In two different studies, the effect of phenolic compounds in beer on the formation of PAHs in charcoal-grilled chicken wings and pork was investigated, and a 67% inhibition of eight PAHs (BaA, CHR, BbF, BkFL, BaP, DBahA, BghiP and IP) was shown for the former [18], while a 30–68% inhibition of the same eight PAHs depending on the beer variety shown for the latter [19], with both studies showing a significant correlation between phenolic compounds in beer, antioxidant activity and PAHs inhibition. However, a poor correlation was shown between the antioxidant activity of added sugarcane molasses extract and inhibition of HAs formation in deep-fried chicken wings [20], as well as the antioxidant activity of green tea phenolic compounds and inhibition of PAHs formation in charcoal-grilled chicken wings [21], implying a complex inhibition mechanism involved in the reduction of HAs and PAHs formation. To evaluate the inhibitory effects of fruit extracts on HAs inhibition in roasted yellow croaker fish, Li et al. [22] recently reported that 1% blueberry extract, acerola cherry extract and grape seed extract could reduce the total HAs content with blueberry extract exhibiting the highest inhibitory effect and the inhibition mechanism can be associated with antioxidant activity of fruit extracts, generation of HAs precursors and lipid/protein oxidation during meat processing. Thus, it is necessary to explore some other natural products in inhibiting the formation of HAs and PAHs in meat products during processing.

Most of the reported studies discussed above were found to determine only a decreasing number of HAs or PAHs and there is a paucity of data dealing with comprehensive evaluation of most HAs and PAHs with simultaneous extraction of both. Nevertheless, in our recent study, Lai et al. [23] reported that by employing a QuEChERS (quick, easy, cheap, effective, rugged and safe) method, along with an optimal solvent system for extraction, a high accuracy and precision was attained for the simultaneous extraction of 20 HAs and 23 PAHs from pork jerky for the subsequent analysis by ultra-high-performance liquid chromatograph-tandem mass spectrometer (UPLC-MS/MS) and gas chromatograph-tandem mass spectrometer (GC-MS/MS), respectively. Thus, this study aims to adopt the above QuEChERS method for the simultaneous extraction of HAs and PAHs from marinated pork and juice, a popular meat commodity in Taiwan, for further analysis by UPLC-MS/MS and GC-MS/MS, and explore the effects of marinating time length and different flavorings including soy sauce, sugar, cinnamon powder and green tea powder on formation and inhibition of HAs and PAHs in ground pork/juice during marinating.

2. Materials and Methods

2.1. Processing of Marinated Pork

A total of 9 kg of raw ground pork was divided into 18 groups with 500 g each for marinating. The unmarinated juice was prepared by mixing of 10 g (1%) of sugar, 100 g (10%) of soy sauce and 890 g (89%) of deionized water (standard formula). In addition, two more formulas were prepared: standard formula plus 0.5% (weight/volume, w/v) cinnamon powder and standard formula plus 0.5% (w/v) green tea powder. Each unmarinated juice was poured into a stainless-steel saucepan with a lid on the top and heated by gas. After boiling, 500 g of raw ground pork was poured into a saucepan and heated for 2, 4, 8, 12 and 24 h, and a total of 18 saucepans, including 3 saucepans without heating, were used. The temperature was controlled at 90 ± 2 °C, and 200 mL of hot water added to maintain the volume at 1 L every hour during marinating. Before sampling, water was added to the same original volume as that before heating. Then the marinated pork and juice were subjected to analysis of HAs and PAHs by QuEChERS coupled with UPLC-MS/MS and GC-MS/MS, respectively.

2.2. Materials and Chemical Reagents

A total of 9 kg of raw ground pork was purchased from a local supermarket located in New Taipei City, Taiwan, and was stored in a −20 °C freezer prior to use. A total of 21 HA standards including DMIP, IFP, iso-IQ, IQ, MeIQ, IQ [4,5-b], IQx, 8-MeIQx, 7,8-DiMeIQx, 4,8-DiMeIQx, 4,7,8-TriMeIQx (internal standard), Phe-P-1, AαC, Trp-P-2, Trp-P-1, Glu-P-2, Glu-P-1, PhIP, Harman, Norharman and MeAαC were procured from Toronto Research Chemicals (Downsview, Ontario, Canada), while 24 PAH standards including Nap, AcPy, AcP, Flu, Phe, Ant, FL, Pyr, BaA, CHR, BbFL, BaP, IP, DBahA, BghiP, BjF, BcF, CcdP, MCH, DBalP, DBaeP, DBaiP, DBahP and Triphenylene (internal standard) were from Sigma-Aldrich Co. (St. Louis, MO, USA). A list of names and abbreviations of all the 21 HAs and 24 PAHs used in this study are provided at the end of this article.

The HPLC-grade solvents including methanol, acetonitrile, hexane and acetone were from Merck Co. (Darmstadt, Germany), while glacial acetic acid was from Sigma-Aldrich Co. Deionized water was made using a Milli-Q water purification system from Millipore Co. (Bedford, MA, USA). An Acquity UPLC BEH C18 column (100 × 2.1 mm ID., particle size 1.7 μm) was from Waters Co. (Milford, MA, USA), and a DB-5 MS capillary column (30 m × 0.25 mm ID, film thickness 0.25 μm) was from Agilent Technologies Co. (Palo Alto, CA, USA). The QuEChERS extraction (UR-EX) and purification (UR-CLEAN-II) kit was from Yu-Ho Co. (New Taipei City, Taiwan).

2.3. Simultaneous Extraction and Purification of HAs and PAHs in Unmarinated/Marinated Pork and Juice by QuEChERS

A method based on Lai et al. [23] was used for the simultaneous extraction of HAs and PAHs from marinated pork and juice. In brief, a 2-g sample of unmarinated pork, marinated pork or juice was mixed with one ceramic homogenizer in a 50 mL centrifuge tube, followed by adding 10 mL of deionized water and shaking the mixture at 200 rpm for 10 min. Then 10 mL of 100% acetonitrile containing 1% acetic acid was added for shaking at 200 rpm for 10 min, after which the extraction powder ($MgSO_4$ 4 g and CH_3 COONa 1 g) was added for shaking for one min, followed by centrifuging at 4000× *g* for 10 min (4 °C) and collecting the supernatant (6 mL) for purification. Then the supernatant was poured into a centrifuge tube containing 300 mg PSA, 900 mg $MgSO_4$ and 300 mg C18 EC for shaking for 1 min, and then centrifuged at 4000× *g* for 10 min (4 °C). Then the supernatant (1 mL) was collected, evaporated to dryness under nitrogen, dissolved in 0.2 mL of methanol containing 1 ppb of internal standard (4,7,8-TriMeIQx) and filtered through a 0.22-μm membrane filter for HA analysis by UPLC-MS/MS. For PAH analysis, the residue was dissolved in 0.2-mL of hexane containing 10 ppb of internal standard (Triphenylene) and filtered through a 0.22-μm membrane filter for GC-MS/MS analysis.

2.4. Separation and Identification of HAs in Unmarinated/Marinated Pork and Juice by UPLC-MS/MS

A method based on Hsu and Chen [1] was used to analyze HAs in unmarinated pork, marinated pork and juice by UPLC-MS/MS. An ACQUITY BEH C18 column (50 × 2.1 mm ID, particle size 1.7 μm) and a gradient mobile phase of (A) 20 mM ammonium acetate (pH 4.75) and (B) 100% acetonitrile with an initial proportion of 95% A and 5% B, maintained for 2 min, raised to 90% B in 3.3 min, maintained for 4 min and returned to original ratio in 4.1 min for separation of 21 HAs with the column temperature at 30 °C and flow rate at 0.7 mL/min. The various MS parameters and selected reaction monitoring (SRM) mode used for detection of various HAs in unmarinated/marinated pork and juice as well as precursor ions and product ions were the same as those reported by Hsu and Chen [24].

2.5. Separation and Identification of PAHs in Unmarinated/Marinated Pork and Juice by GC-MS/MS

A method based on Lai et al. [23] was used to analyze PAHs in unmarinated pork, marinated pork and juice. A DB-5 MS capillary column (30 m × 0.25 mm ID, film thickness 0.25 μm) with He as carrier gas, flow rate at 1.25 mL/min, injector temperature at 320 °C, injection in splitless mode and MS interface temperature at 280 °C was used to separate 23 PAHs within 78 min with the following temperature programming condition: 80 °C initially, maintained for 1 min, increased to 200 °C at 5 °C/min, maintained for 10 min, raised to 220 °C at 5 °C/min, maintained for 5 min, increased to 230 °C at 1 °C/min, maintained for 10 min, and raised to 320 °C at 10 °C/min and maintained for 10 min.

The SRM mode was used for detection of various PAHs in unmarinated pork, marinated pork and juice samples, and the parameters, such as precursor ions and product ions, were the same as that reported by Lai et al. [23]. Furthermore, an internal standard (Triphenylene) was added to pork sample for quantitation of various PAHs in unmarinated/marinated pork and juice.

2.6. Method Validation of HAs and PAHs

The method validation of HAs and PAHs in freeze-dried raw ground pork was reported in our recent study [23], and the same raw ground pork was used to prepare the marinated pork in this study. Thus, the method validation was not performed separately in this study.

2.7. Determination of HA Precursors

The determination of HA precursors such as reducing sugar, amino acid and creatine/creatinine in unmarinated pork/juice as well as marinated pork/juice was performed by a method reported by Chen et al. [24], TFDA [25] and Gibis and Loeffler [26], respectively. A detailed procedure is provided in the Supplementary Material.

2.8. Determination of Antioxidant Activity

The determination of antioxidant activity of unmarinated pork/juice as well as marinated pork/juice was performed by a method reported by Serpen et al. [27]. A detailed procedure is provided in the Supplementary Material.

2.9. Determination of Bioactive Compounds in Cinnamon Powder by UPLC-MS/MS

The determination of bioactive compounds in cinnamon powder was performed by a method reported by Wang et al. [28]. A detailed procedure is provided in the Supplementary Material.

2.10. Determination of Bioactive Compounds in Green Tea Powder by HPLC-Diode Array Detection (DAD)

The determination of bioactive compounds in green tea powder was performed by a method reported by Lin et al. [29]. A detailed procedure is provided in the Supplementary Material.

2.11. Determination of PAH Precursors

The determination of PAH precursors of unmarinated pork/juice as well as marinated pork/juice was performed by a method reported by Bueno et al. [30]. A detailed procedure is provided in the Supplementary Material.

2.12. Principal Component Analysis (PCA)

Principal component analysis (PCA) was performed on the mean data of triplicate determinations by grouping experimental data of various treatments and transforming a set of correlated variables into a new set of linearly uncorrelated variables based on eigenvalue >1. The possible correlation between formation of HAs and PAHs in marinated pork/juice as affected by different flavorings and marinating time length were elucidated by running PCA with a Kaiser–Meyer–Olkin value of 0.80 and $p < 0.05$ by using Origin® 2019 b version 9.65 (OriginLab Corporation, Northampton, MA, USA).

2.13. Statistical Analysis

All the data were subjected to statistical analysis using the statistical analysis system (SAS) 9.4 software system [31] for analysis of variance (ANOVA) and Duncan's multiple range test for significance in mean comparison ($p < 0.05$).

3. Results and Discussion

3.1. HA Contents in Unmarinated/Marinated Pork and Juice as Affected by Flavorings and Heating Time

Figures 1 and 2 show the UPLC-MS/MS chromatogram of HAs, respectively, in marinated pork and juice after 24 h marinating without and with 0.5% cinnamon powder or 0.5% green tea powder, as detected by SRM mode. The HA contents in unmarinated/marinated pork and juice as affected by flavorings and heating time are shown in Table 1. A total of three HAs including Harman (53.51 ng/g), Norharman (7.92 ng/g) and Glu-P-2 (0.15 ng/g) were present in unmarinated pork (Table 1), while a total of four HAs including Harman (120.42 ng/g), Norharman (16.13 ng/g), Phe-P-1 (2.41 ng/g) and Glu-P-2 (0.30 ng/g) were present in unmarinated juice (Table 1). Apparently, the unmarinated juice contained a higher level of total HAs than that in unmarinated pork. The presence of high level of Harman and Norharman in both unmarinated pork and juice is probably from soy sauce as it was reported that amino acids such as tryptophan may react with pyruvate or acetate to generate Harman and Norharman [32]. In a previous study, Pfau and Skog [32] pointed out that Harman and Norharman were present at a level of 130–250 ng/mL and 5–71 ng/mL in soy sauce, respectively. Moreover, the contents of Harman and Norharman in five commercial soy sauces were shown to be from 111.47–301.30 ng/g and 80.76–199.27 ng/g, respectively [33].

In Table 1, the HA contents were shown to follow a time-dependent rise during pork marinating, and a total of nine HAs were detected after 24-h marinating, with Harman present at the highest level (2456.96 ng/g), followed by Norharman (335.49 ng/g), Phe-P-1 (189.12 ng/g), IQ [4,5-b] (1.53 ng/g), Glu-P-2 (1.01 ng/g), PhIP (0.95 ng/g), DMIP (0.86 ng/g), IQx (0.47 ng/g) and 8-MeIQx (0.10 ng/g). However, after incorporation of 0.5% cinnamon powder or 0.5% green tea powder, the total HA contents were reduced to 2641.90 and 2528.75 ng/g, respectively, following 24-h marinating, implying that green tea powder was more effective than cinnamon powder in minimizing HA formation in marinated pork.

Like marinated pork, the HA contents also followed a time-dependent increase during juice marinating. Following 24-h marinating, a total of nine HAs were produced with Harman present at the highest level (1481.22 ng/g), followed by Norharman (204.82 ng/g), Phe-P-1 (102.72 ng/g), DMIP (1.33 ng/g), Glu-P-2 (0.96 ng/g), PhIP (0.48 ng/g), IQx (0.37 ng/g), 8-MeIQx (0.18 ng/g) and IQ [4,5-b] (trace) (Table 1). However, following the addition of 0.5% cinnamon powder or 0.5% green tea powder, the total HA contents were reduced to 1282.32 and 1398.56 ng/g, respectively, after 24-h marinating, revealing that cinnamon powder was more efficient than green tea powder in minimizing HA formation in marinated juice. Furthermore, in both marinated pork and juice, the non-polar HAs such as Harman, Norharman, Phe-P-1 and Glu-P-2 were more susceptible to formation than the polar HAs during marinating. This outcome is in accordance with a report by Gibis [34], showing that the non-polar HAs were more readily formed at a lower temperature. As both Harman and Norharman were shown to enhance mutagenicity of other toxic compounds such as Trp-p-1, BaP, N-2-fluorenylacetamide and 4-dimethyl aminoazobenzene [35], their presence at high levels in marinated pork and juice cannot be ignored.

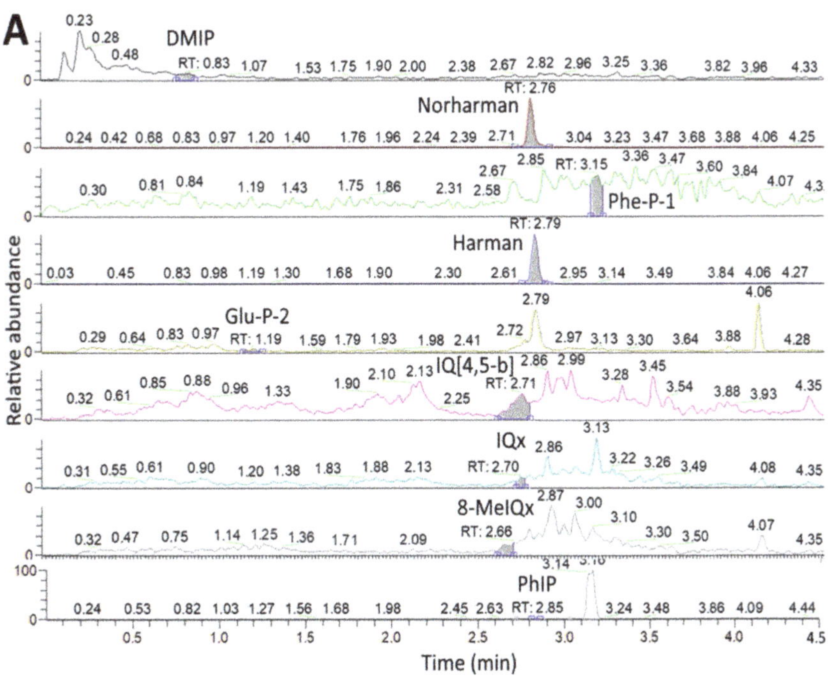

Figure 1. *Cont.*

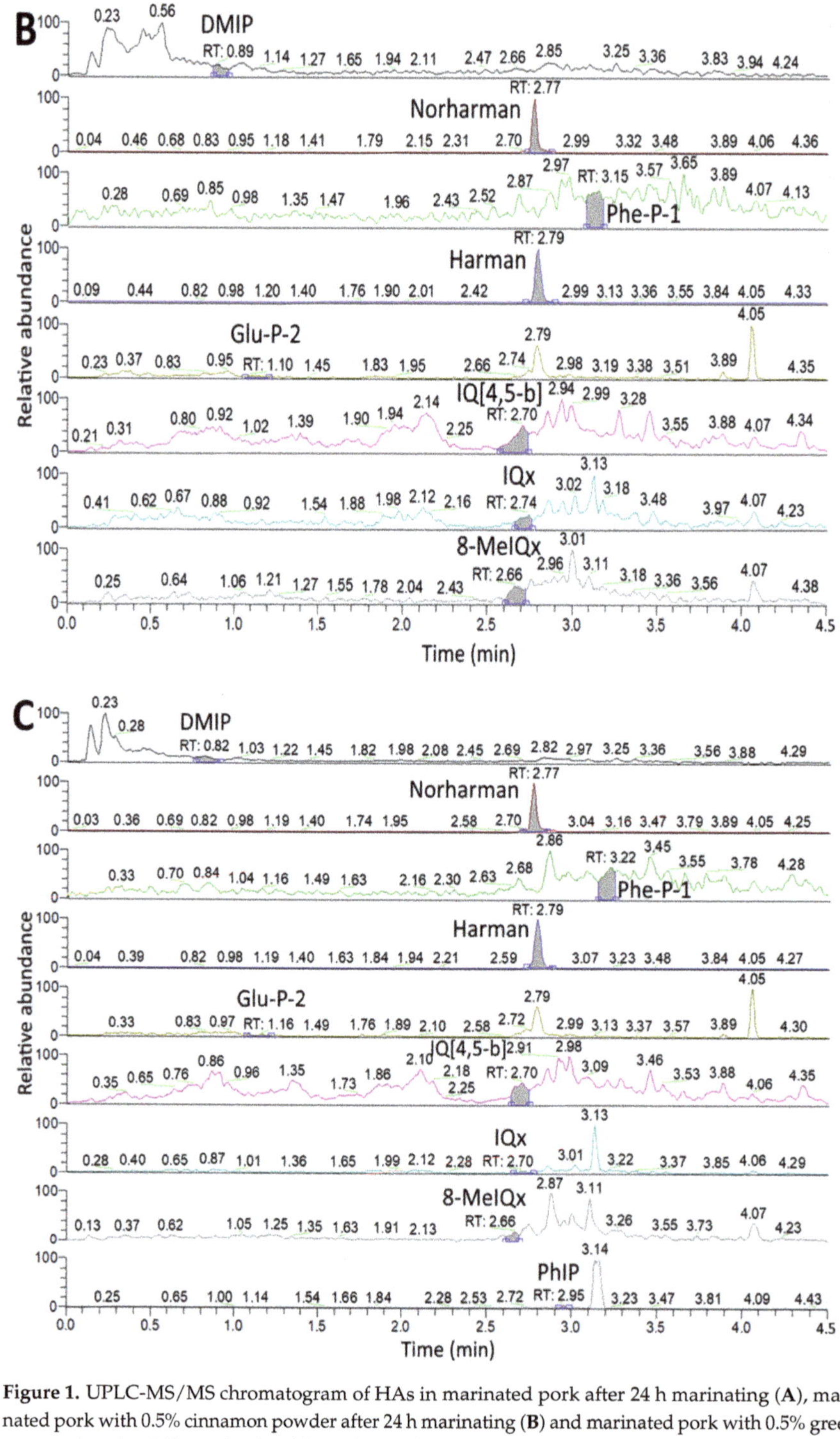

Figure 1. UPLC-MS/MS chromatogram of HAs in marinated pork after 24 h marinating (**A**), marinated pork with 0.5% cinnamon powder after 24 h marinating (**B**) and marinated pork with 0.5% green tea powder after 24 h marinating (**C**), as detected by SRM mode.

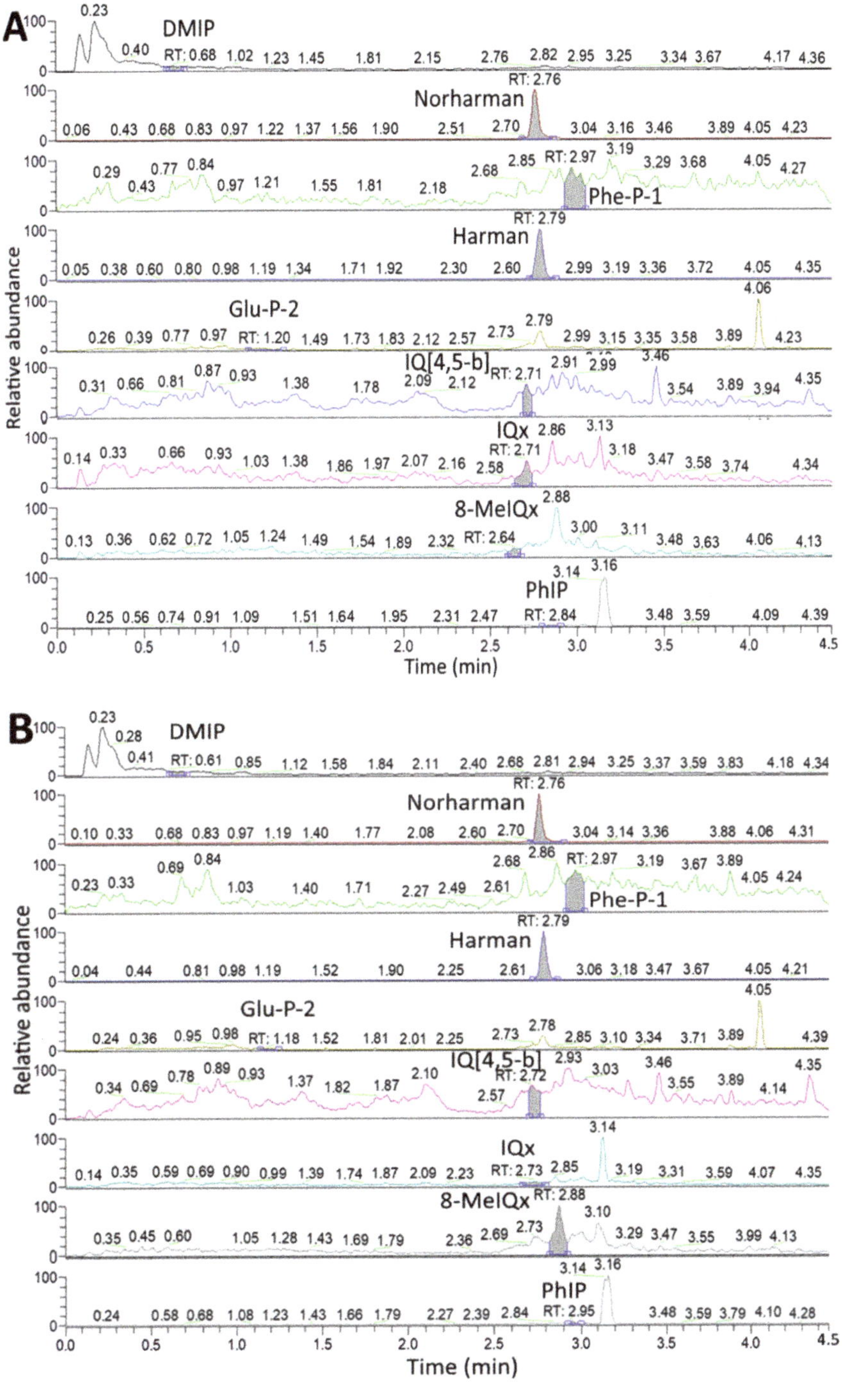

Figure 2. *Cont.*

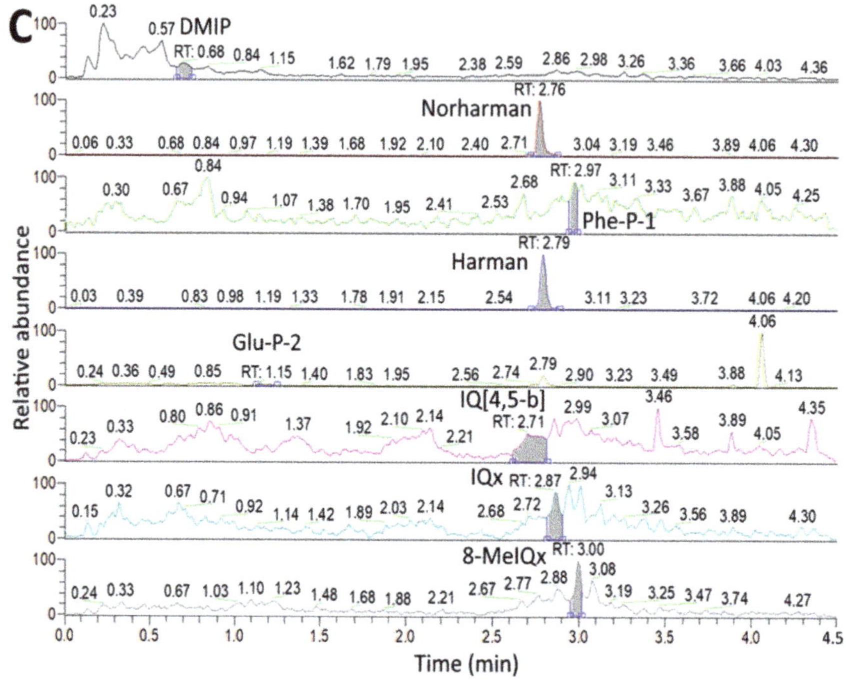

Figure 2. UPLC-MS/MS chromatogram of HAs in marinated juice (**A**), marinated juice with 0.5% cinnamon powder after 24 h marinating (**B**) and marinated juice with 0.5% green tea powder after 24 h marinating (**C**), as detected by SRM mode.

Table 1. HAs contents (ng/g) in marinated pork and juice as affected different flavorings and time length.

HAs	0 h	2 h	4 h	8 h	12 h	24 h
			Unmarinated pork/marinated pork [1]			
DMIP	nd [2]	nd	nd	0.21 ± 0.02 [c]	0.367 ± 0.04 [b]	0.86 ± 0.05 [a]
Norharman	7.92 ± 0.30 [f]	79.66 ± 2.27 [e]	120.75 ± 5.37 [d]	212.19 ± 9.04 [c]	243.57 ± 9.41 [b]	335.49 ± 15.04 [a]
Phe-P-1	nd	23.98 ± 1.93 [e]	30.57 ± 1.68 [d]	64.66 ± 4.07 [c]	118.57 ± 6.81 [b]	189.12 ± 4.67 [a]
Harman	53.51 ± 4.48 [f]	403.18 ± 6.46 [e]	686.35 ± 12.27 [d]	1243.92 ± 11.53 [c]	1624.68 ± 18.25 [b]	2456.96 ± 41.45 [a]
Glu-P-2	0.15 ± 0.01 [c]	0.58 ± 0.09 [b]	0.63 ± 0.10 [b]	0.70 ± 0.08 [b]	0.99 ± 0.10 [a]	1.01 ± 0.16 [a]
IQ [4,5-b]	nd	nd	nd	nd	0.15 ± 0.02 [b]	1.53 ± 0.10 [a]
IQx	nd	nd	0.21 ± 0.01 [b]	0.24 ± 0.04 [b]	0.28 ± 0.02 [b]	0.47 ± 0.04 [a]
8-MeIQx	nd	nd	nd	nd	trace [3]	0.10 ± 0.02 [a]
PhIP	nd	nd	0.24 ± 0.03 [d]	0.32 ± 0.05 [c]	0.68 ± 0.05 [b]	0.95 ± 0.12 [a]
Total	61.58 ± 4.69 [f]	507.39 ± 10.69 [e]	838.75 ± 15.97 [d]	1522.23 ± 14.34 [c]	1989.28 ± 33.09 [b]	2986.47 ± 60.52 [a]
			Unmarinated pork/marinated pork + 0.5% cinnamon powder [1]			
DMIP	nd	nd	0.20 ± 0.08 [c]	0.37 ± 0.05 [b]	0.42 ± 0.12 [b]	0.98 ± 0.04 [a]
Norharman	8.44 ± 0.61 [f]	64.32 ± 3.47 [e]	110.40 ± 8.48 [d]	137.97 ± 6.25 [c]	180.39 ± 3.75 [b]	279.45 ± 8.89 [a]
Phe-P-1	nd	4.27 ± 0.03 [e]	9.87 ± 0.81 [d]	17.78 ± 0.75 [c]	31.31 ± 1.75 [b]	69.67 ± 1.93 [a]
Harman	50.18 ± 3.50 [f]	285.69 ± 7.25 [e]	607.12 ± 11.64 [d]	891.23 ± 21.56 [c]	1376.23 ± 26.68 [b]	2287.77 ± 42.26 [a]
Glu-P-2	0.20 ± 0.03 [e]	0.50 ± 0.05 [d]	0.65 ± 0.10 [d]	0.87 ± 0.12 [c]	1.47 ± 0.10 [b]	1.73 ± 0.08 [a]
IQ [4,5-b]	nd	nd	nd	nd	nd	1.62 ± 0.18 [a]
IQx	nd	0.19 ± 0.01 [c]	0.25 ± 0.03 [c]	0.32 ± 0.05 [b]	0.36 ± 0.05 [b]	0.54 ± 0.02 [a]
8-MeIQx	nd	nd	trace	trace	0.06 ± 0.01 [b]	0.15 ± 0.04 [a]
Total	58.82 ± 4.10 [f]	354.98 ± 10.47 [e]	728.48 ± 20.86 [d]	1048.53 ± 28.13 [c]	1590.24 ± 30.21 [b]	2641.90 ± 44.61 [a]

Table 1. *Cont.*

HAs	0 h	2 h	4 h	8 h	12 h	24 h
			Unmarinated pork/marinated pork + 0.5% green tea powder [1]			
DMIP	nd	nd	nd	0.25 ± 0.12 [b]	0.34 ± 0.04 [b]	0.50 ± 0.08 [a]
Norharman	8.54 ± 0.53 [e]	69.32 ± 3.60 [d]	107.53 ± 5.06 [c]	177.52 ± 6.34 [b]	238.32 ± 19.22 [a]	263.84 ± 28.33 [a]
Phe-P-1	nd	16.35 ± 0.93 [e]	29.40 ± 1.27 [d]	61.43 ± 2.66 [c]	88.39 ± 6.68 [b]	174.72 ± 9.83 [a]
Harman	52.03 ± 1.52 [f]	346.31 ± 29.96 [e]	629.19 ± 17.17 [d]	1051.16 ± 29.38 [c]	1606.89 ± 21.77 [b]	2086.39 ± 24.34 [a]
Glu-P-2	0.18 ± 0.06 [c]	0.52 ± 0.15 [b]	0.90 ± 0.07 [a]	0.92 ± 0.09 [a]	0.97 ± 0.15 [a]	0.99 ± 0.18 [a]
IQ [4,5-b]	nd	nd	nd	nd	nd	0.87 ± 0.10 [a]
IQx	nd	nd	nd	0.25 ± 0.011 [b]	0.28 ± 0.05 [b]	0.45 ± 0.05 [a]
8-MeIQx	nd	nd	nd	trace	trace	0.16 ± 0.03 [a]
PhIP	nd	nd	nd	0.18 ± 0.02 [c]	0.30 ± 0.09 [b]	0.85 ± 0.05 [a]
Total	60.76 ± 1.69 [f]	432.50 ± 32.36 [e]	767.02 ± 22.01 [d]	1291.71 ± 34.05 [c]	1935.50± 40.63 [b]	2528.75 ± 57.76 [a]
			Unmarinated juice/marinated juice [1]			
DMIP	nd [2]	nd	nd	nd	0.27 ± 0.02 [b]	1.33 ± 0.04 [a]
Norharman	16.13 ± 0.68 [f]	45.50 ± 0.626 [e]	59.25 ± 2.26 [d]	94.23 ± 3.06 [c]	134.02 ± 5.93 [b]	204.82 ± 8.24 [a]
Phe-P-1	2.41 ± 0.17 [d]	11.11 ± 0.49 [c]	11.65 ± 0.85 [c]	13.14 ± 0.55 [c]	54.08 ± 2.70 [b]	102.72 ± 4.67
Harman	120.42 ± 1.44 [f]	304.32 ± 10.70 [e]	373.96 ± 7.65 [d]	598.07 ± 4.40 [c]	1016.41 ± 9.09 [b]	1481.22 ± 41.84 [a]
Glu-P-2	0.30 ± 0.05 [c]	0.35 ± 0.01 [c]	0.57 ± 0.05 [b]	0.59 ± 0.01 [b]	0.59 ± 0.04 [b]	0.96 ± 0.23 [a]
IQ [4,5-b]	nd	nd	nd	nd	nd	trace [3]
IQx	nd	0.18 ± 0.09 [c]	0.26 ± 0.06 [b]	0.26 ± 0.07 [b]	0.36 ± 0.05 [a]	0.37 ± 0.04 [a]
8-MeIQx	nd	nd	nd	nd	0.06 ± 0.01 [b]	0.18 ± 0.02 [a]
PhIP	nd	nd	nd	0.17 ± 0.01 [c]	0.22 ± 0.03 [b]	0.48 ± 0.04 [a]
Total	139.27 ± 2.16 [f]	361.46 ± 11.02 [e]	445.68 ± 10.17 [d]	706.46 ± 4.68 [c]	1206.01 ± 13.55 [b]	1792.08 ± 54.65 [a]
HAs (ng/g)			Unmarinated juice/marinated juice + 0.5% cinnamon powder [1]			
DMIP	nd	nd	nd	nd	nd	0.51 ± 0.03 [a]
Norharman	16.91 ± 0.38 [e]	45.54 ± 2.20 [d]	48.39 ± 2.70	77.62 ± 5.76 [c]	116.80 ± 3.50 [b]	141.38 ± 5.40 [a]
Phe-P-1	1.72 ± 0.14 [e]	9.95 ± 0.19 [d]	18.95 ± 0.93 [c]	19.53 ± 1.22 [c]	23.64 ± 1.53 [b]	73.03 ± 3.93 [a]
Harman	116.23 ± 3.35 [f]	273.02 ± 6.80 [e]	309.93 ± 5.27 [d]	538.04 ± 7.62 [c]	900.94 ± 14.20 [b]	1065.64 ± 22.30 [a]
Glu-P-2	0.38 ± 0.06 [c]	0.36 ± 0.01 [c]	0.53 ± 0.08 [b]	0.55 ± 0.05 [b]	0.61 ± 0.03 [b]	0.71 ± 0.04 [a]
IQ [4,5-b]	nd	nd	nd	nd	nd	0.22 ± 0.04 [a]
IQx	nd	0.21 ± 0.01 [c]	0.22 ± 0.06 [c]	0.36 ± 0.04 [b]	0.34 ± 0.03 [b]	0.42 ± 0.06 [a]
8-MeIQx	nd	nd	nd	trace	trace	0.10 ± 0.03 [a]
PhIP	nd	nd	nd	nd	0.13 ± 0.02 [b]	0.32 ± 0.02 [a]
Total	135.23 ± 3.47 [f]	329.08 ± 9.15 [e]	378.02 ± 7.84 [d]	636.10 ± 11.33 [c]	1042.46 ± 19.24 [b]	1282.32 ± 30.81 [a]
HAs (ng/g)			Unmarinated juice/marinated juice + 0.5% green tea powder [1]			
DMIP	nd	nd	nd	nd	nd	0.98 ± 0.05 [a]
Norharman	18.23 ± 0.88 [e]	23.21 ± 2.05 [e]	32.39 ± 2.26 [d]	50.15 ± 2.89 [c]	92.06 ± 3.47 [b]	130.70 ± 5.45 [a]
Phe-P-1	0.32 ± 0.07 [d]	0.50 ± 0.04 [d]	0.91 ± 0.17 [cd]	1.63 ± 0.26 [c]	7.47 ± 0.41 [b]	15.83 ± 1.41 [a]
Harman	118.63 ± 2.33 [f]	137.13 ± 7.52 [e]	202.93 ± 4.54 [d]	385.25 ± 25.68 [c]	778.41 ± 35.98 [b]	1249.16 ± 31.11 [a]
Glu-P-2	0.35 ± 0.05 [d]	0.47 ± 0.04 [cd]	0.56 ± 0.06 [c]	0.70 ± 0.01 [b]	0.70 ± 0.09 [b]	0.96 ± 0.13 [a]
IQ [4,5-b]	nd	nd	nd	nd	nd	0.45 ± 0.05 [a]
IQx	nd	nd	0.21 ± 0.03 [b]	0.27 ± 0.02 [b]	0.27 ± 0.06 [b]	0.37 ± 0.07 [a]
8-MeIQx	nd	nd	nd	trace	trace	0.10 ± 0.02 [a]
Total	137.52 ± 3.24 [e]	161.30 ± 9.24 [e]	237.001 ± 6.431 [d]	438.00 ± 28.570 [c]	878.91 ± 39.72 [b]	1398.56 ± 38.23 [a]

[1] Data are presented as mean ± standard deviation of triplicate determinations and data with different small letters (a–f) in the same row are significantly different ($p < 0.05$); [2] not detected; [3] LOQ ≥ PAHs levels ≥ LOD.

3.1.1. Amino Acid Change in Unmarinated/Marinated Pork and Juice

Table 2 shows changes of individual and total amino acid contents in unmarinated/marinated pork and juice. Only a minor change in total amino acid contents was shown over a 24-h marinating period for marinated pork, marinated pork plus 0.5% cinnamon powder and marinated pork plus 0.5% green tea powder, indicating that amino acid may play a less important role in HA formation. A similar phenomenon was found for the treatments of marinated juice, marinated juice plus 0.5% cinnamon powder and marinated juice plus 0.5% green tea powder (Table 2).

Table 2. Individual and total amino acid (AA) contents in unmarinated/marinated pork and juice as affected by different flavorings and time length.

AA (mg/g)	0 h	2 h	4 h	8 h	12 h	24 h
			marinated pork [1]			
Asp	0.41 ± 0.04	0.40 ± 0.59	0.36 ± 0.01	0.40 ± 0.01	0.28 ± 0.02	0.41 ± 0.02
Glu	0.94 ± 0.05	0.77 ± 0.00	0.76 ± 0.04	0.82 ± 0.03	0.69 ± 0.01	0.77 ± 0.01
Ser	0.23 ± 0.01	0.21 ± 0.00	0.20 ± 0.00	0.21 ± 0.00	0.19 ± 0.00	0.21 ± 0.00
His	0.06 ± 0.01	0.05 ± 0.01	0.04 ± 0.01	0.05 ± 0.00	0.08 ± 0.01	0.11 ± 0.00
Gly	0.50 ± 0.02	0.46 ± 0.03	0.48 ± 0.04	0.38 ± 0.01	0.46 ± 0.05	0.43 ± 0.01
Thr	0.09 ± 0.01	0.09 ± 0.00	0.08 ± 0.01	0.10 ± 0.00	0.14 ± 0.01	0.16 ± 0.02
Arg	0.49 ± 0.01	0.42 ± 0.01	0.42 ± 0.02	0.40 ± 0.01	0.42 ± 0.01	0.42 ± 0.00
Ala	0.47 ± 0.01	0.43 ± 0.00	0.43 ± 0.03	0.41 ± 0.01	0.39 ± 0.00	0.42 ± 0.00
Tyr	0.22 ± 0.00 [2]	0.06 ± 0.01	0.05 ± 0.02	0.05 ± 0.01	0.30 ± 0.02	0.16 ± 0.00
Cys	0.15 ± 0.01	0.11 ± 0.00	0.11 ± 0.00	0.11 ± 0.00	0.03 ± 0.00	0.05 ± 0.00
Val	0.23 ± 0.04	0.24 ± 0.00	0.20 ± 0.06	0.28 ± 0.04	0.17 ± 0.01	0.19 ± 0.02
Met	0.23 ± 0.14	0.14 ± 0.00	0.15 ± 0.01	0.23 ± 0.07	0.14 ± 0.00	0.16 ± 0.01
Phe	0.21 ± 0.01	0.23 ± 0.00	0.23 ± 0.01	0.20 ± 0.03	0.17 ± 0.01	0.20 ± 0.00
Ile	0.24 ± 0.06	0.28 ± 0.00	0.27 ± 0.01	0.23 ± 0.06	0.30 ± 0.01	0.27 ± 0.01
Leu	0.54 ± 0.20	0.66 ± 0.01	0.69 ± 0.03	0.67 ± 0.02	0.50 ± 0.01	0.55 ± 0.01
Lys	0.53 ± 0.05	0.55 ± 0.01	0.59 ± 0.04	0.61 ± 0.02	0.55 ± 0.04	0.53 ± 0.02
Pro	0.26 ± 0.00	0.40 ± 0.02	0.41 ± 0.04	0.45 ± 0.01	0.43 ± 0.03	0.42 ± 0.01
Total	5.81 ± 0.14	5.83 ± 0.63	5.45 ± 0.10	5.61 ± 0.03	5.24± 0.02	5.45 ± 0.04
			marinated pork + 0.5% cinnamon powder [1]			
Asp	0.42 ± 0.01	0.40 ± 0.02	0.36 ± 0.01	0.38 ± 0.01	0.41 ± 0.01	0.36 ± 0.01
Glu	0.78 ± 0.02	0.72 ± 0.02	0.68 ± 0.00	0.74 ± 0.00	0.78 ± 0.00	0.80 ± 0.01
Ser	0.21 ± 0.00	0.22 ± 0.00	0.21 ± 0.00	0.20 ± 0.00	0.22 ± 0.00	0.21 ± 0.00
His	0.03 ± 0.01	0.08 ± 0.00	0.06 ± 0.00	0.09 ± 0.00	0.04 ± 0.00	0.04 ± 0.00
Gly	0.58 ± 0.02	0.55 ± 0.04	0.13 ± 0.01	0.13 ± 0.01	0.14 ± 0.01	0.10 ± 0.00
Thr	0.14 ± 0.01	0.15 ± 0.00	0.42 ± 0.01	0.43 ± 0.00	0.44 ± 0.00	0.49 ± 0.00
Arg	0.53 ± 0.00	0.48 ± 0.00	0.96 ± 0.00	0.98 ± 0.01	1.02 ± 0.01	1.18 ± 0.00
Ala	0.44 ± 0.00	0.44 ± 0.00	0.40 ± 0.00	0.41 ± 0.00	0.43 ± 0.01	0.49 ± 0.00
Tyr	0.13 ± 0.00	0.14 ± 0.00	0.13 ± 0.00	0.15 ± 0.00	0.16 ± 0.00	0.14 ± 0.00
Cys	0.14 ± 0.01	0.14 ± 0.00	0.04 ± 0.01	0.06 ± 0.00	0.06 ± 0.00	0.04 ± 0.00
Val	0.18 ± 0.00	0.16 ± 0.01	0.16 ± 0.00	0.18 ± 0.00	0.20 ± 0.00	0.17 ± 0.01
Met	0.16 ± 0.01	0.15 ± 0.00	0.14 ± 0.00	0.15 ± 0.00	0.16 ± 0.00	0.13 ± 0.00
Phe	0.23 ± 0.07	0.19 ± 0.01	0.17 ± 0.00	0.19 ± 0.00	0.21 ± 0.00	0.21 ± 0.01
Ile	0.36 ± 0.16	0.25 ± 0.01	0.28 ± 0.01	0.25 ± 0.01	0.27 ± 0.01	0.27 ± 0.01
Leu	0.43 ± 0.09	0.56 ± 0.00	0.49 ± 0.01	0.53 ± 0.01	0.56 ± 0.02	0.65 ± 0.01
Lys	0.45 ± 0.06	0.58 ± 0.04	0.60 ± 0.01	0.56 ± 0.01	0.55 ± 0.01	0.44 ± 0.01
Pro	0.29 ± 0.02	0.56 ± 0.00	0.44 ± 0.01	0.43 ± 0.01	0.43 ± 0.01	0.38 ± 0.00
Total	5.49 ± 0.11	5.79 ± 0.10	5.56 ± 0.02	5.85 ± 0.06	6.07 ± 0.04	6.10 ± 0.07
			marinated pork + 0.5% green tea powder [1]			
Asp	0.53 ± 0.05	0.43 ± 0.09	0.41 ± 0.08	0.65 ± 0.37	0.36 ± 0.06	0.34 ± 0.08
Glu	0.94 ± 0.08	0.81 ± 0.21	0.99 ± 0.23	1.04 ± 0.27	1.04 ± 0.22	1.06 ± 0.26
Ser	0.23 ± 0.02	0.24 ± 0.05	0.23 ± 0.05	0.23 ± 0.05	0.24 ± 0.05	0.23 ± 0.06
His	0.04 ± 0.00	0.04 ± 0.00	0.03 ± 0.01	0.04 ± 0.00	0.03 ± 0.01	0.03 ± 0.01
Gly	0.12 ± 0.01	0.14 ± 0.02	0.18 ± 0.06	0.17 ± 0.06	0.19 ± 0.07	0.17 ± 0.07
Thr	0.44 ± 0.01	0.46 ± 0.08	0.53 ± 0.08	0.48 ± 0.08	0.57 ± 0.08	0.56 ± 0.08
Arg	1.13 ± 0.06	1.10 ± 0.20	1.19 ± 0.23	1.24 ± 0.25	1.33 ± 0.23	1.32 ± 0.23
Ala	0.48 ± 0.03	0.46 ± 0.09	0.50 ± 0.10	0.52 ± 0.11	0.56 ± 0.10	0.56 ± 0.10
Tyr	0.01 ± 0.00	0.01 ± 0.00	0.14 ± 0.03	0.01 ± 0.00	0.09 ± 0.02	0.20 ± 0.04
Cys	0.01 ± 0.00	0.07 ± 0.04	0.08 ± 0.00	0.06 ± 0.00	0.09 ± 0.01	0.08 ± 0.01
Val	0.23 ± 0.02	0.19 ± 0.03	0.12 ± 0.04	0.24 ± 0.05	0.24 ± 0.05	0.22 ± 0.06
Met	0.14 ± 0.02	0.17 ± 0.03	0.21 ± 0.07	0.17 ± 0.05	0.22 ± 0.06	0.22 ± 0.07

Table 2. *Cont.*

AA (mg/g)	0 h	2 h	4 h	8 h	12 h	24 h
			marinated pork + 0.5% green tea powder [1]			
Phe	0.16 ± 0.01	0.18 ± 0.03	0.24 ± 0.05	0.20 ± 0.03	0.24 ± 0.03	0.24 ± 0.04
Ile	0.36 ± 0.04	0.34 ± 0.09	0.23 ± 0.01	0.38 ± 0.11	0.43 ± 0.13	0.36 ± 0.09
Leu	0.52 ± 0.03	0.57 ± 0.11	0.45 ± 0.08	0.60 ± 0.19	0.70 ± 0.13	0.72 ± 0.13
Lys	0.50 ± 0.15	0.71 ± 0.20	0.71 ± 0.18	0.57 ± 0.14	0.64 ± 0.18	0.57 ± 0.13
Pro	0.31 ± 0.01	0.42 ± 0.04	0.38 ± 0.02	0.27 ± 0.02	0.35 ± 0.02	0.35 ± 0.01
Total	6.16 ± 0.51	6.36 ± 1.14	6.54 ± 1.00	6.63 ± 1.41	7.25 ± 1.32	7.23 ± 1.42
			marinated juice [1]			
Asp	0.41 ± 0.04	0.71 ± 0.59	0.36 ± 0.01	0.40 ± 0.01	0.28 ± 0.02	0.41 ± 0.02
Glu	0.94 ± 0.05	0.77 ± 0.00	0.76 ± 0.04	0.82 ± 0.03	0.69 ± 0.01	0.77 ± 0.01
Ser	0.23 ± 0.01	0.21 ± 0.00	0.20 ± 0.00	0.21 ± 0.00	0.19 ± 0.00	0.21 ± 0.00
His	0.45 ± 0.27	0.79 ± 0.50	0.83 ± 0.54	0.73 ± 0.50	0.71 ± 0.49	0.69 ± 0.45
Gly	0.50 ± 0.02	0.46 ± 0.03	0.48 ± 0.04	0.38 ± 0.01	0.46 ± 0.05	0.43 ± 0.01
Thr	0.09 ± 0.01	0.09 ± 0.00	0.08 ± 0.01	0.10 ± 0.00	0.14 ± 0.01	0.16 ± 0.02
Arg	0.49 ± 0.01	0.42 ± 0.01	0.42 ± 0.02	0.40 ± 0.01	0.42 ± 0.01	0.42 ± 0.00
Ala	0.47 ± 0.01	0.43 ± 0.00	0.43 ± 0.03	0.41 ± 0.01	0.39 ± 0.00	0.42 ± 0.00
Tyr	0.22 ± 0.00	0.06 ± 0.01	0.05 ± 0.02	0.05 ± 0.01	0.30 ± 0.02	0.16 ± 0.00
Cys	0.15 ± 0.01	0.11 ± 0.00	0.11 ± 0.00	0.11 ± 0.00	0.03 ± 0.00	0.05 ± 0.00
Val	0.23 ± 0.04	0.24 ± 0.00	0.20 ± 0.06	0.28 ± 0.04	0.17 ± 0.01	0.19 ± 0.02
Met	0.23 ± 0.14	0.14 ± 0.00	0.15 ± 0.01	0.23 ± 0.07	0.14 ± 0.00	0.16 ± 0.01
Phe	0.21 ± 0.01	0.23 ± 0.00	0.23 ± 0.01	0.20 ± 0.03	0.17 ± 0.01	0.20 ± 0.00
Ile	0.24 ± 0.06	0.28 ± 0.00	0.27 ± 0.01	0.23 ± 0.06	0.30 ± 0.01	0.27 ± 0.01
Leu	0.54 ± 0.20	0.66 ± 0.01	0.69 ± 0.03	0.67 ± 0.02	0.50 ± 0.01	0.55 ± 0.01
Lys	0.53 ± 0.05	0.55 ± 0.01	0.59 ± 0.04	0.61 ± 0.02	0.55 ± 0.04	0.53 ± 0.02
Pro	0.26 ± 0.00	0.40 ± 0.02	0.41 ± 0.04	0.45 ± 0.01	0.43 ± 0.03	0.42 ± 0.01
Total	12.44 ± 1.64	13.11 ± 2.06	14.26 ± 3.04	14.94 ± 3.61	14.17 ± 3.05	13.86 ± 2.95
			marinated juice + 0.5% cinnamon powder [1]			
Asp	0.42 ± 0.01	0.40 ± 0.02	0.36 ± 0.01	0.38 ± 0.01	0.41 ± 0.01	0.36 ± 0.01
Glu	0.78 ± 0.02	0.72 ± 0.02	0.68 ± 0.00	0.74 ± 0.00	0.78 ± 0.00	0.80 ± 0.01
Ser	0.21 ± 0.00	0.22 ± 0.00	0.21 ± 0.00	0.20 ± 0.00	0.22 ± 0.00	0.21 ± 0.00
His	0.55 ± 0.36	0.70 ± 0.45	0.71 ± 0.52	0.62 ± 0.46	0.58 ± 0.38	0.53 ± 0.34
Gly	0.58 ± 0.02	0.55 ± 0.04	0.13 ± 0.01	0.13 ± 0.01	0.14 ± 0.01	0.10 ± 0.00
Thr	0.14 ± 0.01	0.15 ± 0.00	0.42 ± 0.01	0.43 ± 0.00	0.44 ± 0.00	0.49 ± 0.00
Arg	0.53 ± 0.00	0.48 ± 0.00	0.96 ± 0.00	0.98 ± 0.01	1.02 ± 0.02	1.18 ± 0.00
Ala	0.44 ± 0.00	0.44 ± 0.00	0.40 ± 0.00	0.41 ± 0.00	0.43 ± 0.01	0.49 ± 0.00
Tyr	0.13 ± 0.00	0.14 ± 0.00	0.07 ± 0.02	0.09 ± 0.03	0.09 ± 0.03	0.10 ± 0.03
Cys	0.14 ± 0.01	0.14 ± 0.00	0.16 ± 0.13	0.18 ± 0.16	0.16 ± 0.01	0.12 ± 0.04
Val	0.18 ± 0.00	0.16 ± 0.01	0.16 ± 0.00	0.18 ± 0.00	0.20 ± 0.00	0.17 ± 0.01
Met	0.16 ± 0.01	0.15 ± 0.00	0.14 ± 0.00	0.15 ± 0.00	0.16 ± 0.00	0.13 ± 0.00
Phe	0.23 ± 0.07	0.19 ± 0.01	0.17 ± 0.00	0.19 ± 0.00	0.21 ± 0.00	0.21 ± 0.01
Ile	0.36 ± 0.16	0.25 ± 0.01	0.28 ± 0.01	0.25 ± 0.01	0.27 ± 0.01	0.27 ± 0.01
Leu	0.43 ± 0.09	0.56 ± 0.00	0.49 ± 0.01	0.53 ± 0.01	0.56 ± 0.02	0.65 ± 0.01
Lys	0.45 ± 0.06	0.58 ± 0.04	0.60 ± 0.01	0.56 ± 0.01	0.55 ± 0.01	0.44 ± 0.01
Pro	0.29 ± 0.02	0.56 ± 0.00	0.44 ± 0.01	0.43 ± 0.01	0.43 ± 0.01	0.38 ± 0.00
Total	13.47 ± 2.56	13.61 ± 1.25	14.41 ± 3.46	14.85 ± 4.18	15.19 ± 4.37	15.25 ± 3.98
			marinated juice + 0.5% green tea powder [1]			
Asp	0.53 ± 0.05	0.43 ± 0.09	0.41 ± 0.08	0.65 ± 0.37	0.36 ± 0.06	0.34 ± 0.08
Glu	0.94 ± 0.08	0.81 ± 0.21	0.99 ± 0.23	1.04 ± 0.27	1.04 ± 0.22	1.06 ± 0.26
Ser	0.23 ± 0.02	0.24 ± 0.05	0.23 ± 0.05	0.23 ± 0.05	0.24 ± 0.05	0.23 ± 0.06
His	0.37 ± 0.20	0.49 ± 0.30	0.44 ± 0.31	0.30 ± 0.18	0.30 ± 0.19	0.28 ± 0.15
Gly	0.12 ± 0.01	0.14 ± 0.02	0.18 ± 0.06	0.17 ± 0.06	0.19 ± 0.07	0.17 ± 0.07
Thr	0.44 ± 0.01	0.46 ± 0.08	0.53 ± 0.08	0.48 ± 0.08	0.57 ± 0.08	0.56 ± 0.08
Arg	1.13 ± 0.06	1.10 ± 0.20	1.19 ± 0.23	1.24 ± 0.25	1.33 ± 0.23	1.32 ± 0.23
Ala	0.48 ± 0.03	0.46 ± 0.09	0.50 ± 0.10	0.52 ± 0.11	0.56 ± 0.10	0.56 ± 0.10
Tyr	0.15 ± 0.01	0.07 ± 0.02	0.09 ± 0.03	0.12 ± 0.05	0.09 ± 0.02	0.20 ± 0.04

Table 2. *Cont.*

AA (mg/g)	0 h	2 h	4 h	8 h	12 h	24 h
			marinated juice + 0.5% green tea powder [1]			
Cys	0.14 ± 0.03	0.36 ± 0.23	0.24 ± 0.09	0.05 ± 0.01	0.09 ± 0.01	0.08 ± 0.01
Val	0.23 ± 0.02	0.19 ± 0.03	0.12 ± 0.04	0.24 ± 0.05	0.24 ± 0.05	0.22 ± 0.06
Met	0.14 ± 0.02	0.17 ± 0.03	0.21 ± 0.07	0.17 ± 0.05	0.22 ± 0.06	0.22 ± 0.07
Phe	0.16 ± 0.01	0.18 ± 0.03	0.24 ± 0.05	0.20 ± 0.03	0.24 ± 0.03	0.24 ± 0.04
Ile	0.36 ± 0.04	0.34 ± 0.09	0.23 ± 0.01	0.38 ± 0.11	0.43 ± 0.13	0.36 ± 0.09
Leu	0.52 ± 0.03	0.57 ± 0.11	0.45 ± 0.08	0.60 ± 0.19	0.70 ± 0.13	0.72 ± 0.13
Lys	0.50 ± 0.15	0.71 ± 0.20	0.71 ± 0.18	0.57 ± 0.14	0.64 ± 0.18	0.57 ± 0.13
Pro	0.31 ± 0.01	0.42 ± 0.04	0.38 ± 0.02	0.27 ± 0.02	0.35 ± 0.02	0.35 ± 0.01
Total	11.88 ± 0.95	13.66 ± 2.60	14.93 ± 3.97	15.58 ± 2.23	16.38 ± 5.75	16.65 ± 5.50

[1] Data are presented as mean ± standard deviation of triplicate determinations. [2] Data with the standard deviation values <0.005 are shown as ±0.00 to maintain 2 significant digits.

3.1.2. Reducing Sugar Content Change in Unmarinated/Mar-inated Pork and Juice

Conversely, reducing sugar should play a more important role than amino acid in HA formation in marinated pork, marinated pork plus 0.5% cinnamon powder and marinated pork plus 0.5% green tea powder, as evidenced by a time-dependent rise in reducing sugar content being shown for all the treatments during pork marinating (Table 3). Specifically, following 24-h marinating, the reducing sugar content was raised from 0.09 mg/g in unmarinated pork to 4.03 mg/g in marinated pork. Similarly, after addition of 0.5% cinnamon powder or 0.5% green tea powder, the reducing sugar contents further increased to 5.33 mg/g and 6.50 mg/g, respectively, following 24-h marinating, which can be attributed to the decomposition of sucrose into glucose and fructose during heating. Compared to marinated pork, a higher level of reducing sugar in the treatments of 0.5% cinnamon powder and 0.5% green tea powder implied that a less amount of reducing sugar participated in the Maillard browning reaction, leading to a reduction in total HAs (Table 1). However, a different trend in reducing sugar content change was shown in marinated juice (Table 3), as a plateau was reached after 2-h marinating for marinated juice (12.18 mg/g), marinated juice plus 0.5% cinnamon powder (11.01 mg/g) and marinated juice plus 0.5% green tea powder (8.92 mg/g), followed by a decline to 7.89, 9.74 and 8.57 mg/g, respectively, after 24-h marinating. It may be postulated that the decrease in reducing sugar content during marinating is probably due to the reaction rate of Maillard browning reaction being higher than the decomposition rate of sucrose. Compared to 0.5% green tea powder, the addition of 0.5% cinnamon powder resulted in a more reduction of the reducing sugar content during 24-h marinating, revealing that a higher amount of reducing sugar participated in the Maillard browning reaction, leading to a further reduction of total HAs for the latter treatment (Table 1).

3.1.3. Creatine and Creatinine Change in Unmarinated/Marinated Pork and Juice

Creatine, commonly present in creatine phosphate form in animal muscle tissue as energy source, can be hydrolyzed to creatinine during heating, a precursor of imidazole formation [36,37]. Moreover, the formation of creatine/creatinine in meat products during heating can be correlated positively with mutagenicity [38]. In Table 3, the creatine content followed a time-dependent decrease in marinated pork, marinated pork plus 0.5% cinnamon powder and marinated pork plus 0.5% green tea powder and declined to 6.72, 7.27 and 7.62 mg/100 g after 24-h marinating, respectively, which can be attributed to the conversion of creatine to creatinine during heating. A similar trend was observed for the creatine content change in marinated juice, marinated juice plus 0.5% cinnamon powder and marinated juice plus 0.5% green tea powder, with the creatine content being reduced to 9.30, 10.26 and 10.79 mg/100 g, respectively, after 24-h marinating (Table 3). Comparatively, unmarinated pork contained a much higher level of creatine than unmarinated juice, however, following marinating for 4 h and above, marinated juice was shown to contain a higher level of creatine than marinated pork, which may partially due to leaching of creatine from pork into juice. Both treatments of 0.5% cinnamon powder and 0.5% green tea powder showed the same phenomenon. For the creatinine content change, it followed a time-dependent rise over a 8-h marinating period and a maximum was reached for marinated pork (76.43 mg/100 g), marinated pork plus 0.5% cinnamon powder (56.60 mg/100 g) and marinated pork plus 0.5% green tea powder (51.65 mg/100 g), followed by a decline to 44.47, 45.43 and 49.31 mg/100 g, respectively, after 24-h marinating (Table 3), probably due to participation of creatinine in the Maillard browning reaction. It is also possible that the formation rate of creatinine from creatine was slower than the Maillard browning reaction rate after prolonged marinating for ≥8 h.

Table 3. Changes of HA precursor contents in unmarinated/marinated pork and juice as affected by different flavorings and time length.

Time Length	0 h	2 h	4 h	8 h	12 h	24 h
			Reducing sugar content (mg/g) [1]			
marinated pork (MP)	$0.09 \pm 0.25^{e,A}$	$1.37 \pm 0.46^{d,B}$	$1.12 \pm 0.08^{d,B}$	$2.32 \pm 0.28^{c,B}$	$3.50 \pm 0.02^{b,C}$	$4.03 \pm 0.05^{a,C}$
MP + 0.5% cinnamon powder	$0.08 \pm 0.48^{c,A}$	$1.56 \pm 1.70^{b,A}$	$1.03 \pm 0.44^{b,B}$	$1.83 \pm 0.37^{b,C}$	$4.73 \pm 0.10^{a,B}$	$5.33 \pm 0.66^{a,B}$
MP + 0.5% green tea powder	$0.07 \pm 0.40^{d,A}$	$1.22 \pm 0.76^{c,B}$	$4.08 \pm 1.65^{b,A}$	$3.75 \pm 0.47^{b,A}$	$6.73 \pm 1.81^{a,A}$	$6.50 \pm 0.21^{a,A}$
			Creatine content (mg/100 g) [1]			
marinated pork (MP)	$57.85 \pm 0.21^{a,A}$	$25.70 \pm 0.1^{b,A}$	$20.43 \pm 0.14^{c,A}$	$15.94 \pm 0.20^{d,A}$	$11.34 \pm 0.08^{e,A}$	$6.72 \pm 0.07^{f,C}$
MP + 0.5% cinnamon powder	$56.99 \pm 0.61^{a,A}$	$22.98 \pm 0.13^{b,B}$	$14.12 \pm 0.10^{c,B}$	$11.82 \pm 0.17^{d,B}$	$8.96 \pm 0.08^{e,B}$	$7.27 \pm 0.08^{f,B}$
MP + 0.5% green tea powder	$57.14 \pm 0.37^{a,A}$	$21.16 \pm 0.12^{b,C}$	$13.46 \pm 0.56^{c,B}$	$10.66 \pm 0.08^{d,C}$	$8.72 \pm 0.09^{e,C}$	$7.62 \pm 0.06^{f,A}$
			Creatinine content (mg/100 g) [1]			
marinated pork (MP)	$5.97 \pm 0.15^{e,A}$	$34.57 \pm 0.54^{d,B}$	$60.33 \pm 1.23^{b,A}$	$76.43 \pm 2.74^{a,A}$	$60.26 \pm 1.55^{b,A}$	$44.47 \pm 0.92^{c,B}$
MP + 0.5% cinnamon powder	$5.84 \pm 0.29^{e,A}$	$37.07 \pm 0.06^{d,A}$	$46.38 \pm 0.17^{c,C}$	$56.60 \pm 0.64^{a,B}$	$50.22 \pm 0.26^{b,C}$	$45.43 \pm 1.72^{c,B}$
MP + 0.5% green tea powder	$5.11 \pm 0.47^{d,A}$	$37.17 \pm 0.26^{c,A}$	$49.03 \pm 1.12^{b,B}$	$51.65 \pm 0.47^{a,C}$	$51.57 \pm 0.14^{a,B}$	$49.31 \pm 1.25^{b,A}$
			Reducing sugar content (mg/g) [1]			
marinated juice (MJ)	$0.15 \pm 0.06^{d,A}$	$12.18 \pm 0.13^{a,A}$	$9.30 \pm 0.22^{b,B}$	$9.95 \pm 0.29^{b,B}$	$7.74 \pm 0.17^{c,C}$	$7.89 \pm 0.11^{c,C}$
MJ + 0.5% cinnamon powder	$0.10 \pm 0.04^{e,A}$	$11.01 \pm 0.07^{b,B}$	$10.91 \pm 0.14^{b,A}$	$11.61 \pm 0.19^{a,A}$	$9.43 \pm 0.24^{d,A}$	$9.74 \pm 0.17^{c,A}$
MJ + 0.5% green tea powder	$0.13 \pm 0.02^{e,A}$	$8.92 \pm 0.06^{a,C}$	$7.96 \pm 0.08^{d,C}$	$7.87 \pm 0.09^{d,C}$	$8.76 \pm 0.10^{b,B}$	$8.57 \pm 0.09^{c,B}$
			Creatine content (mg/100 g) [1]			
marinated juice (MJ)	$23.30 \pm 0.18^{a,A}$	$22.95 \pm 0.07^{b,A}$	$22.28 \pm 0.06^{c,A}$	$16.57 \pm 0.24^{d,A}$	$10.42 \pm 0.15^{e,B}$	$9.30 \pm 0.14^{f,,B}$
MJ + 0.5% cinnamon powder	$23.21 \pm 0.03^{a,A}$	$22.46 \pm 0.02^{b,B}$	$18.18 \pm 0.08^{c,B}$	$14.44 \pm 0.49^{d,B}$	$13.38 \pm 0.13^{e,A}$	$10.26 \pm 0.02^{f,A}$
MJ + 0.5% green tea powder	$23.01 \pm 0.21^{a,A}$	$20.43 \pm 1.22^{b,C}$	$15.84 \pm 0.38^{c,C}$	$13.14 \pm 0.15^{d,C}$	$10.21 \pm 0.10^{e,B}$	$10.79 \pm 0.72^{e,A}$
			Creatinine content (mg/100 g) [1]			
marinated juice (MJ)	nd [2]	$6.47 \pm 0.10^{e,B}$	$11.23 \pm 0.23^{b,A}$	$14.44 \pm 0.52^{a,A}$	$10.07 \pm 0.26^{c,A}$	$8.20 \pm 0.17^{d,B}$
MJ + 0.5% cinnamon powder	nd	$6.77 \pm 0.01^{d,A}$	$8.51 \pm 0.03^{c,C}$	$10.74 \pm 0.12^{a,B}$	$9.17 \pm 0.05^{b,C}$	$8.29 \pm 0.31^{c,B}$
MJ + 0.5% green tea powder	nd	$6.83 \pm 0.05^{d,A}$	$9.17 \pm 0.21^{c,B}$	$9.76 \pm 0.09^{a,C}$	$9.51 \pm 0.03^{ab,B}$	$9.37 \pm 0.23^{bc,A}$

[1] Data are presented as mean ± standard deviation of triplicate determinations and data with different small letters (a–f) in the same row and capital letters (A–C) in the same column are significantly different ($p < 0.05$).
[2] not detected.

A similar trend was observed for the creatinine content change in marinated juice, marinated juice plus 0.5% cinnamon powder and marinated juice plus 0.5% green tea powder, with a peak of 14.44, 10.74 and 9.76 mg/100 g being attained, respectively, after 8-h marinating, accompanied by a drop to 8.20, 8.29 and 9.37 mg/100 g, respectively, after prolonged marinating for 24 h (Table 3). Like marinated pork, the formation rate of creatinine from creatine should be slower than the Maillard browning reaction rate for all the three marinated juice treatments.

3.1.4. Antioxidant Capacity of Unmarinated/Marinated Pork and Juice

Table 4 shows the antioxidant capacity of unmarinated pork, marinated pork and juice. Interestingly, only a minor change in DPPH value was shown for both marinated pork and juice over a 24-h marinating period. However, for FRAP, both marinated pork and marinated pork plus 0.5% cinnamon powder followed a time-dependent increase during marinating for 24 h. For the treatments of marinated juice and marinated pork plus 0.5% green tea powder, the highest FRAP was observed after 8-h marinating, followed by a decline afterwards. Nevertheless, the addition of 0.5% cinnamon powder or 0.5% green tea powder was shown to increase FRAP in both marinated pork and juice, which can be attributed to the presence of high level of cinnamon aldehyde (2.34 mg/g) in cinnamon powder and total catechin (114.64 mg/g) in green tea powder (Table 5).

Table 4. Antioxidant capacity as well as Maillard browning index of marinated pork and juice as affected by different flavorings and time length.

Time Length	0 h	2 h	4 h	8 h	12 h	24 h
			FRAP (TAC, mmol Trolox equivalent/kg) [1]			
marinated pork (MP)	11.75 ± 0.17 [f,B]	13.84 ± 0.21 [e,C]	18.98 ± 0.23 [d,C]	24.77 ± 0.14 [c,C]	25.51 ± 0.20 [b,C]	28.90 ± 0.70 [a,C]
MP + 0.5% cinnamon powder	11.96 ± 0.19 [f,B]	20.13 ± 0.20 [e,B]	27.32 ± 0.64 [d,B]	35.71 ± 0.69 [c,B]	37.22 ± 0.49 [b,B]	42.74 ± 0.34 [a,B]
MP + 0.5% green tea powder	21.33 ± 0.29 [f,A]	49.97 ± 0.52 [c,A]	43.24 ± 0.77 [e,A]	54.27 ± 0.17 [a,A]	51.08 ± 0.05 [b,A]	47.09 ± 0.58 [d,A]
			DPPH (TAC, mmol Trolox equivalent/kg) [1]			
marinated pork (MP)	28.94 ± 0.11 [c,C]	27.10 ± 0.22 [e,C]	27.37 ± 0.13 [d,B]	30.13 ± 0.07 [b,C]	30.28 ± 0.08 [b,B]	30.63 ± 0.09 [a,B]
MP + 0.5% cinnamon powder	29.26 ± 0.19 [d,B]	29.47 ± 0.07 [c,B]	30.85 ± 0.02 [ab,A]	30.78 ± 0.02 [b,B]	30.83 ± 0.09 [ab,A]	30.97 ± 0.09 [a,A]
MP + 0.5% green tea powder	30.58 ± 0.07 [c,A]	31.02 ± 0.04 [a,A]	30.84 ± 0.06 [ab,A]	31.01 ± 0.05 [a,A]	30.75 ± 0.33 [bc,A]	30.93 ± 0.02 [ab,A]
			FRAP (TAC, mmol Trolox equivalent/kg) [1]			
marinated juice (MJ)	54.33 ± 0.96 [a,C]	49.30 ± 0.47 [c,C]	50.76 ± 0.19 [b,C]	54.87 ± 0.64 [a,C]	44.04 ± 0.91 [e,C]	45.99 ± 1.09 [d,C]
MJ + 0.5% cinnamon powder	56.27 ± 1.11 [b,B]	52.87 ± 0.83 [c,B]	53.50 ± 0.37 [c,B]	57.62 ± 0.32 [a,B]	58.31 ± 0.36 [a,B]	51.36 ± 0.51 [d,B]
MJ + 0.5% green tea powder	67.13 ± 0.14 [b,A]	69.64 ± 0.15 [a,A]	69.37 ± 0.46 [a,A]	66.18 ± 0.22 [c,A]	62.98 ± 0.81 [d,A]	62.98 ± 0.81 [d,A]
			DPPH (TAC, mmol Trolox equivalent/kg) [1]			
marinated juice (MJ)	30.32 ± 0.03 [a,A]	29.94 ± 0.52 [ab,A]	30.13 ± 0.03 [a,A]	30.06 ± 0.05 [a,A]	30.12 ± 0.11 [a,A]	29.66 ± 0.07 [b,A]
MJ + 0.5% cinnamon powder	30.41 ± 0.02 [a,A]	30.23 ± 0.18 [ab,A]	30.02 ± 0.25 [b,A]	30.02 ± 0.05 [b,A]	29.71 ± 0.06 [c,B]	29.65 ± 0.02 [c,A]
MJ + 0.5% green tea powder	30.18 ± 0.05 [b,A]	30.34 ± 0.09 [a,A]	30.11 ± 0.02 [b,A]	29.95 ± 0.09 [c,A]	29.68 ± 0.02 [d,B]	29.50 ± 0.03 [e,A]
Maillard browning index [1]	0 h	2 h	4 h	8 h	12 h	24 h
marinated juice (MJ)	0.21 ± 0.01 [d,B]	0.26 ± 0.00 [c,A,2]	0.30 ± 0.01 [b,A]	0.31 ± 0.01 [b,A]	0.31 ± 0.01 [b,B]	0.38 ± 0.01 [a,B]
MJ + 0.5% cinnamon powder	0.23 ± 0.01 [d,A]	0.26 ± 0.02 [d,A]	0.26 ± 0.03 [d,A]	0.33 ± 0.01 [c,A]	0.38 ± 0.01 [b,B]	0.49 ± 0.02 [a,AB]
MJ + 0.5% green tea powder	0.26 ± 0.01 [b,A]	0.25 ± 0.02 [b,A]	0.26 ± 0.05 [b,A]	0.35 ± 0.07 [b,A]	0.53 ± 0.07 [a,A]	0.56 ± 0.11 [a,A]

[1] Data are presented as mean ± standard deviation of triplicate determinations and data with different small letters (a–f) in the same row and capital letters (A–C) in the same column are significantly different ($p < 0.05$).
[2] Data with the standard deviation values <0.005 are shown as ±0.00 to maintain 2 significant digits.

Table 5. Analysis of the bioactive compounds (mg/g) in curcuma powder and cinnamon powder.

Cinnamon Powder										
Compound	**Neochlorogenic Acid**	**Benzoic Acid**	**Caffeic Acid**	**Hyperoside + Isoquecetin**	**Coumarin**	**Quercetin**	**Cinnamic Acid**	**Cinnam-Aldehyde**	**Eugenol**	**p-Coumaric Acid**
Content (mg/g) [1]	0.017 ± 0.002	0.038 ± 0.006	0.004 ± 0.000	0.007 ± 0.001	0.001 ± 0.000	0.004 ± 0.000	0.323 ± 0.018	2.338 ± 0.012	0.008 ± 0.001	0.001 ± 0.000

Green Tea Powder					
Compound	**Epicatechin (EC)**	**Epigallocatechin Gallate (EGCG)**	**Gallocatechin Gallate (GCG)**	**Epicatechin Gallate (ECG)**	**Total**
Content (mg/g) [1]	60.41 ± 0.15	35.90 ± 0.10	1.30 ± 0.01	17.04 ± 0.10	114.64 ± 0.35

[1] Mean of duplicate determinations.

In several previous studies, Unal et al. [39] reported that the addition of 0.2% clove or 0.5% cinnamon was effective in inhibiting HA formation in barbecued sucuk, while the HA formation in meat and seafood was retarded by incorporation of cinnamon aldehyde [40]. Likewise, the incorporation of 0.5% garlic, onion, red chilli, paprika, ginger or black pepper was shown to reduce total HAs content, but not PAHs in beef and chicken meatballs fried at 180 °C, and a high negative correlation was shown between TEAC/ORAC antioxidant activity of these spices and total HAs content [17]. By using an optimum ratio of turmeric to lemon grass at 52.4:47.6, Sepahpour et al. [41] demonstrated a 94.7% inhibition of total HAs content in grilled beef with a reduction of IQ, PhIP, Harman, Norharman and AαC by 52, 92, 99, 97 and 98%, respectively, while an inhibition of 82, 61, 28 and 79% was shown for PhIP, IQx, MeIQx and 4,8-DiMeIQx, respectively, in grilled beef marinated with 0.5% Sichaun pepper and 0.01% sanshoamide extract [16]. In a later study, Teng et al. [42] explored the inhibitory effects of *Sonchus olearleu* herb extract on HAs formation in roasted pork patties and reported that a significant reduction in IQ, Harman and Norharman was correlated negatively with total phenolic content and antioxidant potential of *S. olearleu* herb extract. More recently, Lai et al. [23] demonstrated that the addition of 0.5% cinnamon powder was efficient in inhibiting formation in pork jerky during roasting. Similar findings were shown for HA inhibition in pan-fried beef marinated with green tea, and the longer the marinating time, the less the formation of HAs in beef during pan-frying [43]. In addition to the antioxidant capacity of cinnamon powder and green tea powder, the effect of Maillard browning reaction products (MRPs) on inhibition of HAs cannot be ignored. As shown in Table 4, the Maillard browning reaction index followed a time-dependent rise for all three marinated juice treatments, with green tea powder showing a higher Maillard browning reaction index than cinnamon powder and marinated juice after prolonged marinating, leading to a reduction in total HAs in marinated juice (Table 1). A similar phenomenon was observed by Lai et al. [23], reporting that the formation of MRPs was effective in inhibiting HA formation in roasted pork jerky.

Taken together, the addition of 0.5% cinnamon powder was more effective than 0.5% green tea powder in inhibiting HA formation in marinated pork, while in marinated juice, the incorporation of 0.5% green tea powder was more efficient than 0.5% cinnamon powder. This difference may be accounted for by the presence of the low-polar cinnamon aldehyde from cinnamon powder mainly in marinated pork, while the polar catechin in green tea powder may be mainly present in marinated juice.

3.2. Analysis of PAHs in Unmarinated/Marinated Pork and Juice

Figures 3 and 4 show the GC-MS/MS chromatogram of PAHs in marinated pork and juice, respectively, after 24 h marinating without and with addition of 0.5% cinnamon powder or 0.5% green tea powder. The PAH contents in marinated pork as affected by time length are shown in Table 6. A total of five PAHs were found in unmarinated pork, with pyrene being present in the largest amount (31.00 ng/g), followed by DBahA (1.54 ng/g), DBalP (0.72 ng/g), DBaeP (0.71 ng/g) and Phe (0.60 ng/g). However, a time-dependent increase in PAH contents was shown during marinating for 24 h, with six PAHs being generated at the highest level including Pyr (50.51 ng/g), Phe (1.64 ng/g), DBahA (1.54 ng/g), BghiP (0.81 ng/g), DBalP (0.72 ng/g) and DBaeP (0.71 ng/g). The same tendency was observed for both treatments of marinated pork plus 0.5% cinnamon powder and marinated pork plus 0.5% green tea powder, with the total PAH contents of 58.76 and 58.04 ng/g being produced, respectively. By comparison at the same time length (2–12 h), the addition of 0.5% cinnamon powder or 0.5% green tea powder was effective in reducing PAH formation. However, after prolonged marinating for 24 h, the addition of 0.5% cinnamon powder or 0.5% green tea powder failed to reduce PAH formation, probably due to degradation of bioactive compounds such as cinnamaldehyde in cinnamon powder or catechin in green tea powder.

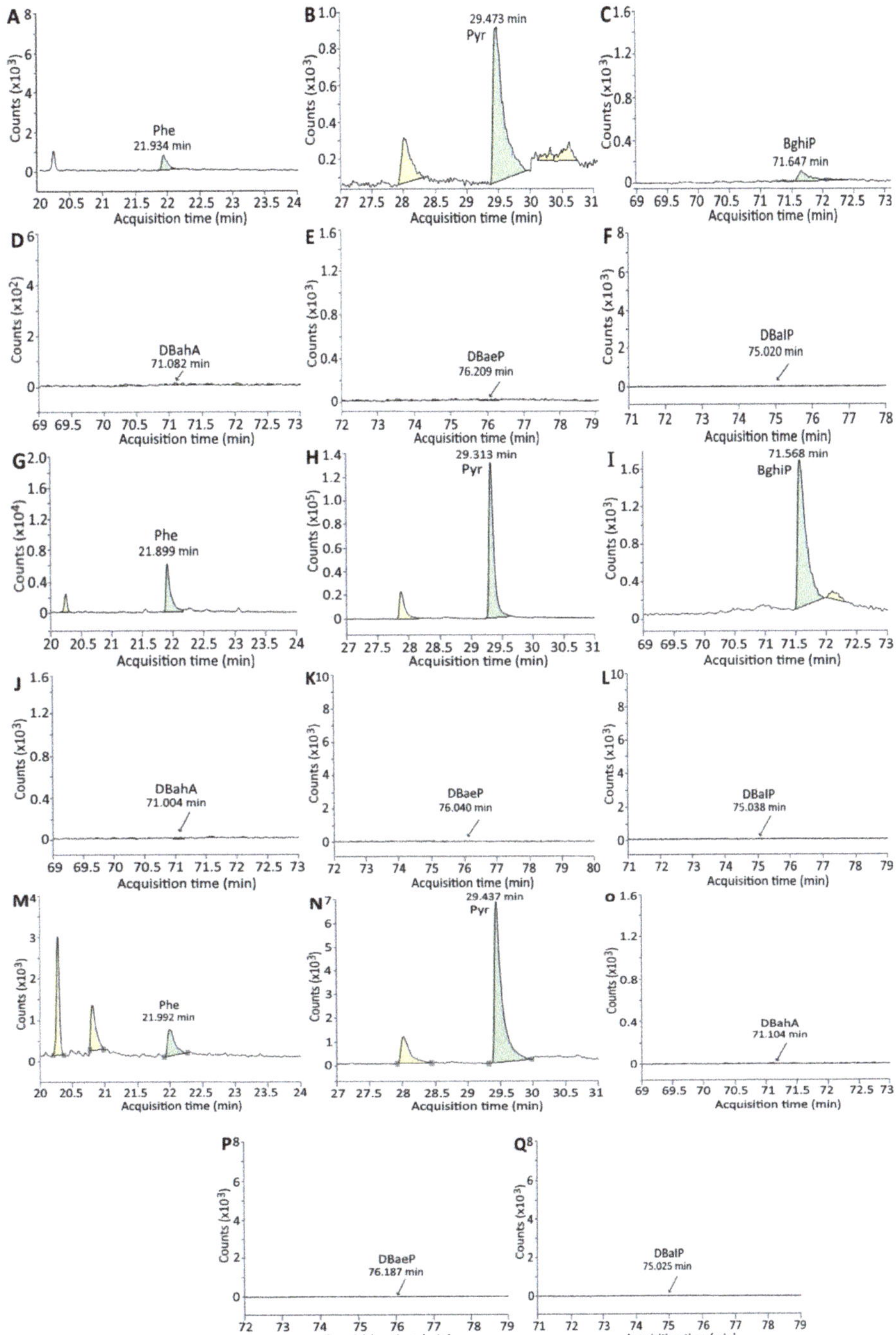

Figure 3. GC-MS/MS chromatogram of PAHs in marinated pork after 24 h marinating (**A–F**), marinated pork with 0.5% cinnamon powder after 24 h marinating (**G–L**) and marinated pork with 0.5% green tea powder after 24 h marinating (**M–Q**), as detected by SRM mode. (**A,G,M**): Phe; (**B,H,N**): Pyr; (**C,I**): BghiP; (**D,J,O**): DBahA; (**E,K,P**): DBaeP; (**F,L,Q**): DBalP.

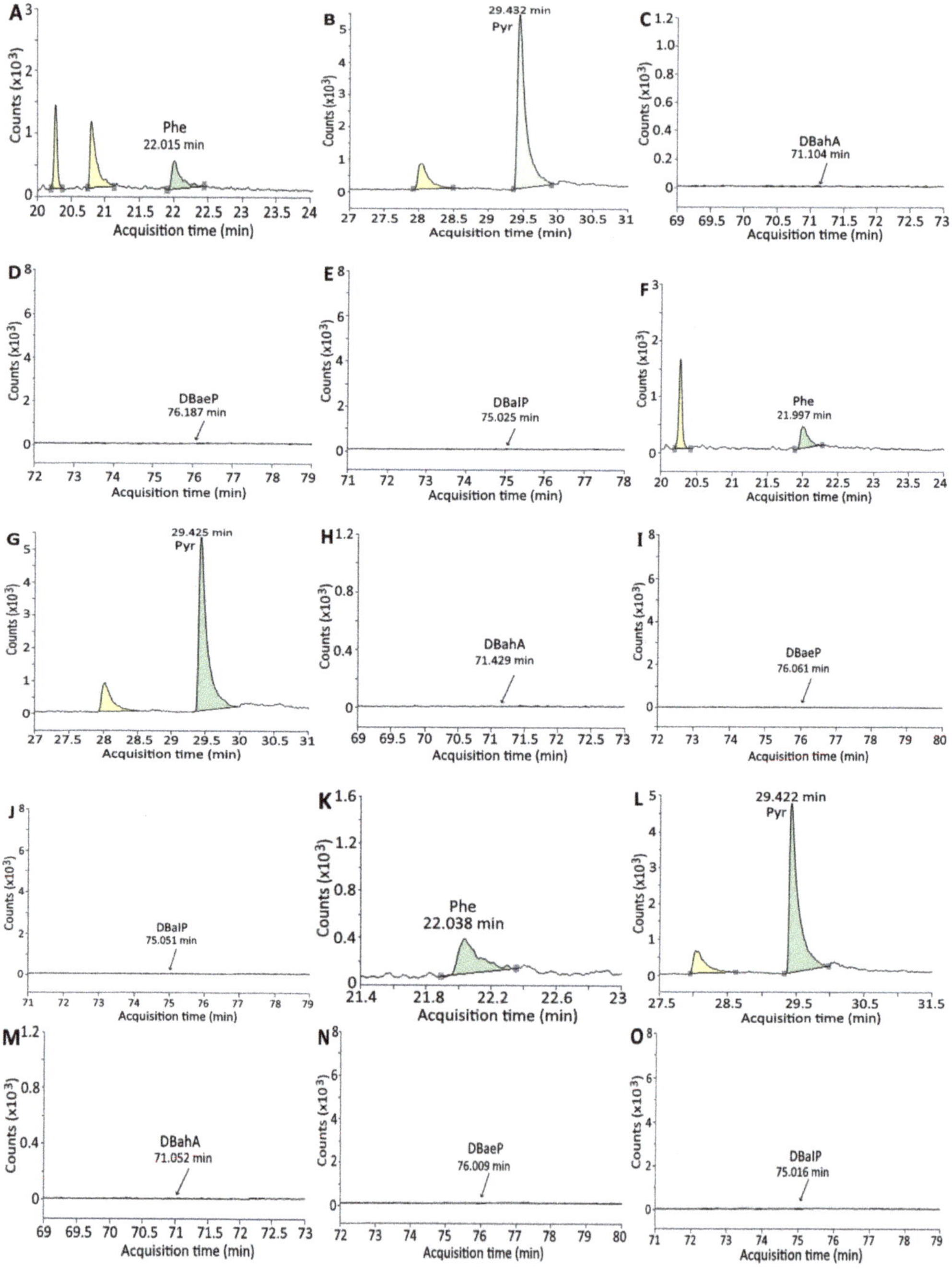

Figure 4. GC-MS/MS chromatogram of PAHs in marinated juice after 24 h marinating (**A–E**), marinated juice with 0.5% cinnamon powder after 24 h marinating (**F–J**) and marinated juice with 0.5% green tea powder after 24 h marinating (**K–O**), as detected by SRM mode. (**A,F,K**): Phe; (**B,G,L**): Pyr; (**C,H,M**): DBahA; (**D,I,N**): DBaeP; (**E,J,O**): DBalP.

Table 6. PAH contents (ng/g) in marinated pork and juice as affected by different flavorings and time length.

PAHs	0 h	2 h	4 h	8 h	12 h	24 h
			marinated pork [1,2]			
Pheneanthrene (Phe)	0.60 ± 0.01 e	0.63 ± 0.05 e	0.78 ± 0.04 d	1.24 ± 0.16 c	1.47 ± 0.04 b	1.64 ± 0.09 a
Pyrene (Pyr)	31.00 ± 0.15 f	38.81 ± 0.72 e	41.79 ± 0.79 d	43.87 ± 0.85 c	46.94 ± 0.98 b	50.51 ± 0.64 a
Benzo[ghi]perylene (BghiP)	nd [3]	trace [4]	0.34 ± 0.01 d	0.43 ± 0.02 c	0.56 ± 0.04 b	0.81 ± 0.01 a
Dibenz[a,h]anthracene (DBahA)	1.54 ± 0.01 a	1.53 ± 0.01 a	1.53 ± 0.01 a	1.53 ± 0.01 a	1.53 ± 0.01 a	1.54 ± 0.01 a
Dienzo[a,e]pyrene (DBaeP)	0.71 ± 0.01 a	0.71 ± 0.00 a	0.71 ± 0.00 a	0.71 ± 0.00 a	0.71 ± 0.00 a	0.71 ± 0.01 a
Dienzo[a,l]pyrene (DBalP)	0.72 ± 0.00 b	0.72 ± 0.00 ab	0.72 ± 0.00 ab	0.72 ± 0.00 a	0.72 ± 0.00 a	0.72 ± 0.01 a
Total	34.56 ± 0.17 f	42.40 ± 0.77 e	45.87 ± 0.76 d	48.49 ± 0.999 c	51.92 ± 0.94 b	55.93 ± 0.73 a
			marinated pork + 0.5% cinnamon powder [1,2]			
Pheneanthrene (Phe)	trace	trace	trace	0.44 ± 0.04 c	4.83 ± 0.25 b	7.29 ± 0.40 a
Pyrene (Pyr)	30.61 ± 1.10 e	36.08 ± 0.84 d	37.15 ± 1.11 cd	37.86 ± 1.01 bc	39.66 ± 0.60 b	45.02 ± 1.31 a
Benzo[ghi]perylene (BghiP)	nd	nd	trace	0.45 ± 0.04 c	1.23 ± 0.09 b	3.46 ± 0.15 a
Dibenz[a,h]anthracene (DBahA)	1.53 ± 0.01 ab	1.53 ± 0.01 ab	1.52 ± 0.00 ab	1.52 ± 0.00 b	1.52 ± 0.00 b	1.55 ± 0.03 a
Dienzo[a,e]pyrene (DBaeP)	0.71 ± 0.01 a	0.71 ± 0.01 a	0.71 ± 0.00 a	0.71 ± 0.01 a	0.71 ± 0.00 a	0.72 ± 0.01 a
Dienzo[a,l]pyrene (DBalP)	0.72 ± 0.01 ab	0.72 ± 0.00 b	0.72 ± 0.00 b	0.72 ± 0.00 b	0.72 ± 0.00 b	0.73 ± 0.01 a
Total	33.57 ± 1.09 e	39.04 ± 0.86 de	40.10 ± 1.14 cd	41.70 ± 1.06 c	48.74 ± 0.50 b	58.76 ± 1.87 a
			marinated pork + 0.5% green tea powder [1,2]			
Pheneanthrene (Phe)	0.33 ± 0.01 d	0.78 ± 0.06 c	0.79 ± 0.06 c	0.84 ± 0.04 c	1.05 ± 0.13 b	1.77 ± 0.09 a
Pyrene (Pyr)	28.80 ± 1.50 d	30.94 ± 1.12 d	35.05 ± 0.80 c	35.22 ± 1.79 bc	37.35 ± 0.82 b	49.77 ± 0.76 a
Benzo[ghi]perylene (BghiP)	nd	nd	nd	nd	nd	nd
Dibenz[a,h]anthracene (DBahA)	1.53 ± 0.01 a	1.57 ± 0.05 a	1.54 ± 0.01 a	1.54 ± 0.01 a	1.55 ± 0.03 a	1.55 ± 0.05 a
Dienzo[a,e]pyrene (DBaeP)	0.71 ± 0.00 c	0.71 ± 0.00 c	0.71 ± 0.00 c	0.72 ± 0.01 bc	0.72 ± 0.00 b	0.73 ± 0.00 a
Dienzo[a,l]pyrene (DBalP)	0.72 ± 0.00 c	0.73 ± 0.01 b	0.73 ± 0.00 b	0.73 ± 0.08 b	0.73 ± 0.01 b	0.76 ± 0.01 a
Total	32.09 ± 1.50 d	34.72 ± 1.11 d	38.82 ± 2.73 c	39.05 ± 0.46 c	41.40 ± 0.95 b	54.59± 0.54 a
			marinated juice [1,2]			
Pheneanthrene (Phe)	0.34 ± 0.01 e	0.41 ± 0.06 d	0.45 ± 0.03 d	0.63 ± 0.03 c	0.71 ± 0.01 b	0.80 ± 0.04 a
Pyrene (Pyr)	23.56 ± 0.87 e	23.70 ± 0.26 e	26.00 ± 1.78 d	33.07 ± 0.54 c	37.33 ± 1.62 b	40.36 ± 1.09 a
Dibenz[a,h]anthracene (DBahA)	1.53 ± 0.00 b	1.53 ± 0.00 b	1.52 ± 0.00 b	1.52 ± 0.01 b	1.53 ± 0.01 b	1.55 ± 0.00 a
Dienzo[a,e]pyrene (DBaeP)	0.71 ± 0.00 c	0.71 ± 0.00 c	0.71 ± 0.00 bc	0.71 ± 0.00 bc	0.72 ± 0.00 b	0.73 ± 0.01 a
Dienzo[a,l]pyrene (DBalP)	0.72 ± 0.00 d	0.72 ± 0.00 d	0.72 ± 0.00 cd	0.73 ± 0.00 bc	0.73 ± 0.01 b	0.74 ± 0.00 a
Total	26.85 ± 0.86 e	27.06 ± 0.24 e	29.41 ± 1.75 d	36.67 ± 0.53 c	41.01 ± 1.62 b	44.17 ± 1.05 a
			marinated juice + 0.5% cinnamon powder [1,2]			
Pheneanthrene (Phe)	trace [3]	0.38 ± 0.02 b	0.42 ± 0.02 b	0.70 ± 0.29 a	0.77 ± 0.03 a	0.85 ± 0.04 a
Pyrene (Pyr)	24.19 ± 0.64 f	34.96 ± 0.36 e	42.76 ± 1.27 d	46.41 ± 1.74 c	49.13 ± 1.11 b	54.52 ± 1.20 a
Dibenz[a,h]anthracene (DBahA)	1.52 ± 0.00 b	1.53 ± 0.00 b	1.52 ± 0.00 b	1.53 ± 0.01 b	1.53 ± 0.01 b	1.55 ± 0.01 a
Dienzo[a,e]pyrene (DBaeP)	0.71 ± 0.00 b	0.71 ± 0.00 b	0.71 ± 0.00 b	0.71 ± 0.01 b	0.71 ± 0.00 b	0.73 ± 0.01 a
Dienzo[a,l]pyrene (DBalP)	0.72 ± 0.00 c	0.72 ± 0.00 c	0.73 ± 0.00 b	0.73 ± 0.00 b	0.73 ± 0.01 ab	0.74 ± 0.00 a
Total	27.14 ± 0.63 f	38.29 ± 0.34 e	46.14 ± 1.26 d	50.06 ± 1.58 c	52.87 ± 1.12 b	58.39 ± 1.17 a
			marinated juice + 0.5% green tea powder [1,2]			
Pheneanthrene (Phe)	0.40 ± 0.03 f	0.54 ± 0.01 e	0.58 ± 0.01 d	0.76 ± 0.03 c	0.83 ± 0.02 b	0.97 ± 0.03 a
Pyrene (Pyr)	22.64 ± 1.05 f	26.57 ± 2.12 e	31.60 ± 0.84 d	37.77 ± 0.54 c	40.72 ± 0.89 b	47.69 ± 1.81 a
Dibenz[a,h]anthracene (DBahA)	1.53 ± 0.01 a	1.53 ± 0.00 a	1.53 ± 0.01 a	1.53 ± 0.01 a	1.54 ± 0.01 a	1.54 ± 0.03 a
Dienzo[a,e]pyrene (DBaeP)	0.71 ± 0.00 c	0.71 ± 0.00 c	0.71 ± 0.00 bc	0.71 ± 0.00 ab	0.71 ± 0.00 a	0.72 ± 0.00 a
Dienzo[a,l]pyrene (DBalP)	0.72 ± 0.00 d	0.73 ± 0.00 d	0.73 ± 0.00 cd	0.73 ± 0.00 bc	0.74 ± 0.00 b	0.76 ± 0.00 a
Total	26.00 ± 1.09 f	30.07 ± 2.11 e	35.14 ± 0.83 d	41.50 ± 0.56 c	44.54 ± 0.87 b	51.68 ± 1.81 a

[1] Data are presented as mean ± standard deviation of triplicate determinations and data with different small letters (a–f) in the same row are significantly different ($p < 0.05$); [2] Data with the standard deviation values <0.005 are shown as ±0.00 to maintain 2 significant digits; [3] not detected; [4] LOQ ≥ PAHs levels ≥ LOD.

Similar to unmarinated pork, a total of five PAHs including Pyr (23.56 ng/g), DBahA (1.53 ng/g), DBalP (0.72 ng/g), DBaeP (0.71 ng/g) and Phe (0.34 ng/g) were detected in unmarinated juice (Table 6). A time-dependent rise in PAH contents was also shown over a 24-h marinating period, with Pyr present at the highest level (40.36 ng/g), followed by DBahA (1.55 ng/g), Phe (0.80 ng/g), DBalP (0.74 ng/g) and DBaeP (0.73 ng/g). Both treatments of marinated juice plus 0.5% cinnamon powder and marinated juice plus 0.5% green tea powder showed the same trend; however, by comparison at the same time length, the incorporation of 0.5% cinnamon powder generated the highest level of total PAHs (58.39 ng/g) followed by 0.5% green tea powder (51.68 ng/g) and marinated juice (44.17 ng/g). This outcome implied that both 0.5% cinnamon powder and 0.5% green tea powder failed to inhibit PAH formation in marinated juice, probably due to degradation of bioactive compounds during heating as indicated above. Nevertheless, the bioactive compounds in cinnamon powder and green tea powder may be more susceptible to degradation in marinated juice when compared to marinated pork. Furthermore, the presence of five PAHs in unmarinated juice may be originally from soy sauce as 3,4-benzopyrene was detected in Japanese soy sauce [44]. In a study dealing with the effect of curing on PAH formation in poultry meat, a small amount of PAHs were detected in raw meat, while the total PAH was produced at a higher level for 24-h curing than for 12-h curing [3]. This result is similar to our finding that a small amount of PAHs were present in both unmarinated pork and juice.

The PAH formation and inhibition in model system and food products during heating has been controversial. For instance, Britt et al. [45] studied PAH formation in three model systems by heating at 840 °C for 10 sec, a model system of glucose and proline was shown to generate the highest amount of PAHs (Phe and Ant), followed by the model systems of glucose and proline. Obviously, the formation of MRPs between glucose and proline during heating can promote PAH formation. In a study dealing with the effect of small molecular weight aldose and basic amino acid on PAH formation in grilled pork sausage, Nie et al. [46] pointed out that glucose was more susceptible to PAH formation than fructose, probably due to formation of small aldehyde compounds from the former for the subsequent polymerization or condensation to produce PAHs. Furthermore, the addition of basic amino acids such as lysine and arginine was shown to generate more PAHs than acidic amino acids, probably caused by a rise in pH from the former to accelerate Maillard reaction for the subsequent formation of 1-amino-1-deoxy-2-ketosaccharide and then PAHs [46]. More specifically, sugar can be dehydrated to form 5-HMF, methylglyoxal and acetone compounds under acidic condition, followed by indene formation from 5-HMF through Diels–Alder reaction, dehydrogenation and acetylene addition to generate FL [47–49]. In meat products, the PAH formation in thin slices of dried pork as affected by flavorings and roasting conditions was studied by Hung et al. [2], reporting that a flavoring of 16% sugar and 8% soy sauce or 16% sugar and 4% soy sauce generated some more varieties of PAHs at 200 °C. Furthermore, the AcP oxidation during roasting of thin slices of dried pork may lead to AcPy formation. In another study the addition of elderberry was shown to inhibit PAH formation in charcoal-grilled pork by 82%, followed by white wine vinegar (79%), red wine vinegar (66%), apple cider vinegar (66%) and fruit vinegar with raspberry juice (55%) [50]. Sinaga et al. [51] further reported that following marinating in andaliman fruit juice for 60 min for the subsequent grilling, the BaP content in charcoal-grilled duck was reduced by 2.6-fold.

Analysis of PAH Precursors in Unmarinated/Marinated Pork and Juice

Table 7 shows the content changes of PAH precursors including 2-cyclohexene-1-one, benzaldehyde and trans,trans-2,4-decadienal in marinated pork and juice. In most treatments the total amount of PAH precursors followed a time-dependent increase and resulted in a rise of total PAH contents. Of the three precursors, only 2-cyclohexene-1-one and trans,trans-2,4-decadienal followed a time-dependent increment for all the marinated pork and juice treatments during marinating, while an inconsistent change was observed for the benzaldehyde, probably due to its volatility and instability. Comparatively, benzaldehyde was produced at a higher level than 2-cyclohexene-1-one and trans,trans-2,4-decadienal in marinated pork and juice, and should play a more important role in PAH formation as reported by Chen and Chen [7]. Furthermore, the addition of 0.5% cinnamon powder may promote benzaldehyde formation in both marinated pork and juice, as it was reported that cinnamaldehyde could be converted to benzaldehyde during heating [52]. This may explain why a maximum level of benzaldehyde (922.47 ng/g) was produced in pork after 2-h marinating. Moreover, the addition of 0.5% green tea powder may induce 2-cyclohexene-1-one formation as it was detected in green tea [53]. Conversely, the addition of 0.5% cinnamon powder or 0.5% green tea powder was effective in inhibiting formation of trans,trans-2,4-decadienal, a degradation product of linoleic acid oxidation during heating. In a previous study, Chen and Chen [7] studied the PAH formation mechanism in food lipids, and postulated that benzaldehyde could be generated through cyclohexene oxidation, while trans,trans-2,4-decadienal and 2-cyclohexene-1-one produced through degradation of linoleic acid and linolenic acid during heating, respectively.

Table 7. Changes of PAH precursor contents (ng/g) in unmarinated/marinated pork and juice as affected by different flavorings and time length.

Time Length	0 h	2 h	4 h	8 h	12 h	24 h
2-cyclohexene-1-one						
marinated pork (MP)	5.30	9.71	10.88	12.90	28.43	43.19
MP + 0.5% cinnamon powder	3.72	16.30	17.79	28.32	40.51	151.24
MP + 0.5% green tea powder	nd [1]	11.25	21.46	45.09	54.02	181.89
benzaldehyde						
marinated pork (MP)	140.93	155.59	191.86	226.28	289.67	240.47
MP + 0.5% cinnamon powder	183.88	922.47	522.05	338.95	294.31	253.05
MP + 0.5% green tea powder	200.81	121.63	177.61	155.51	176.41	180.34
trans,trans-2,4-decadienal						
marinated pork (MP)	15.06	80.45	152.74	224.86	233.75	248.04
MP + 0.5% cinnamon powder	16.35	42.03	62.44	126.85	205.47	296.07
MP + 0.5% green tea powder	25.12	40.69	48.64	71.42	123.28	218.44
Total (2-cyclohexene-1-one + benzaldehyde + trans,trans-2,4-decadienal)						
marinated pork (MP)	161.29	245.75	355.47	464.04	551.85	531.71
MP + 0.5% cinnamon powder	203.95	980.80	602.27	494.12	540.29	700.36
MP + 0.5% green tea powder	225.93	173.57	247.71	272.02	353.71	580.67
2-cyclohexene-1-one						
marinated juice (MJ)	nd	2.30	4.68	4.56	15.81	21.46
MJ + 0.5% cinnamon powder	nd	3.14	4.25	5.03	8.34	24.30
MJ + 0.5% green tea powder	nd	5.13	9.31	8.86	7.11	9.32
benzaldehyde						
marinated juice (MJ)	24.25	39.23	37.19	37.79	36.90	13.77
MJ + 0.5% cinnamon powder	30.98	45.98	53.57	51.35	40.14	36.16
MJ + 0.5% green tea powder	18.38	24.36	24.95	32.15	34.96	33.80
trans,trans-2,4-decadienal						
marinated juice (MJ)	5.92	6.10	7.22	44.38	102.00	93.32
MJ + 0.5% cinnamon powder	nd	6.43	5.16	9.84	51.90	81.73
MJ + 0.5% green tea powder	4.73	4.91	5.96	10.49	13.23	51.09
Total (2-cyclohexene-1-one + benzaldehyde + trans,trans-2,4-decadienal)						
marinated juice (MJ)	30.17	47.63	49.08	86.73	154.72	128.55
MJ + 0.5% cinnamon powder	30.98	55.55	62.98	66.22	100.38	142.20
MJ + 0.5% green tea powder	23.11	34.40	40.23	51.50	55.30	94.21

[1] not detected.

The inhibition of PAHs in meat products by incorporation of antioxidants has been well documented. For instance, the addition of onion (30 g/100 g pork) or garlic (15 g/100 g pork) was effective in inhibiting six PAHs (BbF, BkFL, BaA, BaP, DBahA and BghiP) in fried pork by 60% and 54%, respectively [54,55]. Wang et al. [56] reported that 57% of eight PAHs including BaA, CHR, BbF, BkFL, BaP, DBahA, BghiP and IP in charcoal-grilled chicken wings was inhibited by green tea, which can be attributed to the presence of phenolic compounds such as catechin. In two different studies, the beer-marinated charcoal-grilled chicken wings were shown to reduce the formation of PAHs with both phenolic content and antioxidant activity of beer playing a significant role [18,19]. More recently, Cao et al. [57] compared the effect of apple polyphenol and three antioxidants (ascorbic acid, TBHQ and BHT) on PAH formation in barbecued pork and reported that the addition of 0.2% apple polyphenol possessed the most pronounced inhibition effect which correlated well with the antioxidant activity, peroxide and thiobarbituric acid values. For phenolic acids, protocatechuic acid was shown to be the most effective in inhibiting PAH formation in charcoal-grilled chicken wings, followed by gallic acid and ferulic acid [58]. However, for the DPPH scavenging activity, gallic acid was the most efficient, followed by protocatechuic acid and ferulic acid. Likewise, a total of eight phenolic compounds present in green tea extracts including epigallocatechin gallate, gallocatechin, catechin, epicatechin gallate, catechin gallate, eriodictyol, naringenin and quinic acid were found to inhibit the formation of eight PAHs (BaA, CHR, BbF, BkFL, BaP, DBahA, BghiP and IP) by 15–55% in charcoal-grilled chicken wings, with the phenolic compounds with weak antioxidant activity (quinic acid and naringenin) showing the highest inhibitory effects [21]. This outcome demonstrated that there is no direct relationship between free radical scavenging activity and PAH inhibition. In other words, some other factors may also involve in retarding PAH formation in meat products during heating.

Collectively, in this study we demonstrated that the addition of 0.5% cinnamon powder or 0.5% green tea powder was effective in inhibiting PAH formation in marinated pork through the antioxidant activity. However, no inhibition effect of PAH was observed in marinated juice, which may be due to degradation of phenolic compounds during heating to generate free radicals for the subsequent PAH formation [59]. Of the various PAHs formed in pork and juice during 24-h marinating, only DBahA and DBalP were classified as Group 2A [13]. This should pose no risk to human health as both PAHs were present in small amounts.

3.3. Principal Component Analysis

Figure 5 shows the PCA analysis of HAs and PAHs formation as affected by different flavorings (unmarinated pork/juice, marinated pork/juice, marinated pork/juice plus 0.5% cinnamon powder and marinated pork/juice plus 0.5% green tea powder) and marinating time length (2–24 h). Based on an eigen value >1, a total of four components (PCs 1–4) were shown to describe all the above treatments, with the first two components including PC 1 with 47.64% and PC 2 with 34.35% predominantly contributing to the maximum total variation of 81.99% in HAs or PAHs formation as affected by different flavorings and marinating time length. Analysis of score plot in Figure 5A revealed that the experimental data can be grouped into five groups with Groups 1 and 2 representing, respectively, various HAs and PAHs formed at 36 different treatments including unmarinated pork and juice (p, j), unmarinated pork and juice with 0.5% cinnamon powder/0.5% green tea powder (p+cp/gp, j+cp/gp), marinated pork and juice processed at five different time length (mp2-24 h, mj2-24 h), marinated pork and juice with 0.5% cinnamon powder/0.5% green tea powder at five different time length (mp+cp/gp2-24 h, mj+cp/gp2-24 h). Likewise, regardless of flavoring type, the total contents of HAs and PAHs formed at different marinating time length appeared together in serial line clusters in Group 3, confirming a time-dependent rise in the formation of both HAs and PAHs. On the other hand, for marinated pork/juice without and with 0.5% cinnamon powder/0.5% green tea powder, the total contents of HAs and PAHs formed were grouped as Group 4 and Group 5, respectively. All the five groups (Group 1–5) were well separated suggesting that both HAs and PAHs are formed by different mechanisms with their formation being significantly affected by various flavorings and marinating time length. Furthermore, the score plots in Group 1 were relatively more dispersed compared to the highly overlapping score plots in Group 2, affirming that HAs are more susceptible to formation than PAHs at different flavorings and marinating time length.

To gain more insight on the correlation between the formation of HAs and PAHs, as affected by different flavorings and marinating time length, the loading plots in Figure 5B were analyzed based on inclination of Groups 1–5 towards PC 1 or PC 2 and the angle between them. The loading plots of formation of various PAHs at different flavoring and marinating time length (Group 2) and the contents of total HAs and total PAHs at different marinating time length (Group 3) as well as the contents of total PAHs at different flavorings (Group 5) were inclined to the same direction as PC 1 (47.64%); while that of various HAs formation at different flavorings and marinating time length (Group 1) and the contents of total HAs at different flavorings (Group 4) were inclined to the same direction as PC 2 (34.35%) implying their respective influence on PC 1 and PC 2. Moreover, the loading plots of various HAs and PAHs formed at different flavoring and marinating time length were diverged by a large degree of angle confirming again that their formation by different mechanism and a large variation in their contents. Likewise, comparatively the loading plots of total contents of HAs and PAHs as affected by marinating time length (Group 3) were diverged by a larger degree of angle from the loading plots of various HAs formed (Group 1) than that from various PAHs formed (Group 2), indicating that the formation of HAs is more favored than PAHs with increasing marinating time length. For different flavoring treatments, regardless of marinating time length, the loading plots of total contents of HAs and PAHs diverged by a reasonable angle to appear as two distinct groups (Group 4 and Group 5), revealing a significant variation in the impact of different flavorings on HAs and PAHs formation. Obviously, the incorporation of 0.5% cinnamon powder or 0.5% green tea powder was effective in inhibiting HAs formation, with the former showing a more pronounced inhibition effect for marinated pork and the latter showing a more efficient inhibition for marinated juice (Table 1). However, the addition of 0.5% cinnamon powder or 0.5% green tea powder was only effective in inhibiting PAH formation in marinated pork (Table 6). The aforementioned observations were well conformed with the discussion on HAs and PAHs formation as affected by different flavorings and marinating time length in Sections 3.1 and 3.2 as well as the data tabulated in Tables 1 and 6.

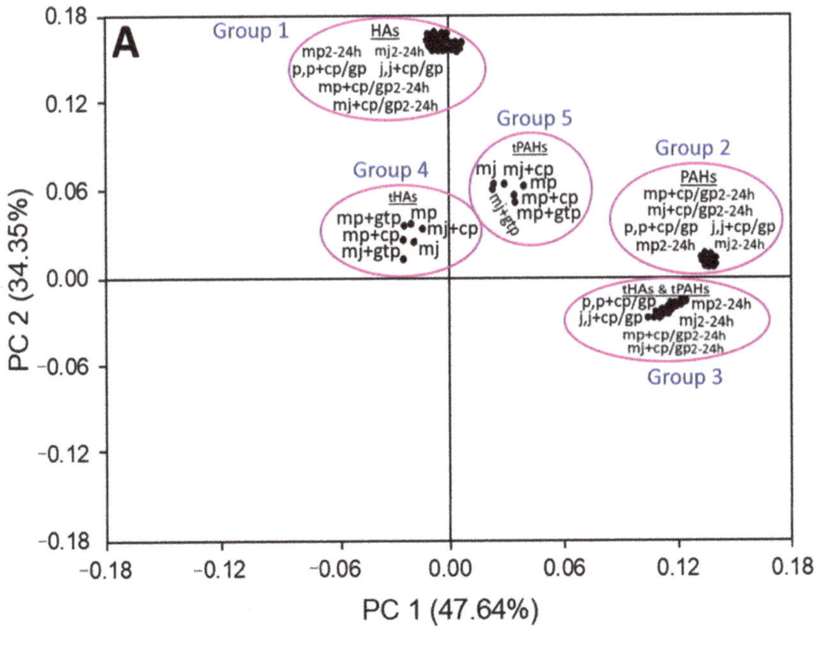

Figure 5. *Cont.*

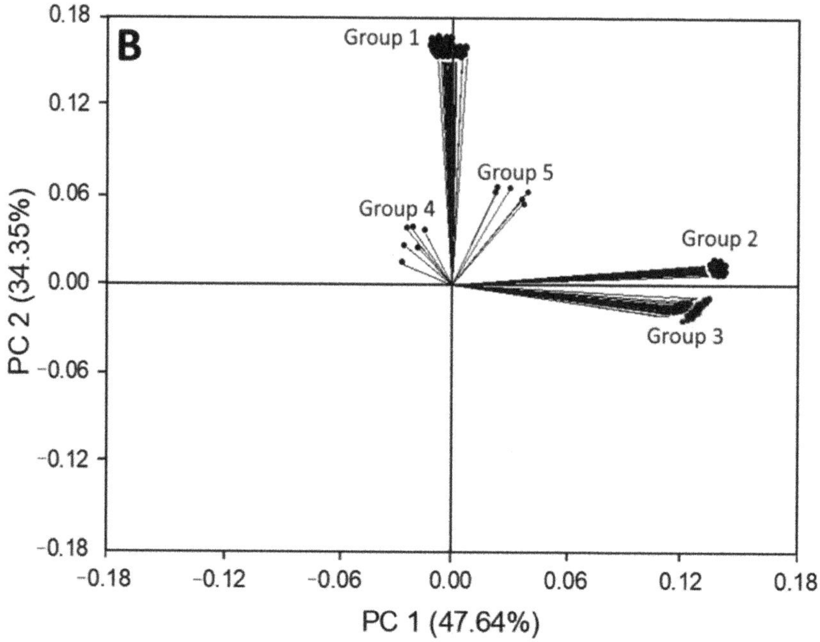

Figure 5. Principle component analysis showing score plot (**A**) and loading plot (**B**) for HAs and PAHs formation as affected by different flavorings and processing time length. Ground pork with different flavoring treatments at various marinating time length was performed by initially preparing the unmarinated juice by mixing 1% of sugar, 10% of soy sauce and 89% of deionized water (standard formula) without or with 0.5% cinnamon powder or 0.5% green tea powder, followed by adding 500 g of raw ground pork and marinating for 2, 4, 8, 12 and 24 h with the temperature controlled at 90 ± 2 °C. Mean values of triplicate determinations were used for PCA analysis. HAs, heterocyclic amines; tHAs, total HAs content; PAHs, polycyclic aromatic hydrocarbons; tPAHs, total PAHs content; p, unmarinated pork; j, unmarinated juice; p+cp/gp, unmarinated pork flavored with 0.5% cinnamon powder or 0.5% green tea powder; j+cp/gp, unmarinated juice flavored with 0.5% cinnamon powder or 0.5% green tea powder; mp2-24 h, marinated pork cooked at 90 ± 2 °C for different time length (2, 4, 8, 12 and 24 h); mj2-24 h, marinated juice cooked at 90 ± 2 °C for different time length (2, 4, 8, 12 and 24 h); mp+cp/gp2-24 h, marinated pork plus 0.5% cinnamon powder or 0.5% green tea powder cooked at 90 ± 2 °C for different time length (2, 4, 8, 12 and 24 h); mj+cp/gp2-24 h, marinated juice plus 0.5% cinnamon powder or 0.5% green tea powder cooked at 90 ± 2 °C for different time length (2, 4, 8, 12 and 24 h).

4. Conclusions

In conclusion, both HAs and PAHs were simultaneously extracted from pork by the QuEChERS method and subsequently analyzed UPLC-MS/MS and GC-MS/MS to study their formation and inhibition in ground pork and juice as affected by different flavorings and time length during marinating. Comparatively, HAs were more prone to formation than PAHs in marinated pork and juice. Following a rise in marinating time length, a time-dependent increase in HA and PAH contents in pork and juice were shown, which can be attributed to elevation in reducing sugar and creatinine levels for the former as well as benzaldehyde, 2-cyclohexene-1-one and trans,trans-2,4-decadienal levels for the latter during marinating. Addition of 0.5% cinnamon powder or 0.5% green tea powder effectively inhibited HA formation in both marinated pork and juice, whereas they were only effective in inhibiting PAH formation in marinated pork. Furthermore, the PCA analysis confirmed that both HAs and PAHs are formed by different mechanisms with their formation being significantly affected by different flavorings and marinating time length.

Supplementary Materials: The following supporting information can be downloaded at: https://www.mdpi.com/article/10.3390/foods11193080/s1, Methods for determination of HA precursors (reducing sugar, amino acid, creatine/creatinine), antioxidant activity, bioactive compounds in cinnamon powder by UPLC-MS/MS, bioactive compounds in green tea powder by HPLC-DAD and PAH precursors.

Author Contributions: Conceptualization, B.-H.C.; Methodology, Y.-W.L. and Y.-T.L.; Formal analysis, Y.-W.L. and Y.-T.L.; Investigation, Y.-W.L. and Y.-T.L.; Data curation, Y.-W.L., Y.-T.L. and B.S.I.; Validation, Y.-W.L., Y.-T.L. and B.S.I.; Software, Y.-W.L., Y.-T.L. and B.S.I.; Visualization, Y.-W.L., Y.-T.L. and B.S.I.; Writing—original draft, B.-H.C. and B.S.I.; Writing—review and editing, B.-H.C. and B.S.I.; Funding acquisition, B.-H.C.; Resources, B.-H.C.; Project administration, B.-H.C.; Supervision, B.-H.C. All authors have read and agreed to the published version of the manuscript.

Funding: This research was funded by a grant (MOST-111-2327-B-030-001) from Ministry of Science and Technology, Taiwan.

Institutional Review Board Statement: Ethical approval was not sought for the present study.

Informed Consent Statement: No informed consent was required for experiments in the present study.

Data Availability Statement: The data that support the findings of this study are available within the manuscript.

Conflicts of Interest: The authors declare no conflict of interest.

Abbreviations

List of names and abbreviations of 21 HAs and 24 PAHs used in this study.

Abbreviation	Full Names
HAs	
DMIP	2-amino-1,6-dimethylimidazo [4,5-b]-pyridine
IFP	2-amino-1,6-dimethyl-furo [3,2-e]imidazo [4,5-b]-pyridine
iso-IQ	2-amino-1-methyl-imidazo [4,5-f]-quinoline
IQ	2-amino-3-methyl-imidazo [4,5-f]-quinoline
MeIQ	2-amino-3,4-dimethyl-imidazo [4,5-f]- quinolin
IQ [4,5-b]	2-amino-1-methyl-imidazo [4,5-b]-quinoline
IQx	2-amino-3-methyl-imidazo [4,5-f]-quinoxaline
8-MeIQx	2-amino-3,8-dimethyl-imidazo [4,5-f]- quinoxaline
7,8-DiMeIQx	2-amino-3,7,8-dimethyl-imidazo [4,5-f]-quinoxaline
4,8-DiMeIQx	2-amino-3,4,8-dimethyl-imidazo [4,5-f]-quinoxaline
4,7,8-TriMeIQx	2-amino-3,4,7,8-dimethyl-imidazo [4,5-f]- quinoxaline (internal standard)
Phe-P-1	2-amino-5-phenylpyridine
AαC	2-amino-9 H-pyrido [2,3-b]indole
Trp-P-2	3-amino-1-methyl-5 H-pyrido [4,3-b]indole
Trp-P-1	3-amino-1,4-methyl-5 H-pyrido [4,3-b]indole
GIu-P-2	2-aminodipyrido-[1,2-a:3′,2′-d]imidazole
Glu-P-1	2-amino-6-methyldipyrido-[1,2-a:3′,2′-d]imidazole
PhIP	2-amino-1-methyl-6-phenylimidazo [4,5-b]-pyridine
Harman	1-methyl-9 H-pyrido [3,4-b]indole
Norharman	9 H-pyrido [3,4-b]indole
MeAαC	2-Amino-3-methyl-9 H-pyrido [2,3-b]indole
PAHs	
Nap	naphthalene
AcPy	acenaphylene
AcP	acenaphthene
Flu	fluorene
Phe	phenanthrene
Ant	anthracene

FL	fluoranthene
Pyr	pyrene
BaA	benzo[a]anthracene
CHR	chrysene
BbFL	benzo[b]fluoranthene
BaP	benzo[a]pyrene
IP	indeno [1,2,3-c,d]pyrene
DBahA	dibenzo[a,h]anthracene
BghiP	benzo[g,h,i]perylene
BjF	benzo[j]fluoranthene
BcF	benzo[c]fluorene
CcdP	cyclopenta[c,d]pyrene
MCH	5-methylchrysene
DBalP	dibenzo[a,l]pyrene
DBaeP	dibenzo[a,e]pyrene
DBaiP	dibenzo[a,i]pyrene
DBahP	dibenzo[a,h]pyrene
Triphenylene	Triphenylene (internal standard)

References

1. Hsu, K.Y.; Chen, B.H. Analysis and reduction of heterocyclic amines and cholesterol oxidation products in chicken by controlling flavorings and roasting condition. *Food Res. Int.* **2020**, *131*, 109004. [CrossRef] [PubMed]
2. Hung, Y.T.; Lee, Y.T.; Inbaraj, B.S.; Sridhar, K.; Chen, B.H. Analysis and formation of polycyclic aromatic hydrocarbons and cholesterol oxidation products in thin slices of dried pork during processing. *Food Chem.* **2021**, *353*, 129474. [CrossRef] [PubMed]
3. Kao, T.H.; Chen, S.; Chen, C.J.; Huang, C.W.; Chen, B.H. Evaluation of analysis of polycyclic aromatic hydrocarbons by the QuEChERS method and gas chromatography–mass spectrometry and their formation in poultry meat as affected by marinating and frying. *J. Agric. Food Chem.* **2012**, *60*, 1380–1389. [CrossRef]
4. Alaejos, M.S.; Afonso, A.M. Factors that affect the content of heterocyclic aromatic amines in foods. *Comp. Rev. Food Sci. Food Saf.* **2011**, *10*, 52–108. [CrossRef]
5. Nagao, M.; Fujita, Y.; Wakabayashi, K.; Sugimura, T. Ultimate forms of mutagenic and carcinogenic heterocyclic amines produced by pyrolysis. *Biochem. Biophys. Res. Commun.* **1983**, *114*, 626–631. [CrossRef]
6. Ziegenhagen, R.; Boczek, P.; Viell, B. Formation of the comutagenic beta-carboline norharman in a simple tryptophan-containing model system at low temperature (40 °C–80 °C). *Adv. Exp. Med. Biol.* **1999**, *467*, 693–696.
7. Chen, B.H.; Chen, Y.C. Formation of polycyclic aromatic hydrocarbons in the smoke from heated model lipids and food lipids. *J. Agric. Food Chem.* **2001**, *49*, 5238–5243. [CrossRef]
8. IARC, International Agency for Research on Cancer. *Agents Classified by the IARC Monographs*; IARC: Lyon, France, 2012; pp. 1–104.
9. IARC. *Some Naturally Occurring and Synthetic Food Components, Furocoumarins and Ultraviolet Radiation*; IARC: Lyon, France, 1985.
10. IARC, International Agency for Research on Cancer. Some food additives, feed additives and naturally occurring substances. *IARC Monogr. Eval. Carcinog. Risks Hum.* **1987**, *31*, 314.
11. IARC, World Health Organization International Agency for Research on Cancer. Some naturally occurring substances: Food items and constituents, heterocyclic aromatic amines and mycotoxins. *IARC Monogr. Eval. Carcinog. Risk Chem. Hum.* **1993**, *56*, 165–242.
12. Cheng, K.W.; Chen, F.; Wang, M. Heterocyclic amines: Chemistry and health. *Mol. Nutr. Food Res.* **2006**, *50*, 1150–1170. [CrossRef]
13. IARC. IARC working group on the evaluation of carcinogenic risks to humans. Some non-heterocyclic polycyclic aromatic hydrocarbons and some related exposures. *IARC Monogr. Eval. Carcinog. Risks Hum.* **2010**, *92*, 1–853.
14. Hsu, K.Y.; Inbaraj, B.S.; Chen, B.H. Evaluation of analysis of cholesterol oxidation products and heterocyclic amines in duck and their formation as affected by roasting methods. *J. Food Drug Anal.* **2020**, *28*, 128–142. [CrossRef] [PubMed]
15. Chen, S.; Kao, T.H.; Chen, C.J.; Huang, C.W.; Chen, B.H. Reduction of carcinogenic polycyclic aromatic hydrocarbons in meat by sugar-smoking and dietary exposure assessment in Taiwan. *J. Agric. Food Chem.* **2013**, *61*, 7645–7653. [CrossRef] [PubMed]
16. Zeng, M.; Wang, J.; Zhang, M.; Chen, J.; He, Z.; Qin, F.; Xu, Z.; Cao, D.; Chen, J. Inhibitory effects of Sichuan pepper (*Zanthoxylum bungeanum*) and sanshoamide extract on heterocyclic amine formation in grilled ground beef patties. *Food Chem.* **2018**, *239*, 111–118. [CrossRef]
17. Lu, F.; Kuhnle, G.K.; Cheng, Q. The effect of common spices and meat type on the formation of heterocyclic amines and polycyclic aromatic hydrocarbons in deep-fried meatballs. *Food Cont.* **2018**, *92*, 399–411. [CrossRef]
18. Wang, C.; Xie, Y.; Wang, H.; Bai, Y.; Dai, C.; Li, C.; Xu, X.; Zhou, G. Phenolic compounds in beer inhibit formation of polycyclic aromatic hydrocarbons from charcoal-grilled chicken wings. *Food Chem.* **2019**, *294*, 578–586. [CrossRef] [PubMed]
19. Viegas, O.; Yebra-Pimentel, I.; Martinez-Carballo, E.; Simal-Gandara, J.; Ferreira, I.M. Effect of beer marinades on formation of polycyclic aromatic hydrocarbons in charcoal-grilled pork. *J. Agric. Food Chem.* **2014**, *62*, 2638–2643. [CrossRef]

20. Cheng, Y.; Yu, Y.; Wang, C.; Zhu, Z.; Huang, M. Inhibitory effect of sugarcane (*Saccharum officinarum* L.) molasses extract on the formation of heterocyclic amines in deep-fried chicken wings. *Food Cont.* **2021**, *119*, 107490. [CrossRef]
21. Wang, C.; Xie, Y.; Wang, H.; Bai, Y.; Dai, C.; Li, C.; Xu, X.; Zhou, G. The influence of natural antioxidants on polycyclic aromatic hydrocarbon formation in charcoal-grilled chicken wings. *Food Cont.* **2019**, *98*, 34–41. [CrossRef]
22. Li, M.; Lin, S.; Wang, R.; Gao, D.; Bao, Z.; Chen, D.; Tang, Y.; Sun, N.; Zhang, S. Inhibitory effect and mechanism of various fruit extracts on the formation of heterocyclic aromatic amines and flavor changes in roast large yellow croaker (*Pseudosciaena crocea*). *Food Cont.* **2022**, *131*, 108410. [CrossRef]
23. Lai, Y.W.; Lee, Y.T.; Cao, H.; Zhang, H.L.; Chen, B.H. Extraction of heterocyclic amines and polycyclic aromatic hydrocarbons from pork jerky and the effect of flavoring on formation and inhibition. *Food Chem.* **2022**, *402*, 134291. [CrossRef] [PubMed]
24. Chen, Y.; Chien, J.; Inbaraj, B.S.; Chen, B.H. Formation and inhibition of cholesterol oxidation products during marinating of pig feet. *J Agric. Food Chem.* **2012**, *60*, 173–179. [CrossRef] [PubMed]
25. TFDA, Taiwan Food and Drug Administration. *Method of Test for Free Amino Acids, Glucosamine and Taurine in Foods*; TFDA: Taipei, Taiwan, 2017.
26. Gibis, M.; Loeffler, M. Effect of creatine and glucose on formation of heterocyclic amines in grilled chicken breasts. *Foods* **2019**, *8*, 616. [CrossRef]
27. Serpen, A.; Gokmen, V.; Fogliano, V. Total antioxidant capacities of raw and cooked meats. *Meat Sci.* **2012**, *90*, 60–65. [CrossRef]
28. Wang, F.; Pu, C.; Zhou, P.; Wang, P.; Liang, D.; Wang, Q.; Hu, Y.; Li, B.; Hao, X. Cinnamaldehyde prevents endothelial dysfunction induced by high glucose by activating Nrf2. *Cell. Physiol. Biochem.* **2015**, *36*, 315–324. [CrossRef]
29. Lin, Y.H.; Wang, C.C.; Lin, Y.H.; Chen, B.H. Preparation of catechin nanoemulsion from oolong tea leaf waste and its inhibition of prostate cancer cells DU-145 and tumors in mice. *Molecules* **2021**, *26*, 3260. [CrossRef]
30. Bueno, M.; Resconi, V.C.; Campo, M.M.; Ferreira, V.; Escudero, A. Development of a robust HS-SPME-GC-MS method for the analysis of solid food samples. Analysis of volatile compounds in fresh raw beef of differing lipid oxidation degrees. *Food Chem.* **2019**, *281*, 49–56. [CrossRef]
31. SAS, Institute Inc. *SAS®9.4 Output Delivery System: User's Guide*, 5th ed.; SAS Institute Inc.: Cary, NC, USA, 2019.
32. Pfau, W.; Skog, K. Exposure to β-carbolines norharman and harman. *J. Chromatogr. B* **2004**, *802*, 115–126. [CrossRef]
33. Pan, H.; Wang, Z.; Guo, H.; Na, N.; Chen, L.; Zhang, D. Potential sources of β-carbolines harman and norharman in braised sauce meat. *Sci. Agric. Sin.* **2013**, *46*, 3003–3009.
34. Gibis, M. Heterocyclic aromatic amines in cooked meat products: Causes, formation, occurrence, and risk assessment. *Comp. Rev. Food Sci. Food Saf.* **2016**, *15*, 269–302. [CrossRef]
35. Nagao, M.; Yahagi, T.; Sugimura, T. Differences in effects of norharman with various classes of chemical mutagens and amounts of S-9. *Biochem. Biophys. Res. Commun.* **1978**, *83*, 373–378. [CrossRef]
36. Bertram, H.C.; Aaslyng, M.D. Pelvic suspension and fast post-mortem chilling: Effects on technological and sensory quality of pork–A combined NMR and sensory study. *Meat Sci.* **2007**, *76*, 524–535. [CrossRef]
37. Gibis, M.; Thellmann, K.; Thongkaew, C.; Weiss, J. Interaction of polyphenols and multilayered liposomal-encapsulated grape seed extract with native and heat-treated proteins. *Food Hydrocoll.* **2014**, *41*, 119–131. [CrossRef]
38. Dundar, A.; Sarıçoban, C.; Yılmaz, M.T. Response surface optimization of effects of some processing variables on carcinogenic/mutagenic heterocyclic aromatic amine (HAA) content in cooked patties. *Meat Sci.* **2012**, *91*, 325–333. [CrossRef]
39. Unal, K.; Karakaya, M.; Oz, F. The effects of different spices and fat types on the formation of heterocyclic aromatic amines in barbecued sucuk. *J. Sci. Food Agric.* **2018**, *98*, 719–725. [CrossRef]
40. Friedman, M. Chemistry, antimicrobial mechanisms, and antibiotic activities of cinnamaldehyde against pathogenic bacteria in animal feeds and human foods. *J. Agric. Food Chem.* **2017**, *65*, 10406–10423. [CrossRef] [PubMed]
41. Sepahpour, S.; Selamat, J.; Khatib, A.; Manap, M.Y.A.; Abdull Razis, A.F.; Hajeb, P. Inhibitory effect of mixture herbs/spices on formation of heterocyclic amines and mutagenic activity of grilled beef. *Food Addit. Contam. A* **2018**, *35*, 1911–1927. [CrossRef]
42. Teng, H.; Chen, Y.; Lin, X.; Lv, Q.; Chai, T.T.; Wong, F.C.; Chen, L.; Xiao, J. Inhibitory effect of the extract from *Sonchus olearleu* on the formation of carcinogenic heterocyclic aromatic amines during the pork cooking. *Food Chem. Toxicol.* **2019**, *129*, 138–143. [CrossRef] [PubMed]
43. Quelhas, I.; Petisca, C.; Viegas, O.; Melo, A.; Pinho, O.; Ferreira, I.M.P.L.V.O. Effect of green tea marinades on the formation of heterocyclic aromatic amines and sensory quality of pan-fried beef. *Food Chem.* **2010**, *122*, 98–104. [CrossRef]
44. Shiraishi, Y.; Shirotori, T.; Takabatake, E. Determination of polycyclic aromatic hydrocarbons in foods 2) 3, 4-benzopyrene in Japanese daily foods. *Food Hyg. Saf. Sci. (Shokuhin Eiseigaku Zasshi)* **1973**, *14*, 173–178. [CrossRef]
45. Britt, P.F.; Buchanan, A.C.; Owens, C.V.; Todd Skeen, J. Does glucose enhance the formation of nitrogen containing polycyclic aromatic compounds and polycyclic aromatic hydrocarbons in the pyrolysis of proline? *Fuel* **2004**, *83*, 1417–1432. [CrossRef]
46. Nie, W.; Cai, K.Z.; Li, Y.Z.; Zhang, S.; Wang, Y.; Guo, J.; Chen, C.G.; Xu, B.C. Small molecular weight aldose (d-Glucose) and basic amino acids (l-Lysine, l-Arginine) increase the occurrence of PAHs in grilled pork sausages. *Molecules* **2018**, *23*, 3377. [CrossRef] [PubMed]
47. Capuano, E.; Fogliano, V. Acrylamide and 5-hydroxymethylfurfural (HMF): A review on metabolism, toxicity, occurrence in food and mitigation strategies. *LWT-Food Sci. Technol.* **2011**, *44*, 793–810. [CrossRef]
48. Zhou, H.; Wu, C.; Onwudili, J.A.; Meng, A.; Zhang, Y.; Williams, P.T. Polycyclic aromatic hydrocarbons (PAH) formation from the pyrolysis of different municipal solid waste fractions. *Waste Manag.* **2015**, *36*, 136–146. [CrossRef] [PubMed]

49. Lund, M.N.; Ray, C.A. Control of Maillard reactions in foods: Strategies and chemical mechanisms. *J. Agric. Food Chem.* **2017**, *65*, 4537–4552. [CrossRef]
50. Cordeiro, T.; Viegas, O.; Silva, M.; Martins, Z.E.; Fernandes, I.; Ferreira, I.; Pinho, O.; Mateus, N.; Calhau, C. Inhibitory effect of vinegars on the formation of polycyclic aromatic hydrocarbons in charcoal-grilled pork. *Meat Sci.* **2020**, *167*, 108083. [CrossRef]
51. Sinaga, K.; Legowo, A.; Supriatna, E.; Pramono, Y. Reduction of benzo(a)pyrene in charcoal grilled duck meat by marinating with andaliman (Zanthoxylum acanthopodium, DC) fruit juice. *J. Indon. Trop. Anim. Agric.* **2016**, *41*, 204–208. [CrossRef]
52. Friedman, M.; Kozukue, N.; Harden, L.A. Cinnamaldehyde content in foods determined by gas chromatography−mass spectrometry. *J. Agric. Food Chem.* **2000**, *48*, 5702–5709. [CrossRef]
53. Kim, Y.; Goodner, K.L.; Park, J.D.; Choi, J.; Talcott, S.T. Changes in antioxidant phytochemicals and volatile composition of *Camellia sinensis* by oxidation during tea fermentation. *Food Chem.* **2011**, *129*, 1331–1342. [CrossRef]
54. Janoszka, B. HPLC-fluorescence analysis of polycyclic aromatic hydrocarbons (PAHs) in pork meat and its gravy fried without additives and in the presence of onion and garlic. *Food Chem.* **2011**, *126*, 1344–1353. [CrossRef]
55. Sampaio, G.R.; Guizellini, G.M.; da Silva, S.M.; de Almeida, A.P.; Pinaffi-Langley, A.C.C.; Rogero, M.M.; de Camargo, A.C.; Torres, E.A. Polycyclic aromatic hydrocarbons in foods: Biological effects, legislation, occurrence, analytical methods, and strategies to reduce their formation. *Int. J. Mol. Sci.* **2021**, *22*, 6010. [CrossRef]
56. Wang, C.; Xie, Y.; Qi, J.; Yu, Y.; Bai, Y.; Dai, C.; Li, C.; Xu, X.; Zhou, G. Effect of tea marinades on the formation of polycyclic aromatic hydrocarbons in charcoal-grilled chicken wings. *Food Cont.* **2018**, *93*, 325–333. [CrossRef]
57. Cao, J.; Yang, L.; Ye, B.; Chai, Y.; Liu, L. Effect of apple polyphenol and three antioxidants on the formation of polycyclic aromatic hydrocarbon in barbecued pork. *Polycycl. Aromat. Compd.* 2022; *in press*. [CrossRef]
58. Wang, H.; Wang, C.; Li, C.; Xu, X.; Zhou, G. Effects of phenolic acid marinades on the formation of polycyclic aromatic hydrocarbons in charcoal-grilled chicken wings. *J. Food Prot.* **2019**, *82*, 684–690. [CrossRef] [PubMed]
59. Bhushan, B.; Bhat, R.; Sharma, A. Status of free radicals in indian monsooned coffee beans γ-irradiated for disinfestation. *J. Agric Food Chem.* **2003**, *51*, 4960–4964. [CrossRef]

foods

MDPI

Article

Changes in Triacylglycerols Content and Quality Control Implications of Coix Seeds during Processing and Storage

Weiwei Tang [1,†], Jiancheng Wang [1,†], Wei Li [1], Chaojun Zhang [2], Ping Li [1,*] and Jun Chen [3,*]

[1] State Key Laboratory of Natural Medicines, China Pharmaceutical University, Nanjing 210009, China
[2] Zhejiang Kanglaite Pharmaceutical Co., Ltd., Hangzhou 310018, China
[3] Department of Pharmacognosy, School of Traditional Chinese Pharmacy, China Pharmaceutical University, Nanjing 211198, China
* Correspondence: liping@cpu.edu.cn (P.L.); 1020060894@cpu.edu.cn (J.C.)
† These authors contributed equally to this work.

Abstract: Coix seed is a kind of widespread cereal, and it is used as a folk medicine in China. The present work focuses on the analysis of changes in triacylglycerols (TAGs) content and mycotoxins of coix seed during the processing and storage period for its quality control. Using ultra-high-performance liquid chromatography coupled with quadrupole time-of-flight tandem mass spectrometry (UHPLC-Q-TOF-MS) and high-performance liquid chromatography coupled with evaporative light-scattering detector (HPLC-ELSD) techniques, 42 lipid components in coix seeds were identified, and seven molecular species of TAG in coix seeds from different localities in China were measured and compared, respectively. A correlation analysis between the morphological features and TAGs contents revealed the integrity instead of the particle size of the seed, displaying a highly positive correlation with its quality. The higher contents of TAGs in hulled coix seed than in polished coix seed proposed an alternative processing way. During storage, the changes in TAGs contents of seeds indicated that the storage period should be less than 3 months, and the intact seeds could maintain lipid stability better than the powder. Furthermore, the air humidity and temperature should be controlled during coix seed storage to prevent the production of mycotoxins. These results provide significant insight into the effective control of coix seed quality during processing and storage.

Keywords: coix seed; triacylglycerols; quality control; processing; storage

1. Introduction

Coix seed is the seed of *Coix lachryma-jobi* L. var. ma-yuen Stapf and also known as adlay, Job's tears, Coicis Semen, Yiyiren. Coix seed is an important minor cereal with high nutritional and medicinal value [1,2]. It is widely used as food [3], an ingredient for brewing [4], nutraceuticals supplements [5] and for some adjuvant oncology therapies [6]. Lipids, especially triglycerides, are one of the main constituents of coix seed. Previous studies exhibited the lipid contents in polished coix seed ranging from 5.1 to 9.4% [7,8], which is significantly higher than in most common cereals. The highly abundant lipid contents in coix seed inspired the research to explore it for the quality control evaluation of coix seed. A previous study indicated that lipids could potentially be used as a key index to distinguish the geographical origin of coix seeds [9]. Triolein, one of the main triacylglycerols in coix seed, has been investigated and applied as a marker compound for quality control [7]. In addition, the total contents of triacylglycerols (TAGs) are widely used for quality control of coix seeds in most regulations. The US pharmacopeia required that a summation of seven TAGs contents should be more than 3.5% in coix seeds. Both Chinese and European pharmacopeia defined that the content of triolein should not be less than 0.5%. However, the current research mainly focuses on the highly abundant lipids, such as triolein, since there are barely any comprehensive reports related to global mapping lipids in coix seed. Therefore, an in-depth analysis of its lipid profiling should be characterized for a deeper understanding and more use in quality control of coix seed.

The production process of coix seeds generally involves drying, hulling, polishing and storage. During the harvest and processing procedures of coix seeds, the morphological features are first assessed. As described in the US pharmacopeia [10], the macroscopic characteristics, including the shape, color, size and texture, are primarily used for quality evaluation. The appearance of coix seed is one of the most important characteristics for its classification and grading concerning the commercial quality and organoleptic properties [11]. Global evaluation of the morphological features either in a metric or subjective manner is an important quality indicator throughout the whole commercial utilization chain, from the production to the storage, the marketing, and finally, down to the consumer. However, the integrity of coix seed has not been considered for its quality evaluation. During the harvest and processing of coix seeds, both manual and mechanical manipulation may cause the breakage of the seeds. It was supposed that the damage to each grain affects the loss of quality because the oxidation of the ingredients may begin.

Although both morphological features and TAGs content have been used for quality evaluation, the relationships between them have not been clarified yet. Moreover, other features of the seeds, such as the shell, bran and core, may exert influence on the product quality. During the processing, both the shell and the bran are removed during rice milling to produce the coix seed. All cereals tend to have lipids concentrated in the bran, including rice [12]. Thus, it is necessary to elucidate the role of bran in the quality control of coix seeds during processing. Except for processing, the harvested or processed coix seeds are normally stored to achieve a continued supply to the market. During storage, the physicochemical properties of coix seeds may change under the influence of internal and external factors. It is obvious that there are changes of TAGs content during storage, thus causing a decline in quality [13]. Lipids are known to be highly perishable because they are susceptible to oxidation and hydrolysis, which lead to the decomposition of ingredients and lost properties [14]. Therefore, the stability of the coix seeds under different storage times should be evaluated. Moreover, cereal foods could be easily contaminated by mycotoxins under storage due to the growth of some fungi [15]. Aflatoxins B1, B2, G1, G2, the most common toxics in coix seed, are produced by fungi belonging to *Aspergillus flavus*. Aflatoxins are known as group 1 carcinogenic compounds, which could be produced quickly under warm and moist conditions [16]. Zearalenone is another toxic compound commonly found in coix seed, which could cause severe adverse effects, such as hepatotoxicity, neurotoxicity and estrogenic effects [17]. Aflatoxins and zearalenone are considered stable compounds, which cannot be broken down through heating, cooking or brewing operations [18,19]. Thus, it is necessary to monitor the content of aflatoxins and zearalenone in coix seed during storage for safety.

In this study, the comprehensive lipid profiling of coix seeds was analyzed by UHPLC-QTOF-MS techniques. High-performance liquid chromatography coupled with evaporative light-scattering detector (HPLC-ELSD) method was applied to quantify seven TAGs contents to explore the factors that affect the quality of coix seeds during processing and storage. An immunoaffinity assay coupled with LC-MS/MS detection was used for monitoring the aflatoxins and zearalenone in the coix seed, thus providing more reliable suggestions for quality control.

2. Materials and Methods

2.1. Materials and Reagents

A total of 83 batches of coix seeds for the present study were collected in various production regions in China, including Guizhou, Yunnan, Zhejiang, Fujian and Jiangxi Provinces (Table S1). Trilinolein (LLL), 1,2-dilinoleoyl-3-palmitin (LLP), 1,2-dilinoleoyl-3-olein (LLO), 1-palmitoyl-2-oleoyl-3-linolein (POL), 1,2-dioleoyl-3-linolein (OOL), 1,2-dioleoyl-3-palmitin (OOP) and triolein (OOO) were used as reference compounds. OOO was obtained from Shanghai Aladdin Bio-Chem Technology Co., LTD

(Shanghai, China), and other compounds were obtained from Tianjin NNU Biological Technology Co. Ltd. (Tianjin, China). The purity of each reference compound determined by HPLC-ELSD was above 98%. A Milli-Q water purification system from Millipore (Bedford, MA, USA) was utilized for purifying deionized water. Acetonitrile, methanol (MS-grade) and dichloromethane (HPLC grade) were obtained from Merck (Darmstadt, Germany). Formic acid (HPLC-grade) and ammonium formate (HPLC-grade) were obtained from ROE scientific Inc. (Newark, DE, USA) and Sigma-Aldrich, respectively. Standard solution of zearalenone (25 μg/mL) and mixed standard solutions of aflatoxins containing G2 (0.3 μg/mL), B2 (0.3 μg/mL), G1 (1.0 μg/mL) and B1 (1.0 μg/mL) were purchased from Supelco (Bellefonte, PA, USA). The immunoaffinity column kits containing antibodies to both aflatoxins and zearalenone were obtained from Beijing Zhong Ke Hui Ren Technology Co., Ltd. (Beijing, China).

2.2. Sample Preparation

The TAG reference was dissolved in the dichloromethane solvent first to prepare the stock solution. Then, the stock solution was dissolved in methanol to obtain a final concentration of 1 mg/mL. The stock solutions were diluted with methanol to prepare the desired working solutions of TAGs standards. For sample solution preparation, finely powdered coix seeds were prepared using the pulverizer grinding machine, screened through a No.4 sieve (250 μm). An aliquot of 500 mg powder was accurately weighed and then mixed with 50 mL methanol for ultrasonic extraction (200 W, 40 kHz) for 0.5 h at room temperature. After 30 min extraction, the mixture was cooled to room temperature, and methanol was added for lost weight and then filtered with a 0.45 μm filter. All solutions were stored at 4 °C until use. Samples for aflatoxins and zearalenone detection were prepared as follows: 3 g and 5 g of powdered coix seeds were weighed and dissolved in 15 mL of 70% methanol and 90% acetonitrile solutions, respectively. Both solutions were added to 0.6 g NaCl and then vortexed for 1 min. After centrifugation for 5 min under 2500 rpm/min, 5 mL of the supernatant was transferred and diluted to 25 mL with deionized water. Then, the samples were filtered with a 0.45 μm filter and performed on the immunoaffinity column, according to the kits' instructions.

2.3. UHPLC-QTOF-MS Conditions

An Agilent 1290 series LC system coupled with an Agilent 6530 QTOF tandem mass spectrometer (Agilent Technologies, Santa Clara, CA, USA) was implemented for qualitative analysis. The LC system was equipped with an online degasser, a binary pump, an auto-sampler and a thermostatically controlled column compartment. The column oven temperature was set at 25 °C throughout the analytical process. Chromatographic separation was achieved with an Agilent Infinity Lab Poroshell120 EC-C8 column (3.0 mm × 100 mm, 2.7 μm). The mobile phases were water containing 5 mM ammonium formate (A) and methanol containing 5 mM ammonium formate (B). The gradient elution program was set to 15~70% B at 0~20 min, 70~80% B at 20~25 min, 80~100% B at 25~45 min and 100% B at 45~65 min. The flow rate was set at 0.3 mL/min, and the injection volume was 2 μL.

For MS and MS/MS results acquisition, the drying gas (N_2) flow rate was set as 10 L/min at 350 °C, while the sheath gas flow rate was set as 11 L/min at 350 °C. Other parameters were set as follows: nebulizer, 35 psig; capillary voltages, 3000 V; fragmentor, 135 V; collision energy was set at 15 eV, 30 eV and 50 eV. The data were acquired under both positive and negative ion modes. The mass ranges were set at *m/z* 100–2000.

2.4. HPLC-ELSD Conditions and Single Standard to Determine Multi-Components (SSDMC) Method for Quantitation of TAGs

Quantitative analysis was carried out on a Shimadzu LC-20AT Series HPLC system equipped with PL-ELS 1000 evaporative light-scattering detector (Shimadzu, Kyoto, Japan), a quaternary solvent delivery system and a column temperature controller. Chromatographic separation was conducted with the Agilent Infinity Lab Poroshell120 EC-C8 column (2.7 μm, 3.0 mm × 100 mm). The mobile phase was methanol with isocratic elution, and the flow rate was 0.3 mL/min. The column oven temperature was set at 25 °C throughout the analytical process, and the injection volume was 5 μL. The ELSD detector parameters were set as follows: nebulizing gas pressure, 3.4 bar; heating tube temperature, 40 °C; gain value, 11; filter value, 10 s. Shimadzu LC solution software (ver. 1.21SPI) was used for data collection.

For the quantitation of TAGs, the SSDMC method was developed according to a previous study [7]. The OOO was selected as the internal single standard, and the relative conversion factor (RCF, Fi) was displayed in Table S2. Briefly, the RCF was calculated based on the correlation between the chromatographic response and the concentration of the analyte. Due to the similar structures of these seven compounds, the RCFs of these six analytes compared to OOO were finally confirmed as 1.0.

2.5. UHPLC-QQQ-MS Conditions for the Analysis of Aflatoxins and Zearalenone

A Shimadzu LC-30A UHPLC combined with a Shimadzu 8050 QQQ MS (Kyoto, Japan) were used for detection. An Agilent ZORBAX SB-C18 (4.6 × 50 mm, 1.8 μm) column was used for chromatographic separation. The mobile phases were 0.1% formic acid water solution containing 10 mM ammonium acetate (A), acetonitrile (B), water (C) and methanol (D). The liquid chromatographical conditions were set as follows: for aflatoxins, mobile phases A and B were used, and the elution program was set as 0~4.5 min, 35~85% B; 4.5~6 min, 85~100% B; 6~6.5 min, 100~35% B; 6.5~10 min, 35% B at a column temperature of 25 °C. For zearalenone, mobile phases C and D were used, and the elution program was set as 0~5 min, 25~75% D; 5~6 min, 70% D; 6~9 min, 70~25% D at a column temperature of 40 °C. The flow rate and injection volume were set as 0.3 mL/min and 1.0 μL, respectively, for both aflatoxins and zearalenone.

The QQQ-MS were equipped with an ESI source and operated with the following parameters: interface voltage, 5500 V; nebulizer gas (N2) flow rate, 2.0 L/min; drying gas flow rate, 10.0 L/min. The interface temperatures were set as 550 °C and 350 °C for aflatoxins and zearalenone, respectively. The multiple reaction monitoring (MRM) mode was used for quantification, and the parameters are displayed in Table 1.

Table 1. MS/MS parameters for aflatoxins and zearalenone detection.

Analyte	Parent Ion	Product Ion	CE (eV)
Aflatoxin G2	[M + H] + 331.1	**313.1**	33
		245.1	40
Aflatoxin G1	[M + H] + 329.1	**243.1**	35
		311.1	30
Aflatoxin B2	[M + H] + 315.1	**259.1**	35
		287.1	40
Aflatoxin B1	[M + H] + 313.1	241	50
		285.1	40
Zearalenone	[M − H] − 317.1	**174.9**	25

bold ions were used for quantitation analysis.

2.6. Data Analysis

Agilent MassHunter Qualitative Analysis Software (Version B.07.00) was used for qualitative analysis. Graphpad prism 8.0 and Microsoft Excel 2019 were utilized for data analysis. Shimadzu LC solution software (ver. 1.21SPI) was used for quantitation. For LC-MS/MS quantitation, the software of Shimadzu Labsolutions LCMS Version 5.65 (Kyoto, Japan) was used.

3. Results and Discussion

3.1. Overall Lipid Profiling of Coix Seeds

Lipids, especially TAGs and diacylglycerols (DAGs), are considered some of the main constituents of coix seeds [20]. In this study, the comprehensive lipid features of coix seeds, including TAGs, DAGs, phosphatidylcholines (PCs) and free fatty acids (FAs), were profiled using the UHPLC-QTOF-MS method under the positive and negative ion modes. Based on the analysis from the MS total ion chromatogram, a total of 32 peaks and 10 peaks were detected under positive and negative ion modes, respectively (shown in Figure 1A). Among these peaks, seven components were unambiguously authenticated as LLL (T3), LLP (T5), LLO (T6), POL (T8), OOL (T9), OOP (T11) and OOO (T12) by comparing their retention times, *m/z* values and fragment ions with those of the reference compounds. In addition to the above 7 compounds, 16 other compounds (Table S3) were tentatively identified as TAGs by comparison with the MS and MS/MS spectrometry features of the compounds from previous literature works [9,21,22]. Since the molecular structure of TAG contains three acyl chains, the ions of the fatty acyl group and the residues after the eliminated acyl chain were generated under MS/MS detection [23]. The majority of the fatty acyl in coix seeds is unsaturated, thus making palmitic-, stearic-, oleic- and linoleic acyl-triglycerides the main components of lipids [24]. Using LLP (T5) as an example, the typical $[M + NH_4]+$ adduct ion with *m/z* 872.7729 was observed in MS spectrometry owing to the non-polar nature of TAGs. Through the collision-induced dissociation (CID) procedure, the MS/MS pattern was acquired. The typical $[M + NH_4]+$ adduct ion produces abundant fatty-acyl-specific fragment ions (as combined loss of NH3 and fatty acyl), allowing the characterization of TAGs structures. As shown in Table S3, the *m/z* 263.2369 in the product ions list was characterized as the acyl chain of linoleic acid [C18H32O2-OH] fragmented from LLP. In addition, the MS/MS product ions at *m/z* 599.5045, *m/z* 575.5042, *m/z* 337.2751 and *m/z* 313.2714 are derived from the moiety after the loss of acyl chains corresponding to [M + H-palmitic acyl]+, [M + H-linoleic acyl]+, [M + H-palmitic acyl-linoleic acyl + H2O]+ and [M + H-2linoleic acyl + H2O]+, respectively (shown in Figure S1). DAGs were also detected in coix seeds in this study. As shown in Table S4, the retention times of DAGs (marked as D1 to D7) were shorter than TAGs due to their higher polarity. For MS spectrometry, the quasi-molecular ion peaks of DAGs were different from those of TAGs, with the adduct ion of [M + Na]+ easily generated (shown in Table S4). Due to the similar molecular structures, it is obvious that the fragment pathway of TAGs and DAGs should be alike. From TAGs, the MS/MS product ions of the acyl chain and the moiety after the loss of the acyl chain were fragmented easily and could be applied to the identification of DAGs (Figure S1 showed the typical fragment pathway of DAG:LP). Moreover, seven molecular species of PC were also characterized in this study. As shown in Table S5, the retention times of PCs were less than DAGs (marked as P1 to P7) due to the higher polarity of the phosphorylcholine group. The quasi-molecular ion peaks of PCs were displayed as [M + H]+ adduct ions (Table S5). For the fragmentation of PCs, the product ions of the acyl chain or the moiety after the loss of the acyl chain could be characterized and assigned for the identification of the composition of PCs. More importantly, the presence of m/z 184, 86 and 60 was recognized as the MS/MS product ions of the phosphatidylcholine group (as shown in Figure S1), thus providing distinct diagnostic ions for the identification of PCs.

As for the negative mode detection, linoleic acid (F6), palmitic acid (F7), oleic acid (F9) and stearic acid (F10) were elucidated by comparing the MS data with those of the reference compounds (Figure 1B and Table S6). Another six fatty acids were also characterized according to the previous data [25]. Although gas chromatography (GC) was the most commonly used technique for fatty acids analysis, the HPLC-QTOF/MS technique was considered as an alternative assay, as previously reported [26]. Among the peaks assigned as fatty acids in this study, F1, F2 and F3 were detected with the same *m/z* 295 but at different retention times (25.4, 25.6 and 26.1 min, respectively). Compared with the previous report, these three peaks were tentatively proposed as the hydroxy octadecadienoic acid and its isomers with different locations of the hydroxy group or double bonds. Overall, a total of

42 lipids were identified in this study, thus giving a better understanding of the chemical profile of coix seeds.

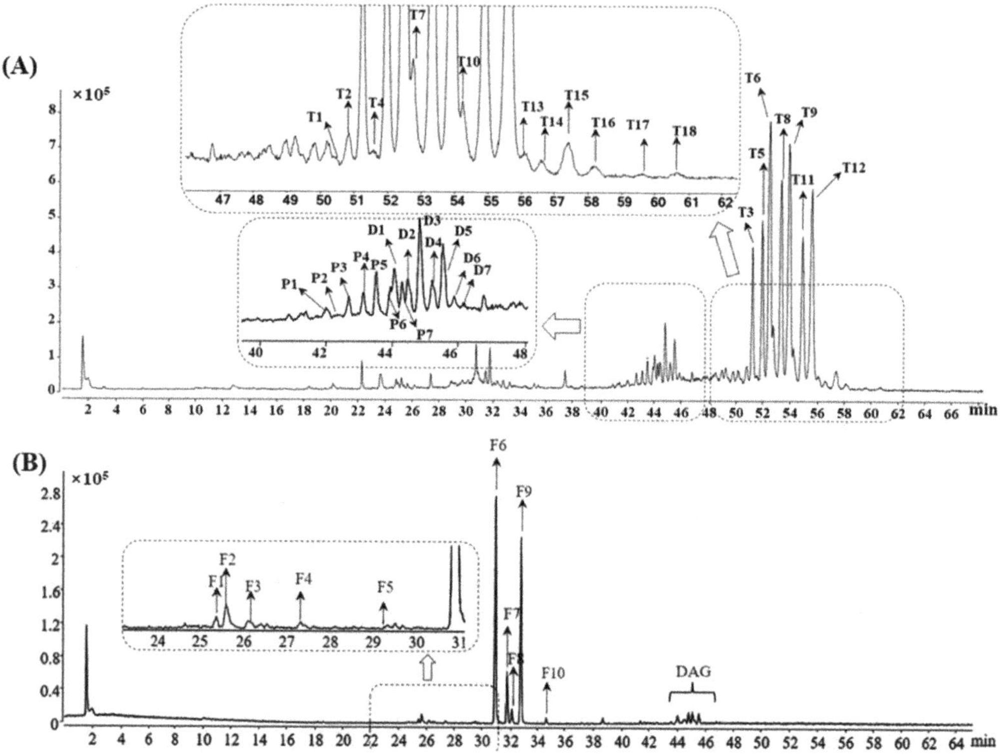

Figure 1. Total ion chromatogram of coix seed sample under positive (A) and negative (B) detection mode by UHPLC-QTOF-MS.

3.2. Quantitation Analysis of Seven TAGs in Coix Seeds

Lipids, among others, are the main active constituents, and according to the above chemical profile, TAGs are the primary constituents of coix seeds. According to previous studies [7,9], seven TAGs, namely LLL (T3), LLP (T5), LLO (T6), POL (T8), OOL (T9), OOP (T11) and OOO (T12), were selected as the mark components of coix seeds. Similarly, due to the high content of OOO in coix seeds and its commercial standard products available, it has been used as the compound mark of coix seeds in Chinese and British pharmacopeias. Moreover, the total content of these seven TAGs is used for quality control in US pharmacopeia. In this study, the SSDMC method using OOO as a reference compound was applied for the quantification of these seven TAGs by using the HPLC-ELSD technique.

A total of 77 batches of coix seed from different geographic areas, namely Fujian, Yunnan, Guizhou, Zhejiang and Jiangxi provinces, were investigated using the SSDMC method. The contents (%) for the percentage of TAGs among the original weight of coix seed were calculated. As shown in Figure 2 and Table S7, OOO, OOL and LLO were displayed as high content, corresponding to the range of 0.32–1.47%, 0.41–1.67% and 0.38–1.47%. The average amount of LLL, LLP, LLO, POL, OOL, OOP, OOO as total TAG was calculated as 5.97%, 8.36%, 21.24%, 14.2%, 21.6%, 10.72%, 17.9%, respectively. A previous study [7] revealed that oleic acid and linolenic acid were the most abundant FAs in coix seed. The TAGs with oleic acid and linolenic acid components were also in high concentration. As one of the most abundant compounds, OOO was marked as a key component required at more than 0.5% in coix seed by the 2020 edition Chinese pharmacopoeia. The results indicated that 94.8% of the 77 batches of coix seed achieved this quality standard. In addition, the total contents of these seven TAGs varied from 2.0 to 7.5% in coix seeds. According to the US pharmacopeia, the total content of TAGs should be more than 3.5%. According to this stated specification, 88.3% of the 77 batches of coix seed qualified in this study.

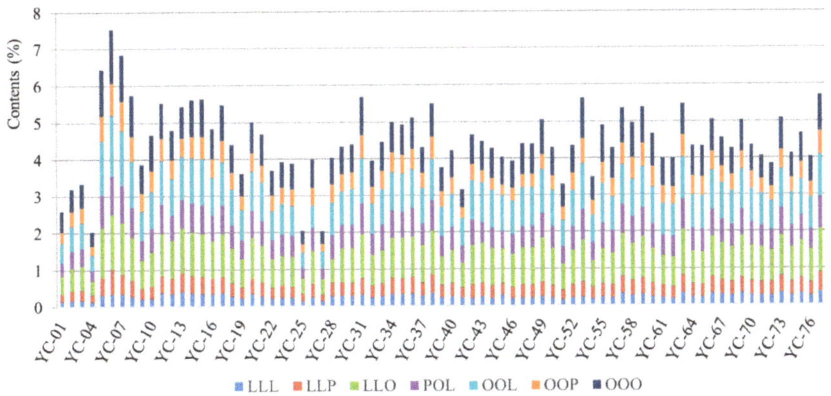

Figure 2. The seven molecular species of TAGs from 77 batches of coix seed (from YC-01 to YC-77). Each color represents different analytes.

3.3. *Evaluation of the Correlations between the Morphological Features of Coix Seed and Its Quality*

The traditional classification of the coix seed depended on its morphological characteristics. During harvest and manufacturing, the morphological features of coix seeds are primarily evaluated for their quality. Those features, such as larger width and full shape of the coix seed, were considered as good quality. However, whether it is reasonable to evaluate the quality of coix seed using these morphological features has not been elucidated scientifically. Therefore, the relationship between the content of these seven TAGs and the morphological characteristics was explored in this section. Since the coix seed with broken features may be selected or used during processing, the integrity of the coix seeds has also been considered. A total of nine batches of coix seed as described above were used for the determination of width, 1000-grain weight, integrity and the content of TAGs. All data were used for the correlation analysis with SPSS Statistics 22 Software (IBM, Chicago, IL, USA). As shown in Table 2, 1000-grain weight displayed high correlations with width (correlation coefficient $r^2 = 0.439$), and integrity showed high correlations with TAGs content (correlation coefficient $r^2 = 0.558$). However, neither the width of the coix seed nor the 1000-grain weight displayed poor correlation with the content of TAGs (correlation coefficient $r^2 = 0.006$ and 0.129, respectively). The results indicated that the width of the coix seed does not affect its quality. Significantly, the integrity of the coix seed may affect the contents of TAGs. To further validate this, samples from the same batches were used. Samples from Guizhou province (YC-42, 45, 52), Yunnan province (YC-56, 60, 65) and Fujian province (YC-67, 70, 74) were categorized into three groups: the broken grain, the small grain with a width of less than 4.5 mm and the large grain with a width over 5.5 mm (Figure 3A). These three groups were divided through suitable sieving and vernier caliper measurement. Using SSDMC for TAGs determination, the results (as shown in Figure 3B) of the total TAGs in broken grains were lower and significantly different to the large and small grain ($p < 0.001$). The result displayed a total TAGs content of over 6.0% for the small and large samples but below 5.0% or even 3.0% for those broken samples. Moreover, the content of each TAGs (namely OOO, OOP, OOL, LLO, LLP, LLL) in the broken, large and small grain also showed a similar trend. The contents of these seven TAGs for small and large seeds were 1.5- to 2.0-fold higher than those of the broken samples. However, there was no significant difference in the content of total TAGs or each TAGs between the large grain and small grain ($p = 0.792$). Previous studies had indicated that the size of the coix seed was considered primarily as a quality control parameter for sample collection and processing. Using traditional morphological features for evaluation, those broken samples but ones with a suitable size would have been processed. Nevertheless, as the results depicted, the broken grains may cause a decrease in the TAGs content, thus providing worse quality and poor yield of TAGs in the final product. Therefore, the integrity of the seed should be taken into consideration for better quality control, and the standardization process of the coix seed should be intensively evaluated to guarantee the quality consistency.

Table 2. Correlation analysis between 1000-grain weight, integrity, width, triglycerides.

Correlation	1000-Grain Weight	Integrity	Width	Triglycerides
1000-grain weight		0.104	0.439 **	0.129
integrity			−0.039	0.558 **
width				0.066
triglycerides				

** Significant difference ($p < 0.01$).

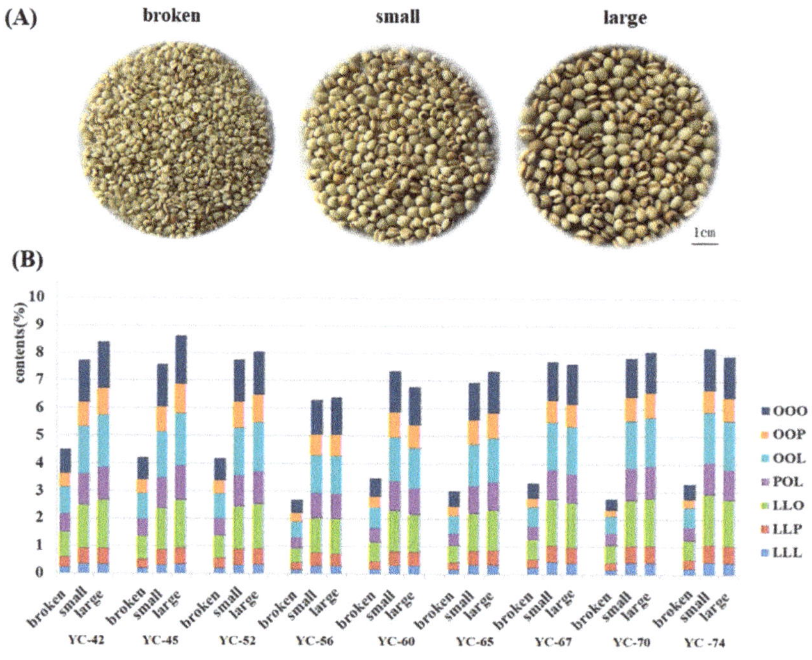

Figure 3. The optical graphs (**A**) and the content of TAGs (**B**) from the broken grain, small grain (width < 4.5 mm) and large grain (width > 5.5 mm).

The coix seed is a ripe caryopsis with a hierarchical structure of the shell, bran and kernel. During processing, the dried coix seed is freed from the shell and then polished to collect the kernel articles. To provide a reasonable basis for processing, the TAGs contents in the different parts of the coix seed were investigated in this section. Three batches of the coix seed (YC-78, YC-79 and YC-80) were acquired and processed with and without the shell. The samples were collected separately and divided into three groups (the hulled coix seed group, the polished coix seed group and the coix seed bran group) for the TAGs content determination. As shown in Figure 4, both the total and each TAGs in the coix seed bran exhibited no significant difference compared with the hulled coix seed. However, the content of TAGs was lower in the polished coix seed than in these two forms. As noted before, the bran usually contains the bulk of the lipids. During processing, the hulled coix seeds are polished to remove the bran. From the quantitation results, the bran appears valuable for obtaining the TAGs and could be preserved during processing. Previous studies have reported that the bran of coix seed exhibited many pharmacological effects, including anti-inflammatory [27–30], thus indicating the importance of the bran. It is therefore recommended that the bran be optional, but the integrity should be maintained during processing to make the quality more controllable.

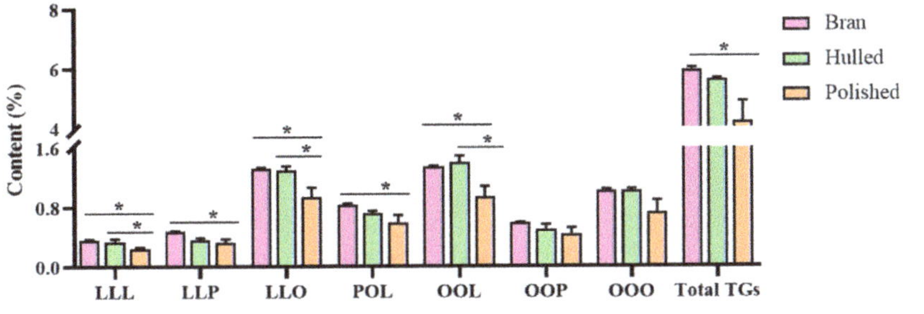

Figure 4. Contents of triglycerides from the hulled, polished coix seed and the bran, * $p < 0.05$.

3.4. Quality Control of Coix Seed during Storage

For large-scale production, a lot of coix seed samples are collected and stored until processing. The long-term stability research of the TAGs content is critical for the quality evaluation during the sample storage. A previous study [13] established changes of coix seed under storage. To evaluate the changes of TAGs content, three batch samples (YC-81, YC-82 and YC-83) were collected and divided into two groups (intact coix seed and coix seed powder) and then stored in a cool and dry place at room temperature. Each sample was then evaluated at three months interval for one year. As shown in Figure 5, both the total and each TAGs content showed little change in the first three months in the intact coix seed. As shown in Table S8, the decline rate of the total TAGs content was in the range from 2.03% to 8.06%. After six months, the decline rate was still below 15%. However, the situation was different in the coix seed powder, which exhibited a 20% decline rate in the first three months. The results in Figure 5 demonstrate that the contents of OOO and total TAGs were below 0.5% and 3.5% in the coix seed powder after 9 months of storage, respectively, while those of the intact coix seed were not. After 12 months of storage, the total TAGs in both the intact coix seed and the coix seed powder were below 50% of the original content. From the result, we suggest that the storage time of coix seed should be less than 3 months for better quality control. In addition, the intact coix seed was shown to be more stable than the coix seed powder, which implied that the bran played an important role in maintaining the integrity of the TAGs.

Similar to other edible grains, coix seed was easily contaminated by aflatoxins and zearalenone due to improper storage conditions. To analyze the trace amounts of aflatoxins and zearalenone in the coix seed, the immunoaffinity column was used for the extraction of analytes and then coupled with the LC-MS/MS technique for detection, according to previous studies [31,32]. The sensitivity and linearity for the detection of aflatoxin G2, G1, B2, B1 and zearalenone are displayed in Table S9. This method was applied to measure the content of aflatoxins and zearalenone in 24 batches of coix seed. As displayed in Table S10, all batches of coix seed were regarded as negative samples of aflatoxin G2 and G1, which were low in coix seed that exceeded the LOD of this study. As for aflatoxin B2 and B1, 21 and 20 batches were regarded as negative samples, respectively. For zearalenone, all samples were shown to be positively detected in the range of 2.31–114.58 μg/kg. According to the China national food safety standards (GB 5009.209-2016), the limit of zearalenone should be below 60 μg/kg in cereal and relevant products. Only one batch of coix seed exceeded this content. To further investigate the changes of aflatoxins and zearalenone under different storage conditions, temperatures ranging from 4 °C to 35 °C and different relative humidity ranges from 50% to 97% were adopted in this study. The storage time was set at four weeks. As displayed in Table 3, aflatoxins G2, G1 and B2 were negatively detected under different conditions. However, 0.61 μg/kg of aflatoxin B1 was detected when the condition changed to 25 °C with 97% humidity. As for zearalenone, it was obviously increased as the temperature and humidity increased. The results reveal that the content of zearalenone was easily affected by humidity, as it was increased over 60 μg/kg when the humidity increased up to 90%. These results suggested that the temperature and humidity should be controlled during storage.

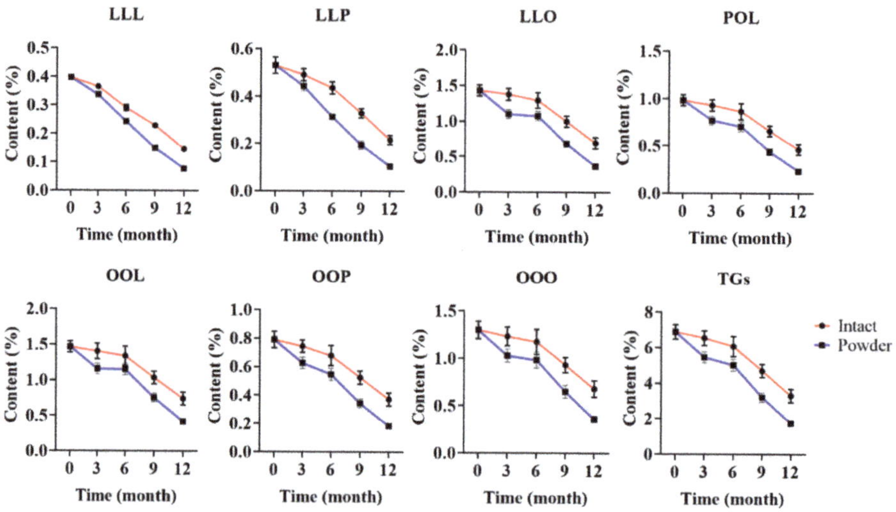

Figure 5. Changes of TAGs content under long-term storage.

Table 3. Changes of aflatoxins and zearalenone under different storage conditions.

Temperature/Relative Humidity (%)	4 °C/50	25 °C/50	35 °C/50	25 °C/97	25 °C/90	25 °C/81
Aflatoxin G2 (μg/kg)	ND	ND	ND	ND	ND	ND
AflatoxinG1 (μg/kg)	ND	ND	ND	ND	ND	ND
AflatoxinB2 (μg/kg)	ND	ND	ND	ND	ND	ND
AflatoxinB1 (μg/kg)	ND	ND	ND	0.61	ND	ND
Zearalenone (μg/kg)	2.61	4.93	30.25	91.71	54.62	81.92

4. Conclusions

In this work, the comprehensive lipid profiling of the coix seed was tentatively characterized by the UHPLC-Q-TOF-MS/MS technique. A total of 42 lipids, thus 18 triglycerides, 7 diglycerides, 7 phospholipids and 10 fatty acids, were identified, hence providing a better understanding of the chemical profiles of the coix seed. For quality control, seven components of TAGs were quantified in this study, especially during the processing and storage procedures of the coix seed. The information from the results indicated that integrity, instead of size, displayed a close correlation with the content of TAGs. The positive correlation between the integrity and content may be useful for assessing the quality during the collection and processing of the coix seed. The quantitative analysis of TAGs from different parts of the coix seed implied that bran was important and could be maintained during processing. The long-term stability research of the TAGs content proved that triglycerides

in the integral coix seed were more stable than those in powdered coix seeds. Similarly, a shorter storage time of less than 3 months does not greatly affect the product quality. Moreover, the temperature and humidity of storage conditions should be strictly controlled to prevent the production of toxic secondary metabolites of fungi. The results obtained from this study would help establish a scientific and rational quality control strategy for the coix seed, especially in the procedures of processing and storage.

Supplementary Materials: The following supporting information can be downloaded at: https://www.mdpi.com/article/10.3390/foods11162462/s1, Table S1: Geographical origins of 83 batches of coix seed; Table S2: The conversion factors of the seven analytes; Table S3: Triacylglycerols identified in coix seed by UHPLC-Q-TOF-MS; Table S4: Diacylglycerols identified in coix seed by UHPLC-Q-TOF-MS; Table S5: Phosphatidylcholines identified in coix seed by UHPLC-Q-TOF-MS; Table S6: Fatty acids identified in coix seed by UHPLC-Q-TOF-MS; Figure S1: Fragmentation pattern of lipid compounds; Table S7: The seven molecular species of TG contents from 77 batches of coix seed (from YC-01 to YC-77); Table S8: The rate of decline in the content of coix seed and coix seed powder; Table S9: The calibration curves, linear range, limit of detection (LOD) and limit of quantification (LOQ) of the aflatoxins and zearalenone; Table S10: Contents of aflatoxin and zearalenone in 24 batches of coix seed.

Author Contributions: Conceptualization, W.T., J.W., P.L. and J.C.; methodology, W.T., J.W., W.L. and C.Z.; formal analysis, W.T. and J.W.; investigation, W.T., J.W., W.L. and C.Z.; data curation, W.T., J.W., W.L., C.Z., P.L. and J.C.; writing—original draft preparation, W.T. and J.W.; writing—review and editing, W.T., J.W., P.L. and J.C.; visualization, W.T., J.W., P.L. and J.C.; supervision, P.L. and J.C.; project administration, P.L. and J.C. All authors have read and agreed to the published version of the manuscript.

Funding: This research was funded by the National Key R&D Program of China (No. 2018YFC1704500) and the National Chinese Medicine Standardization Project (ZYBZH-C-ZJ-61).

Institutional Review Board Statement: Not applicable.

Informed Consent Statement: Not applicable.

Data Availability Statement: All of the data is contained within the article and the Supplementary Materials.

Conflicts of Interest: The authors declare no conflict of interest.

References

1. Zhu, F. Coix: Chemical composition and health effects. *Trends Food Sci. Technol.* **2017**, *61*, 160–175. [CrossRef]
2. Chen, L.C.; Zhang, S.Y.; Zi, Y.; Zhao, H.M.; Wang, H.Y.; Zhang, Y. Functional coix seed protein hydrolysates as a novel agent with potential hepatoprotective effect. *Food Funct.* **2020**, *11*, 9495–9502. [CrossRef]
3. Chaisiricharoenkul, J.; Tongta, S.; Intarapichet, K.-O. Structure and chemical and physicochemical properties of Job's tear (Coix lacryma-jobi L.) kernels and flours. *Suranaree J. Sci. Technol.* **2011**, *18*, 109–122.
4. Wang, C.-Y.; Wu, S.-C.; Ng, C.-C.; Shyu, Y.-T. Effect of Lactobacillus-fermented adlay-based milk on lipid metabolism of hamsters fed cholesterol-enriched diet. *Food Res. Int.* **2010**, *43*, 819–824. [CrossRef]
5. Zhao, M.; Zhu, D.; Sun-Waterhouse, D.; Su, G.; Lin, L.; Wang, X.; Dong, Y. In vitro and in vivo studies on adlay-derived seed extracts: Phenolic profiles, antioxidant activities, serum uric acid suppression, and xanthine oxidase inhibitory effects. *J. Agric. Food Chem.* **2014**, *62*, 7771–7778. [CrossRef] [PubMed]
6. Schwartzberg, L.S.; Arena, F.P.; Bienvenu, B.J.; Kaplan, E.H.; Camacho, L.H.; Campos, L.T.; Waymack, J.P.; Tagliaferri, M.A.; Chen, M.M.; Li, D. A Randomized, Open-Label, Safety and Exploratory Efficacy Study of Kanglaite Injection (KLTi) plus Gemcitabine versus Gemcitabine in Patients with Advanced Pancreatic Cancer. *J. Cancer* **2017**, *8*, 1872–1883. [CrossRef]
7. Hou, J.J.; Guo, J.L.; Cao, C.M.; Yao, S.; Long, H.L.; Cai, L.Y.; Da, J.; Wu, W.Y.; Guo, D.A. Green Quantification Strategy Combined with Chemometric Analysis for Triglycerides in Seeds Used in Traditional Chinese Medicine. *Planta Med.* **2018**, *84*, 457–464. [CrossRef] [PubMed]
8. Devaraj, R.D.; Jeepipalli S.P., K.; Xu, B. Phytochemistry and health promoting effects of Job's tears (Coix lacryma-jobi)—A critical review. *Food Biosci.* **2020**, *34*, 1–13. [CrossRef]
9. Hou, J.J.; Cao, C.M.; Xu, Y.W.; Yao, S.; Cai, L.Y.; Long, H.L.; Bi, Q.R.; Zhen, Y.Y.; Wu, W.Y.; Guo, D.A. Exploring lipid markers of the quality of coix seeds with different geographical origins using supercritical fluid chromatography mass spectrometry and chemometrics. *Phytomedicine* **2018**, *45*, 1–7. [CrossRef] [PubMed]

10. United States Pharmacopeial Convention. Coix Lacryma-Jobi Seed: Herbal Medicines Compendium. 2015. Available online: http://hmc.usp.org/monographs/coix-lacryma-jobi-seed-1-0 (accessed on 4 July 2022).
11. Costa, C.; Antonucci, F.; Pallottino, F.; Aguzzi, J.; Sun, D.-W.; Menesatti, P. Shape Analysis of Agricultural Products: A Review of Recent Research Advances and Potential Application to Computer Vision. *Food Bioprocess Technol.* **2011**, *4*, 673–692. [CrossRef]
12. Wang, N. Effect of variety and crude protein content on dehulling quality and on the resulting chemical composition of red lentil (Lens culinaris). *J. Sci. Food Agric.* **2008**, *88*, 885–890. [CrossRef]
13. Xu, M.; He, D.; Teng, H.; Chen, L.; Song, H.; Huang, Q. Physiological and proteomic analyses of coix seed aging during storage. *Food Chem.* **2018**, *260*, 82–89. [CrossRef] [PubMed]
14. Dussert, S.; Davey, M.W.; Laffargue, A.; Doulbeau, S.; Swennen, R.; Etienne, H. Oxidative stress, phospholipid loss and lipid hydrolysis during drying and storage of intermediate seeds. *Physiol. Plant.* **2006**, *127*, 192–204. [CrossRef]
15. Chen, A.J.; Jiao, X.; Hu, Y.; Lu, X.; Gao, W. Mycobiota and Mycotoxins in Traditional Medicinal Seeds from China. *Toxins* **2015**, *7*, 3858–3875. [CrossRef]
16. Annunziata, L.; Schirone, M.; Visciano, P.; Campana, G.; De Massis, M.R.; Migliorati, G. Determination of aflatoxins, deoxynivalenol, ochratoxin A and zearalenone in organic wheat flour under different storage conditions. *Int. J. Food Sci. Technol.* **2021**, *56*, 4139–4148. [CrossRef]
17. Rogowska, A.; Pomastowski, P.; Sagandykova, G.; Buszewski, B. Zearalenone and its metabolites: Effect on human health, metabolism and neutralisation methods. *Toxicon* **2019**, *162*, 46–56. [CrossRef]
18. Bol, E.K.; Araujo, L.; Veras, F.F.; Welke, J.E. Estimated exposure to zearalenone, ochratoxin A and aflatoxin B1 through the consume of bakery products and pasta considering effects of food processing. *Food Chem. Toxicol.* **2016**, *89*, 85–91. [CrossRef]
19. Mousavi Khaneghah, A.; Fakhri, Y.; Sant'Ana, A.S. Impact of unit operations during processing of cereal-based products on the levels of deoxynivalenol, total aflatoxin, ochratoxin A, and zearalenone: A systematic review and meta-analysis. *Food Chem.* **2018**, *268*, 611–624. [CrossRef]
20. Liu, X.; Zhang, X.; Rong, Y.-Z.; Wu, J.-H.; Yang, Y.-J.; Wang, Z.-W. Rapid Determination of Fat, Protein and Amino Acid Content in Coix Seed Using Near-Infrared Spectroscopy Technique. *Food Anal. Methods* **2014**, *8*, 334–342. [CrossRef]
21. La Nasa, J.; Ghelardi, E.; Degano, I.; Modugno, F.; Colombini, M.P. Core shell stationary phases for a novel separation of triglycerides in plant oils by high performance liquid chromatography with electrospray-quadrupole-time of flight mass spectrometer. *J. Chromatogr. A* **2013**, *1308*, 114–124. [CrossRef]
22. Zhang, X.; Qi, C.; Zhang, Y.; Wei, W.; Jin, Q.; Xu, Z.; Tao, G.; Wang, X. Identification and quantification of triacylglycerols in human milk fat using ultra-performance convergence chromatography and quadrupole time-of-flight mass spectrometery with supercritical carbon dioxide as a mobile phase. *Food Chem.* **2019**, *275*, 712–720. [CrossRef] [PubMed]
23. Indelicato, S.; Bongiorno, D.; Pitonzo, R.; Di Stefano, V.; Calabrese, V.; Indelicato, S.; Avellone, G. Triacylglycerols in edible oils: Determination, characterization, quantitation, chemometric approach and evaluation of adulterations. *J. Chromatogr. A* **2017**, *1515*, 1–16. [CrossRef] [PubMed]
24. Xi, X.J.; Zhu, Y.G.; Tong, Y.P.; Yang, X.L.; Tang, N.N.; Ma, S.M.; Li, S.; Cheng, Z. Assessment of the Genetic Diversity of Different Job's Tears (Coix lacryma-jobi L.) Accessions and the Active Composition and Anticancer Effect of Its Seed Oil. *PLoS ONE* **2016**, *11*, e0153269. [CrossRef]
25. Castro-Gómez, M.P.; Holgado, F.; Rodríguez-Alcalá, L.M.; Montero, O.; Fontecha, J. Comprehensive Study of the Lipid Classes of Krill Oil by Fractionation and Identification of Triacylglycerols, Diacylglycerols, and Phospholipid Molecular Species by Using UPLC/QToF-MS. *Food Anal. Methods* **2015**, *8*, 2568–2580. [CrossRef]
26. He, C.; Cao, J.; Bao, Y.Y.; Sun, Z.X.; Liu, Z.Y.; Li, C. Characterization of lipid profiling in three parts (muscle, head and viscera) of tilapia (Oreochromis niloticus) using lipidomics with UPLC-ESI-Q-TOF-MS. *Food Chem.* **2021**, *347*, 129057. [CrossRef]
27. Chen, H.J.; Hsu, H.Y.; Chiang, W. Allergic immune-regulatory effects of adlay bran on an OVA-immunized mice allergic model. *Food Chem. Toxicol.* **2012**, *50*, 3808–3813. [CrossRef]
28. Chen, H.J.; Chung, C.P.; Chiang, W.; Lin, Y.L. Anti-inflammatory effects and chemical study of a flavonoid-enriched fraction from adlay bran. *Food Chem.* **2011**, *126*, 1741–1748. [CrossRef]
29. Li, S.C.; Chen, C.M.; Lin, S.H.; Chiang, W.; Shih, C.K. Effects of adlay bran and its ethanolic extract and residue on preneoplastic lesions of the colon in rats. *J. Sci. Food Agric.* **2011**, *91*, 547–552. [CrossRef]
30. Chung, C.P.; Hsu, C.Y.; Lin, J.H.; Kuo, Y.H.; Chiang, W.; Lin, Y.L. Antiproliferative lactams and spiroenone from adlay bran in human breast cancer cell lines. *J. Agric. Food Chem.* **2011**, *59*, 1185–1194. [CrossRef]
31. Sun, S.; Yao, K.; Zhao, S.; Zheng, P.; Wang, S.; Zeng, Y.; Liang, D.; Ke, Y.; Jiang, H. Determination of aflatoxin and zearalenone analogs in edible and medicinal herbs using a group-specific immunoaffinity column coupled to ultra-high-performance liquid chromatography with tandem mass spectrometry. *J. Chromatogr. B* **2018**, *1092*, 228–236. [CrossRef]
32. Jedziniak, P.; Panasiuk, L.; Pietruszka, K.; Posyniak, A. Multiple mycotoxins analysis in animal feed with LC-MS/MS: Comparison of extract dilution and immunoaffinity clean-up. *J. Sep. Sci.* **2019**, *42*, 1240–1247. [CrossRef] [PubMed]

foods

MDPI

Article

Insight on a Competitive Nucleophilic Addition Reaction of Nε-(Carboxymethyl) Lysine or Different Amino Acids with 4-Methylbenzoquinone

Zhenhui Zhang [1], Lin Li [1,2], Yuting Li [2], Yi Wu [1], Xia Zhang [1], Haiping Qi [1] and Bing Li [1,*]

[1] Guangdong Province Key Laboratory for Green Processing of Natural Products and Product Safety, Engineering Research Center of Starch and Plant Protein Deep Processing, Ministry of Education, School of Food Science and Engineering, South China University of Technology, 381 Wushan Road, Guangzhou 510640, China; zhang126zzh@126.com (Z.Z.); lilin@dgut.edu.cn (L.L.); mbb_wuyi@163.com (Y.W.); cexzhang@scut.edu.cn (X.Z.); oliveauspicious@163.com (H.Q.)

[2] School of Chemical Engineering and Energy Technology, Dongguan University of Technology, College Road 1, Dongguan 523808, China; liyt@dgut.edu.cn

* Correspondence: lcbingli@scut.edu.cn; Tel.: +86-20-87113252

Abstract: *o*-benzoquinone is a common intermediate which is mostly formed by the oxidation of phenolics or polyphenols containing catechol structure. *o*-benzoquinone has an outstanding nucleophilic ability, while advanced glycation end products (AGEs) are nucleophilic and can undergo a nucleophilic addition reaction with *o*-benzoquinone to mitigate the harmful effects of AGEs on the body. As common nucleophilic substances, amino acids existing in large quantities in food processing and in vivo may bind competitively with *o*-benzoquinone, thus influencing the trapping of *o*-benzoquinone with AGEs. In this study, cyclic voltammetry and coexistence experiments were used to compare the reactivities of Nε-(carboxymethyl) lysine (CML) and amino acids with 4-methylbenzoquinone (4-MBQ). The results showed that CML is more reactive with *o*-benzoquinone than most amino acids, and even in complex systems, *o*-benzoquinone still captured CML. Moreover, almost all adducts were identified by UPLC-QTOF-MS/MS, and their chemical formulas were deduced. Quantum chemistry accurately predicts the efficiency and site of reactions of *o*-benzoquinone and nucleophiles to a certain extent, and found that a secondary amine has a greater reactivity with 4-MBQ than a primary amine in a similar molecular structure. In general, *o*-benzoquinone could capture AGEs, thereby showing potential to reduce the harmfulness of AGEs.

Keywords: advanced glycation end products; *o*-benzoquinone; cyclic voltammetry; coexistence system; binding site; secondary amine; quantum chemistry

1. Introduction

Advanced glycation end products (AGEs) are a type of dangerous chemical formed by the Maillard reaction between reducing sugars and amine groups [1]. AGEs can be divided according to their source into dietary AGEs and endogenous AGEs. Although the medical community has not given maximum limits, there is a clear consensus that reducing dietary AGEs intake and endogenous AGEs production can reduce the risk of diseases such as diabetes, cardiovascular disease and kidney disease [2–4]. There are many targeted studies on methods of inhibiting the physiological damage caused by AGEs through scavenging free radicals and eliminating dicarbonyls [5,6]. These approaches can reduce the production of AGEs to some extent. However, AGEs are inevitably present in foods or in vivo. To date, reports on eliminating existing AGEs have been rare. Unfortunately, long-term accumulation and cross-linking in vivo contribute to the harmful impact of AGEs [7]. Therefore, it is of great significance to eliminate the generated AGEs.

o-benzoquinone is a common reactive intermediate, mostly formed by oxidation of catechol structure, and has high electrophilic properties [8]. It would easily add to nucleophilic groups on amino acids or AGEs [9], as shown in Scheme 1. Epicatechin quinone, which was oxidized by Epicatechin, reacts with the α- or ε-amino group of lysine, thus inhibiting the Maillard reaction [10]. Our previous study found that *o*-benzoquinone also reacts directly with Nε-(carboxymethyl) lysine (CML), a typical representative of AGEs, forming an adduct of *o*-benzoquinone and CML [11], which is deduced to be less absorbable in the intestine than free CML. This discovery provides a new potential way to mitigate or eliminate the health hazards of AGEs. The *o*-benzoquinone structure is ubiquitous in the natural world. For example, catechol, a typical structure found in polyphenols, is easily oxidized into *o*-benzoquinone, which endows polyphenols with outstanding antioxidant ability. In addition, tyrosinase in the body can convert tyrosine or phenylalanine into dopaquinone, which also has an *o*-benzoquinone structure and sequentially polymerizes with cysteine or glutathione to form melanin or converts to dopamine [12]. Evidently, regardless of the food system or inside the body, dietary or endogenous AGEs are very likely to bind to compounds with an *o*-benzoquinone structure, forming AGE-derived adducts that might have different physiochemical properties and biological effects from those of AGEs. However, the binding of *o*-benzoquinone with AGEs is not specific since the reaction can occur between *o*-benzoquinone and any other nucleophilic group. Groups with high nucleophilicity would compete with AGEs to bind *o*-benzoquinone. Among them, amino acids are strong competitors since their amino groups are chemical moieties with strong electron-donating ability and high nucleophilicity. Amino acids are prevalent both in food systems and inside the body, their interference with the binding between *o*-benzoquinone and AGEs is thus unavoidable. For example, cysteine has been confirmed to have very high reaction rates with *o*-benzoquinone derivatives [13]. Therefore, to evaluate the binding ability of *o*-benzoquinone with AGEs and improve the ability of *o*-benzoquinone to mitigate the health hazards of AGEs, it is necessary to investigate the competitive effect of amino acids on the nucleophilic reaction of *o*-benzoquinone and AGEs.

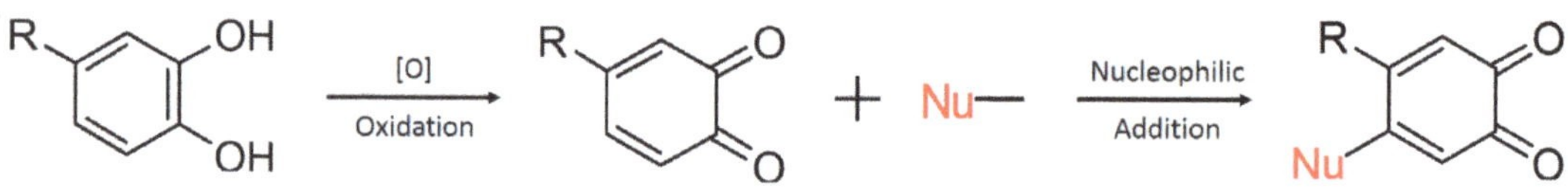

Scheme 1. Nucleophilic addition reaction scheme between catechol structure and nucleophilic groups (Nu) under oxidative conditions.

To date, there have been few reports on the reactivity of amino acids with *o*-benzoquinone. Some studies have investigated the reaction kinetics of CML or various amino acids (cysteine, lysine, methionine, glycine, arginine) with *o*-benzoquinone [11,14–16]. However, these studies investigated only the reactivity of individual nucleophilic amino acid with *o*-benzoquinone, while in food systems or inside the body, AGEs and different amino acids coexist. The binding of *o*-benzoquinone and AGEs would undergo interference through the combined effects of different amino acids. Thus, it is necessary to analyze the reactivity of each amino acid and AGE with *o*-benzoquinone in a competitive environment. Moreover, AGEs and some amino acids have more than a single nucleophilic site, so determining the addition site is helpful to clarify the addition mechanism and provide a reference for further application.

In this study, 4-methylcatechol (4-MC), a representative catechol, was employed to react with CML. The reaction of CML and different amino acids with 4-methylbenzoquinone (4-MBQ) formed after the oxidation of 4-MC was studied by cyclic voltammetry (CV). Ultra Performance Liquid Chromatography tandem quadrupole time-of-flight mass spectrometry (UPLC-QTOF-MS/MS) was used to quantify the residual amount of nucleophile to analyze the reactions of nucleophiles coexisting in mixtures with 4-MBQ. The reactivity of nucleophiles with 4-MBQ and the specific sites of the nucleophilic addition reaction were further investigated by mass spectrometry and quantum chemistry.

2. Materials and Methods

2.1. Materials

4-Methylcatechol (4MC, ≥95%), amino acid mixed standards, and ammonium formate were purchased from Sigma-Aldrich (St. Louis, MO, USA). Twenty amino acids, including L-lysine (Lys, 98%), L-arginine (Arg, 98%), L-alanine (Ala, 98%), L-asparagine (Asn, 98%), L-aspartic acid (Asp, 98%), L-methionine (Met, 99%), glycine (Gly, ≥99%), L-cysteine (Cys, 99%), L-glutamine (Gln, 98%), L-glutamic acid (Glu, 99%), L-histidine (His, 99%), L-isoleucine (Ile, 98%), L-leucine (Leu, 99%), L-phenylalanine (Phe, 98%), L-proline (Pro, 99%), L-serine (Ser, 97%), L-threonine (Thr, 99%), L-tryptophan (Trp, 98%), L-tyrosine (Tyr, 98%), L-valine (Val, 99%), methanol (HPLC grade) and acetonitrile (HPLC grade), were purchased from Aln (Shanghai, China). CML (98%) was purchased from TRC (Toronto, ON, Canada).

2.2. Cyclic Voltammetry

A solution of 1 mM 4MC in pH 7.4 PBS buffer (0.2 M) was prepared, and then CML or one of twenty amino acids was added to obtain a specific final concentration. The solubilities of the substances varied, and the concentrations are indicated in the experimental result chart. CV was performed as described in previous studies with some modifications [17]. By fitting the 4MC IpC1 for reduction and nucleophile concentration with an exponential asymptotic curve correlation, the parameters of the fitting formula were recorded, and the integrated area under the curve was determined. The abscissa interval for the integrated area calculation was standardized as 1–64 mM. The area for an amino acid with a concentration less than 64 mM was calculated according to the maximum value of vertical coordinates.

2.3. Quantitative Analysis of the Reactivity of 4-MBQ with Nucleophiles by UPLC-MS/MS

A 2.5 mM 4-MBQ solution was produced by the bulk electrolysis of 4MC solution as described in a previous study [13]. The concentration of each amino acid in the mixed amino acid standard was 2.5 mM; these were in 0.1 M HCl solution containing 16 amino acids, not including Cys, Asn, Trp, or Gln. In addition, these four amino acids and CML were blended with the amino acid mixed standard product in 0.1 N HCl to obtain 1.25 mM Cys-containing and Cys-free blended solutions, respectively. The 4-MBQ solution and the combined solution were mixed at molar concentration ratios of 0:1 (control group), 1:1 (low group), 5:1 (medium group), and 20:1 (high group), and the solution was diluted with pH 7.4 PBS buffer to give a nucleophile concentration of 25 μM. The diluted solution was allowed to react at room temperature for 1 h, and then an equal volume of 0.2 M HCL solution was added to stop the reaction.

The reaction product solutions were filtered through a 0.22 μM syringe filter. Samples (2 μL) were injected into a UPLC system (AB Sciex Inc., Framingham, MA, USA) equipped with an Agilent Poroshell 120 HILIC-Z column (2.1 × 100 mm, 2.7 μM) and eluted at a flow rate of 0.5 mL/min with eluent A consisting of 10% ammonium formate in aqueous solution (the pH was adjusted to 3 with formic acid) and eluent B consisting of ammonium formate in acetonitrile solution. Gradient elution was performed as follows: 0 min, 100% B; 7 min, 60% B; 9 min, 60% B; and 10 min, 100% B. The column was operated at 30 °C. Mass spectra were obtained by using an X500R QTOF instrument (AB Sciex Inc., Framingham, MA, USA). The IDA model and MS operating parameters adopted were as follows: curtain gas of 30 psi, nebulizer gas of 45 psi, auxiliary gas of 60 psi, source temperature of 400 °C, scan range of m/z 100–800, and ion spray voltage of 5500 V for positive mode analytes. Quantification was performed with by the standard external method.

2.4. Identification of the 4-MBQ-Nucleophile Reaction Products by UPLC-QTOF-MS/MS

4-MBQ was mixed with CML or amino acids at equimolar concentrations, and then the solution was diluted to 25 μM with pH 7.4 PBS buffer and allowed to react at 25 °C for 1 h. The reaction product solutions were filtered through a 0.22 μM syringe filter. Samples

(2 μL) were injected into a UPLC system (AB Sciex Inc., Framingham, MA, USA) equipped with an Agilent InfinityLab Poroshell 120 HILIC-Z column (2.1 × 100 mm, 2.7 μm) and eluted at a flow rate of 0.3 mL/min with eluent A consisting of 0.1% ammonium formic acid and eluent B consisting of methanol. Gradient elution was performed as follows: 0 min, 5% B; 4 min, 5% B; 8 min, 50% B; 10 min, 95% B; 12 min, 95% B; 13 min, 5% B; and 15 min, 5% B. The column was operated at 30 °C. Other mass spectrometry conditions are provided in Section 2.3.

2.5. Quantum Chemistry Calculations of the Nucleophilic and Electrophilic Indexes of Reagents

To obtain stable structures, the geometrical parameters of compounds were optimized in the ground state at the DFT/B3LYP level of theory with a 6-311G+(d,p) basis set. The optimized structural parameters were used in vibrational frequency calculations at DFT levels. The final molecular parameters were calculated at the M062X/def2-TZVP level (for single point energy calculations). The calculations were performed with the Gaussian 09 program package version D.01 by invoking gradient geometry optimization using a personal computer. HOMO–LUMO analysis was carried out to explain charge transfer within the molecule. Global hardness (η), global softness (s), electronegativity (χ), and chemical potential (μ) were calculated using the highest occupied molecular orbital (HOMO) and lowest unoccupied molecular orbital (LUMO).

In 1999, Parr and co-authors [18] propose that a specific property of a chemical species, the square of its electronegativity divided by its chemical hardness, be taken as defining its electrophilicity (ω) index (i.e., global electrophilicity index). In 2008, Domingo and co-authors [19] introduce the nucleophilicity (*N*) index (i.e., global nucleophilicity index) as:

$$N_{(Nu)} = E_{\mathrm{HOMO(Nu)}}\ (\mathrm{eV}) - E_{\mathrm{HOMO(TCE)}}\ (\mathrm{eV}) \tag{1}$$

where this scale of nucleophilicity is referred to tetracyanoethylene (TCE) taken as a reference since it presents the lowest HOMO energy in a large series of molecules already considered. That is, the N scale is simply the negative of the ionization potential calculated by Koopmans' theorem with an arbitrary shifting of the origin. This choice allows us conveniently to handle a nucleophilicity scale of positive values.

Site-specific nucleophilicity was compared to determining the Fukui function (FF) [20] and conceptual density functional theory (CDFT) [21]. The Fukui function describes the degree of change in electron density at each position when the number of electrons in the system changes, and it is strictly defined as:

$$f(r) = \left[\frac{\delta\mu}{\delta v(r)}\right]_N = \left[\frac{\partial\rho(r)}{\partial N}\right]_{v(r)} \tag{2}$$

where μ and N are the chemical potential and the number of electrons in the current system, respectively, and $\rho(r)$ represents the attractive potential of the atomic nucleus for electrons. The Fukui function was condensed to the atom level to obtain the so-called condensed Fukui function (CFF) [22]. Generally, the larger the value of the CFF is at a reaction site, the more likely this site is to be the active site for the corresponding reaction. The CFF describing the nucleophilicity of the specific atom of the molecular is defined as:

$$f_A^- = q_{N-1}^A - q_N^A \tag{3}$$

where q^A is the charge on atom A in the molecule, q_N is the neutral state, and q_{N-1} is the state lacking one electron in the molecule. When calculating CFF, the Hirshfeld method [23] was used to calculate the atomic charge.

Local electrophilicity index (ω^A) and local nucleophilicity index (${N_{Nu}}^A$) are defined as:

$$\omega^A = \omega f_A^- \tag{4}$$

$$N_{Nu}^{A} = N_{Nu} f_{A}^{-} \tag{5}$$

2.6. Statistical Analysis

The experimental results were analyzed using SPSS 22.0, and Origin 9.1 was used for correlation fitting. The data were expressed as the mean ± SD from at least three independent experiments. The F-test was used to assess the homogeneity of variance between groups. A value of $p < 0.05$ was considered significant.

3. Results

3.1. Effects of Nucleophile Concentration on the Reaction of Nucleophiles with o-Benzoquinone

CV is commonly used to analyze the redox characteristics of substances. Compounds with the catechol structure can form *o*-benzoquinone during the positive scan and are reduced to catechols during the negative scan [15]. In the presence of nucleophiles, the current intensity of the catechol cathodic peak in the CV experiment is decreased. Therefore, to analyze the reactivity of CML or amino acids with 4-MBQ, the change in the cathode peak of 4MC (EpC_1 and IpC_1) after adding different concentrations of nucleophiles, including CML and 20 amino acids, was investigated in the CV experiment, and their reaction dynamic parameters were calculated.

The cyclic voltammograms of 4MC with different concentrations of nucleophiles are shown in Figure 1 and Figures S1–S3 in Supplementary Material. With the addition of nucleophiles, the current intensity of the 4MC cathodic peak (IpC_1) showed dose-dependent decreases. These results showed that the interaction of the nucleophile with *o*-benzoquinone which was generated by the oxidation of 4MC occurred, and the reaction efficiency increased with increasing nucleophile concentration. A correlation fitting analysis for reducing the IpC_1 of 4MC with nucleophile concentration showed an exponential asymptotic correlation that conformed to Equation (6):

$$Y = a - b \times c^{x} \tag{6}$$

where the value of a can reflect the amount of 4-MBQ reacting in a cycle detection cycle when the amount of nucleophilic reagent tends to be maximum, while the value of b represents the fluctuation range of the amount of 4-MBQ reacting in the concentration range of the nucleophiles added. The c value is the slope of the fitted curve when the nucleophile is added at the median concentration, and the higher c value indicates that the change of nucleophilic reagent concentration will lead to higher change in the amount of 4-MBQ reaction. This fitting results of CML (Figure 1A) was different from Li, et al.'s [17], where a linear correlation was built up. This difference was possibly due to the wider range of CML concentrations selected in our study. The parameter values in Equation (3) are shown in Table 1. Analyzing the reaction efficiency of nucleophiles and 4-MBQ requires a comprehensive evaluation of these three indicators. At the same time, to make the results more intuitive, this study uses the area under the curve (AUC) as a reference indicator. AUC is a common indicator, which is often used to represent the comprehensive characteristics of the fitted curve [24–26]. In this study, AUC represents the total reaction efficiency within the concentration range, better reflecting the reactivity of nucleophiles with 4-MBQ.

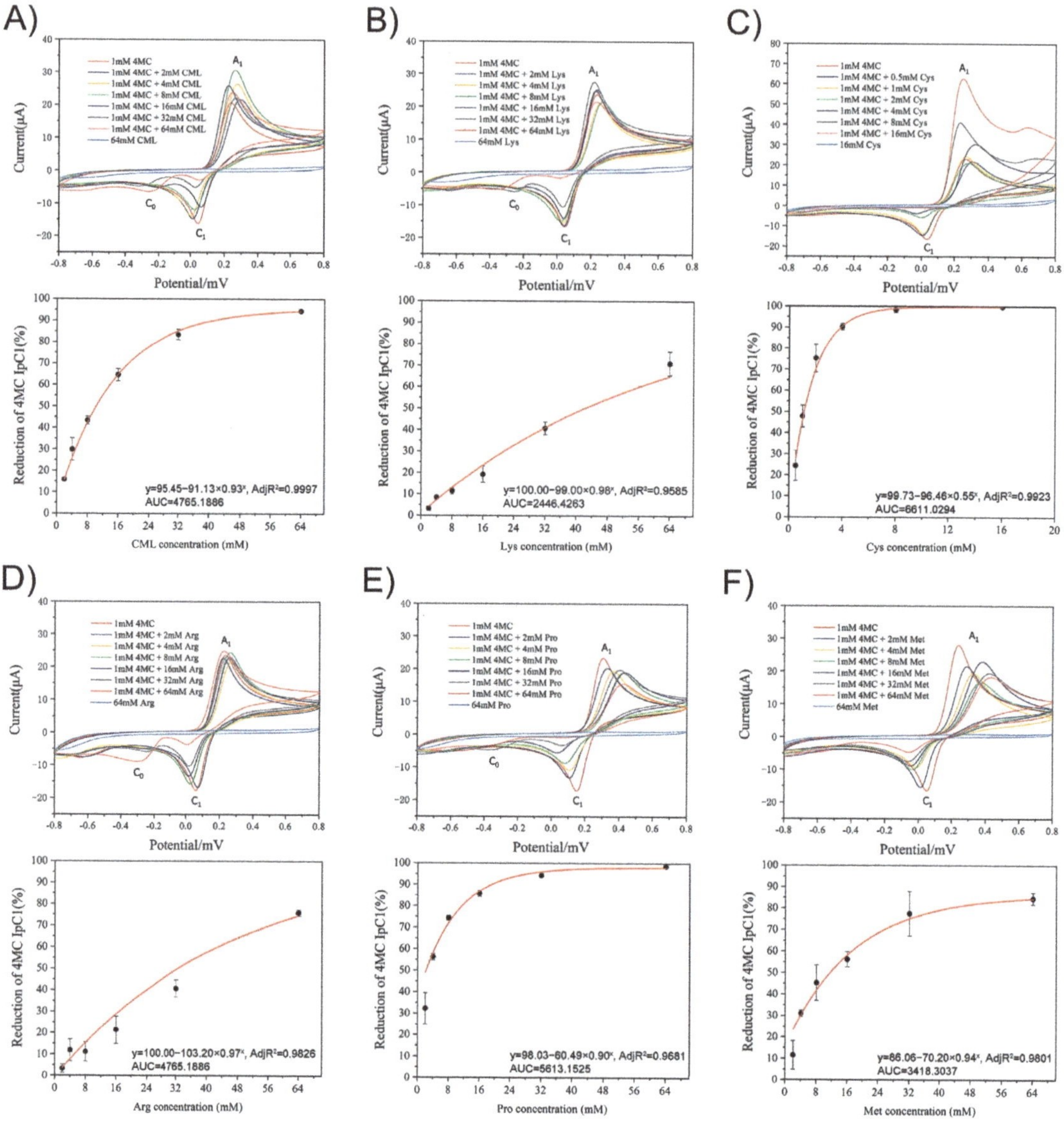

Figure 1. Cyclic voltammograms and fitted curve of 4MC (1 mM) in the presence of different concentrations of nucleophiles ((**A**): CML; (**B**): Lys; (**C**): Cys; (**D**): Arg; (**E**): Pro; (**F**): Met) at the surface of a glass carbon electrode in 0.2 M phosphate buffer (pH 7.4). Scan rate of 10 mV/s. Temperature of 25 ± 1 °C.

Comparing the results for reactions with CML and other amino acids showed that CML has a higher a value than most amino acids, indicating that CML has a higher binding efficiency to 4-MBQ. The b value of CML showed that the change in CML concentration had a greater impact on the cathodic peak current of 4MC than that of most nucleophilic amino acids. In addition, the c value of CML was higher than most amino acids, and the AUC value of CML was second only to that of Cys, Pro, or His, far higher than that of its precursor amino acid Lys, which indicates higher reactivity between CML and 4-MBQ.

The structural difference between CML and Lys lies in the ε-amino group. The reactivity of CML is higher than that of Lys, probably since the ε-amino group is more active after conversion from a primary amine to a secondary amine [27,28]. Therefore, it appears that the attack on *o*-benzoquinone came mainly from the ε-amino group of CML, which may eliminate the physiological hazard of CML. In addition, with the addition of CML, Lys, Arg, Pro, Asp, Gln, Phe, Trp, or Leu, a new weak reduction peak EpC_0 appeared with the reduction peak of 4-MC during the negative scan, which further proved that nucleophiles reacted with the oxidation products of 4-MC to form new complexes.

Table 1. Exponential asymptotic correlation fitting analysis of the 4-MC IpC_1 reduction and nucleophile concentration.

	Asymptote/(a)	Response Range/(b)	Slope/(c)	Adjusted R^2	Area Under Curve/(AUC)
Cys	99.73	96.46	0.55	0.9923	6611.0294
Pro	98.03	60.49	0.90	0.9681	5613.1525
His	99.98	68.95	0.93	0.9777	5256.3847
CML	95.45	91.13	0.93	0.9997	4765.1886
Val	79.88	61.57	0.95	0.9241	3782.85
Ile	88.63	77.40	0.96	0.9943	3737.7741
Met	86.06	70.20	0.94	0.9801	3418.3037
Arg	100.00	103.20	0.97	0.9826	2763.1127
Asn	100.00	90.49	0.98	0.9750	2550.7153
Lys	100.00	99.00	0.98	0.9585	2446.4263
Thr	69.73	78.10	0.96	0.9828	2436.15446
Gly	87.37	80.96	0.98	0.9396	2389.0043
Phe	61.38	65.23	0.96	0.9934	2349.8997
Asp	46.53	32.88	0.92	0.9577	2214.9758
Ala	37.27	57.51	0.75	0.9688	2192.8232
Trp	30.49	21.84	0.76	0.9801	1848.3408
Glu	27.53	30.99	0.47	0.9555	1720.8173
Leu	40.46	35.41	0.96	0.9593	1639.2518
Ser	100.00	98.41	0.99	0.9951	1526.9632
Gln	37.65	33.30	0.93	0.9339	1518.9773
Tyr	-	-	-	-	-

The reactivity of cysteine is much higher than those of CML and other amino acids. Due to the polarizability and electron-rich nature of sulfur, the sulfhydryl group of Cys is highly nucleophilic [29]. Fortunately, however, Cys is the least abundant amino acid and is often found in the form of cysteine [30]. Its sulfhydryl group is involved in multiple reactions, so it may not be highly competitive with CML in the food system and the body. The reactivity of Pro is second only to that of Cys since Pro has a secondary amine structure similar to that of CML. Brotzel, et al. [27], also confirmed this through photometric experiments. The imidazole group of His also contains a secondary amine, so it is slightly more nucleophilic than CML. While studying detoxification of biomass hydrolysates with nucleophilic amino acids, Xie, et al. [31], found that among twenty amino acids, the ability of His to detoxify biomass hydrolysates was second only to that of Cys, which is thought to be since they both contain secondary nucleophilic functional groups in addition to primary amine groups. According to the results of CV experiments, the reactivities of other amino acids decrease in the order: Val > Ile > Met > Arg > Asn > Lys > Thr > Gly > Phe > Asp > Ala > Trp > Glu > Leu > Ser > Gln > Tyr.

The ability of these amino acids to react with 4-MBQ may be related to the ability of side-chain groups to gain or lose electrons. The general rule is that nucleophiles attack electrophiles and that nucleophiles donate the electrons needed to form a covalent bond [32]. If the side chain of the amino acid is an electron-donating group, it will enhance the electron-donating ability of the α-amino group and improve its nucleophilicity. Studies have confirmed that the alkyl side chains of amino acids have electron donating ability [33,34]. The longer or the more branched it the alkyl chain is, the stronger the electron-donating ability [35], and the higher the nucleophilicity of the α-amino group. Therefore, Val and Ile are significantly more reactive than Ala and Gly. However, Leu shows poor reactivity, although the structure of its side chain is similar to that of Ile, which may be due to different steric hindrances for the side chain groups. The branches of Val and Ile are both at the β-carbon, and the steric hindrance is different from that of Leu. In addition, the inductive effects of alkyl groups are still controversial, and their ability to gain or lose electrons is related to their reaction environment and steric hindrance. The pK_a is also an important indicator, and nucleophilic groups with low pK_a values release protons easily [36]. For example, the pK_a of the amino group of Met is less than that of most amino acids, so its reactivity with *o*-benzoquinone is higher. The pK_a of the Tyr amino group is relatively low, and the electrochemical results showed that Tyr hardly reacts with 4-MBQ. However, solubility may interfere with this result, since only 0.045 g of Tyr dissolves in 100 g of water. The reactions of Asp, Glu, Trp, and Gln are also affected by solubility. Despite their similar structures, Glu and Asn have one more amine reaction site than Gln and Asp, respectively, so their reactivities are slightly higher. Similarly, the high reactivities of Arg, Lys, and Trp also result from this feature. Among them, Trp has the largest and most hydrophobic side chain amino acid [37]; its solubility is only slightly higher than that of Tyr, but it has two amino sites, one of which is a more reactive secondary amine.

3.2. The Reactivity of Coexisting Nucleophiles with 4-MBQ

The above results indicate the reactivities of individual amino acids with *o*-benzoquinone. However, in food and in vivo environments, amino acids coexist with each other. Therefore, it is necessary to analyze the reactivity of each amino acid and CML with *o*-benzoquinone in a competitive environment. CV can be used to monitor changes in the 4MBQ content of the system in real time, but for a coexistence study, it cannot be used to compare the efficiencies of different nucleophiles in reactions with 4MBQ. Fortunately, the reaction efficiencies of nucleophiles and *o*-benzoquinone in the coexistence system can be inferred by quantifying the residues of nucleophiles through mass spectrometry. Figure 2A shows that the Cys content was significantly reduced with increasing 4-MBQ addition. At low concentrations, the remaining content of Cys relative to that of the control group was only 21.8%, and at medium and high concentrations, the mass spectrometric signal for Cys could not even be detected; this illustrated the high reactivity of Cys with *o*-benzoquinone. The residual ratio for CML at low, medium and high concentrations were 97.8%, 87.7%, and 68.8% ($p < 0.05$), respectively. These results further confirmed that CML reacted with *o*-benzoquinone in a competitive environment. In addition, the reactivity of CML was greater than that of Lys, which proved that conversion of the ε-amino group of Lys into a secondary amine endows CML with higher reactivity. In addition, with increasing addition of 4-MBQ, the content of Arg also decreased significantly, indicating that the reactivity of Arg with 4-MBQ is higher than those of most amino acids. The guanidine group of Arg contains free amine groups, secondary amine groups and alkyl amine groups and forms a conjugated structure, so it exhibits extremely high nucleophilicity. Strangely, this is not consistent with the results of the CV experiments. Our hypothesis is that while detection by cyclic voltammetry is instantaneous, Arg does not fully react with 4-MBQ in one cycle. Other nucleophiles did not show significant decreases in the low concentration group, and only Pro and Gln showed significant decreases at medium concentrations; this indicated that Pro is more nucleophilic than other amino acids and is more likely to react with *o*-benzoquinone. The secondary amine group may be the key factor in the high reactivity of Pro. However, interestingly, a higher Gln concentration also caused a significant decrease in the reaction efficiency, which may be related to the limited stability of Gln [38]. This is why there was no Gln in the amino acid mixed solution kit. Cys, Asn, and Trp also showed limited stability. In addition, the signal for Gln showed low intensity in the mass spectrum, and the inspection error was significant, which is another important cause of abnormal results. In the high-concentration group, although most amino acids showed a decrease compared with the control group, the contents of His and Ile decreased significantly ($p < 0.05$), and these two amino acids also showed high reactivity in cyclic voltammetry.

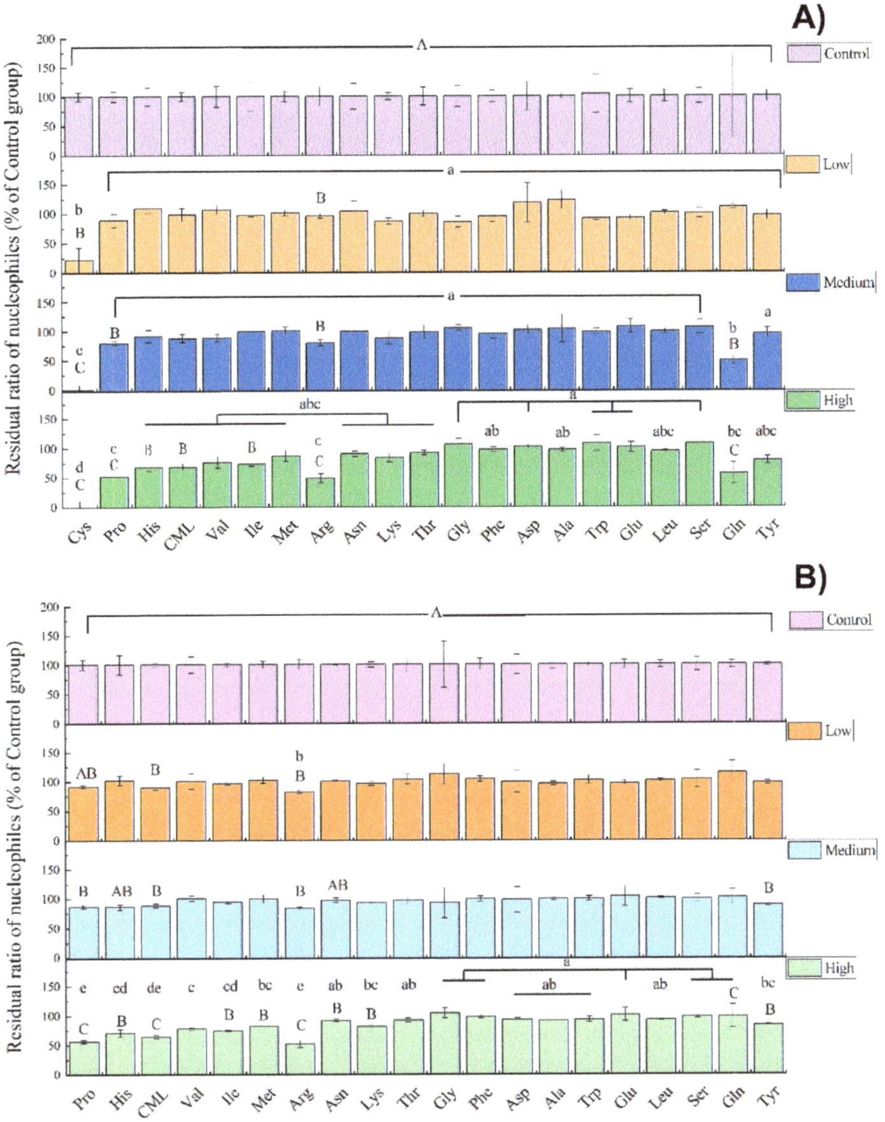

Figure 2. Nucleophilic reaction efficiency of CML and amino acids with 4-MBQ under coexisting conditions. The 4-MBQ solution and the combined solution were mixed at molar concentration ratios of 0:1 (control group), 1:1 (low group), 5:1 (medium group), and 20:1 (high group). (**A**) Full amino acid system, (**B**) full amino acid system except for Cys. The Y axis represented the ratio of the content of nucleophilic reagent after adding 4-MBQ to that without adding 4-MBQ. All data are the averages of triplicate measurements from three separate experiments. Capital letters represent that significant differences ($p < 0.05$) between different groups of the same substance, that is, significant differences in the vertical direction. Lowercase letters represent significant differences between groups ($p < 0.05$), that is, significant differences in the horizontal direction).

Moreover, a comparison between different amino acids at the same concentration showed that there was no significant difference among amino acids in the medium group except for Cys and Gln, which showed that in the system with Cys, low-concentration 4-MBQ reacted preferentially with cysteine, which affected its ability to bind with other nucleophiles. Under high-concentration conditions, the overall trend for the reactions of amino acids was more consistent with the results of the electrochemical experiments. Nevertheless, Tyr showed a higher reactivity than most other amino acids. This was since the solubility of Tyr required testing in the electrochemical experiments with very low concentrations, and its true reaction capacity could not be obtained accurately. Tyr has a low pK_a and a phenolic hydroxyl group, and the hydroxyl group of the phenolic ring is strongly electron-donating [39]. Therefore, the amino group of Tyr is strongly electron-donating, which facilitates nucleophilic addition to *o*-benzoquinone.

The contents of cysteine in food systems and the body are relatively low. Therefore, this study further investigated the ability of each amino acid and CML to react with 4-MBQ in a system without cysteine (shown in Figure 2B). In the low-concentration group, compared with the Cys-containing system, the concentrations of Arg, Pro and CML exhibited more significant reductions, 91.8%, 89.9%, and 82.2%, respectively, than those of other amino acids. In the high concentration group, the residual ratio of the three were reduced to 56.1%, 64.1%, and 51.9%, respectively, consistent with a secondary amine structure with high nucleophilicity. Tyr exhibited good reactivity at medium concentrations, which further showed that its solubility influenced the results of the electrochemical experiments. Under high concentration conditions, compared with the group containing Cys, more amino acids showed significant differences relative to the control group, which implied that the presence of Cys in the system will greatly inhibit binding of o-benzoquinone to CML or other amino acids. In general, the coexistence experiment showed that CML has the ability to bind *o*-benzoquinone and is superior to most amino acids, regardless of whether the system contains Cys.

3.3. *Identification of 4-MBQ-Nucleophile Reaction Products by UPLC-QTOF-MS/MS*

To further confirm the nucleophilic addition reactions of amino acids or CML with *o*-benzoquinone, their adducts with *o*-benzoquinone were detected by mass spectrometry, as shown in Figures S4–S24 in the Supplementary Material and in Table 2. The detected product ions for the CML-4-MBQ adduct were at m/z 206.1187, 152.0717, 130.0874, and 84.0815, and the structural formula was subsequently deduced to prove that the main binding site for 4-MBQ and CML was at the secondary amine of CML. From the product ions, it can be inferred that the binding site of Lys with 4-MBQ should be in the α amino group, which is different from the findings of Li, et al. [16]. The product ions of m/z 178.9184, 136.9303, and 118.9184 were not detected in their spectra. However, these ions are the key to determining the binding sites, and the quantum chemical data in our study (shown in Figure 3) also prove that the α amino group of lysine has higher nucleophilicity than the ε amino group. Another proof occurred with Trp, which also has two amino sites. MS/MS data for the Trp-4MC adduct showed that a product ion with m/z 252.0984 was detected, which proved that binding occurred at the secondary amine site of Trp. His also contains a secondary amine structure, and the presence of a fragment ion with m/z 217.0914 proved that binding occurred with the secondary amine. Moreover, the addition products of L-Ile-4MC and L-Asp-4-MBQ showed a signal only for the precursor ion, and no related product ions were found. This may be related to the instability of the addition product, which would result in low precursor ion abundance and make the product ions difficult to find. For example, the parent ion of a complex formed by 4-MBQ and two Asp was found. The parent ions and product ions of other amino acids were also found, and possible structures and fragmentation methods were deduced, as shown in Figures S4–S24.

Table 2. Mass information for adducts generated by the reactions of 4-MBQ with CML or amino acids.

Adduct	Formula	Precursor Ion/(*m*/*z*)	Product Ion (Relative Intensity %)
L-Cys-4-MC	$C_{10}H_{13}NO_4S$	244.0594	127.0189 (100); 155.0128 (87.4); 86.9896 (6.2) 227.0353 (3.3)
L-Pro-4-MBQ	$C_{12}H_{13}NO_4$	236.0892	192.1011 (100.0); 208.0955 (54.9); 70.0655 (16.1); 162.0908 (12.6)
L-His-4-MBQ	$C_{13}H_{13}N_3O_4$	276.0946	95.0594 (100.0); 232.1043 (53.3); 138.0524 (28.3)
L-His-4MC	$C_{13}H_{15}N_3O_4$	278.1102	232.1066 (100.0); 217.0944 (32.9); 83.0602 (30.4)
CML-4-MBQ	$C_{15}H_{20}N_2O_6$	325.1398	84.0815 (100.0); 130.0874 (30.4); 152.0717 (18.8); 206.1187 (10.2)
L-Val-4-MBQ	$C_{12}H_{15}NO_4$	238.1044	192.0982 (100.0); 164.1049 (86.2); 110.0595 (11.5); 122.0595 (8.0); 210.1104 (4.6); 150.0535 (2.3)
L-Ile-4MC	$C_{11}H_{13}NO_5$	254.1360	Null
L-Met-4-MBQ	$C_{12}H_{15}NO_4S$	270.0761	61.0109 (100.0); 133.0312 (52.8); 176.0695 (43.6); 148.0751 (22.4); 224.0729 (14.4); 194.0795 (14.0)
L-Agr-4-MBQ	$C_{13}H_{18}N_4O_4$	295.1102	70.0642 (100.0); 90.9763 (41.7); 164.1045 (18.3)
L-Asn-4-MBQ	$C_{11}H_{12}N_2O_5$	252.9946	118.9190 (56.9); 168.9032 (52.4); 220.9165 (44.9)
L-Lys-4-MBQ	$C_{13}H_{18}N_2O_4$	267.1315	84.0808 (100.0); 136.9303 (34.3); 118.9184 (23.3); 178.9184 (12.4); 206.1156 (11.4); 222.1148 (8.6)
L-Thr-4-MBQ	$C_{11}H_{13}NO_5$	240.0842	150.0543 (100.0); 122.0597 (70.2); 196.0589 (19.9); 178.0489 (7.5)
Gly-4-MBQ	$C_9H_9NO_4$	196.0568	122.0592 (100.0); 94.0644 (45.2); 150.0538 (37.0)
L-Phe-4-MC	$C_{16}H_{19}NO_4$	288.1207	120.0808 (100.0); 166.0860 (13.5); 149.0572 (2.9)
L-Asp-4-MBQ	$C_{11}H_{11}NO_6$	254.0831	Null
L-Asp-4-MC-L-Asp	$C_{15}H_{18}N_2O_{10}$	387.1258	205.0582 (100.0)
L-Ala-4-MBQ	$C_{10}H_{11}NO_4$	210.0714	136.0747 (100.0); 164.0692 (54.3)
L-Trp-4-MBQ	$C_{18}H_{16}N_2O_4$	325.1146	252.0984 (100.0)
L-Trp-4-MBC	$C_{18}H_{18}N_2O_4$	327.1308	188.0696 (46.6); 146.0591 (34.5); 123.0435 (32.8); 205.0953 (6.0)
L-Glu-4-MBQ	$C_{12}H_{13}NO_6$	268.0787	250.0680 (15.4); 222.0733 (57.3); 204.0631 (25.6); 176.0694 (100.0); 148.0743 (51.3)
L-Leu-4MC	$C_{13}H_{19}NO_4$	254.1363	123.0441 (100.0); 86.0952 (50.0)
L-Ser-4-MBQ	$C_{10}H_{11}NO_5$	226.0680	152.0699 (100.0); 180.0641 (66.3); 123.0672 (32.9); 196.0583 (6.6)
L-Gln-4-MBQ	$C_{12}H_{14}N_2O_5$	267.0928	250.0679 (54.2); 86.0596 (42.9); 222.0734 (40.8); 130.0491 (38.8)
L-Tyr-4-MBQ	$C_{16}H_{15}NO_5$	302.0966	107.0474 (100.0); 150.0529 (41.9); 196.0575 (17.2); 256.0938 (3.7)

3.4. Quantum Chemical Prediction of the Reactivity of 4-MBQ with Nucleophiles

Quantum chemistry is also a method for judging the nucleophilicity and electrophilicity of molecules. The nucleophilicity and electrophilicity of molecules can be calculated with the CFF [20,21]. As shown in Table 3, the global nucleophilic index of Cys was slightly lower than that of Met, which was completely different from the results described above. This is due to the fact that the S atom of Met is saturated and has no site available for a reaction with *o*-benzoquinone. Nevertheless, Met gives a higher calculated result for the global nucleophilic index. The nucleophilic index of Tyr is higher than those of most nucleophiles, which also confirmed that the results of electrochemical experiments were influenced by solubility. In addition, the calculated global nucleophilic index for CML was lower than those of most amino acids, such as lysine, but its global electrophilicity was much higher than those of other amino acids. This does not match the actual efficiency of CML reactions. This may be due to the fact that global nucleophilicity and electrophilicity indicate the overall abilities of molecules to gain or lose electrons and their ability to attract other nucleophilic and electrophilic molecules. Therefore, the high electrophilicity of CML may increase its ability to attract the nucleophilic *o*-benzoquinone. Thereafter, the strong local nucleophilicity exhibited by the ε-amino group of CML (shown in Figure 3) would improve the binding efficiency of CML and *o*-benzoquinone. Therefore, in general, quantum chemical computations can predict the nucleophilicity and electrophilicity of molecules to a certain extent. Nevertheless, the actual efficiency for reactions between molecules are also restricted by many factors.

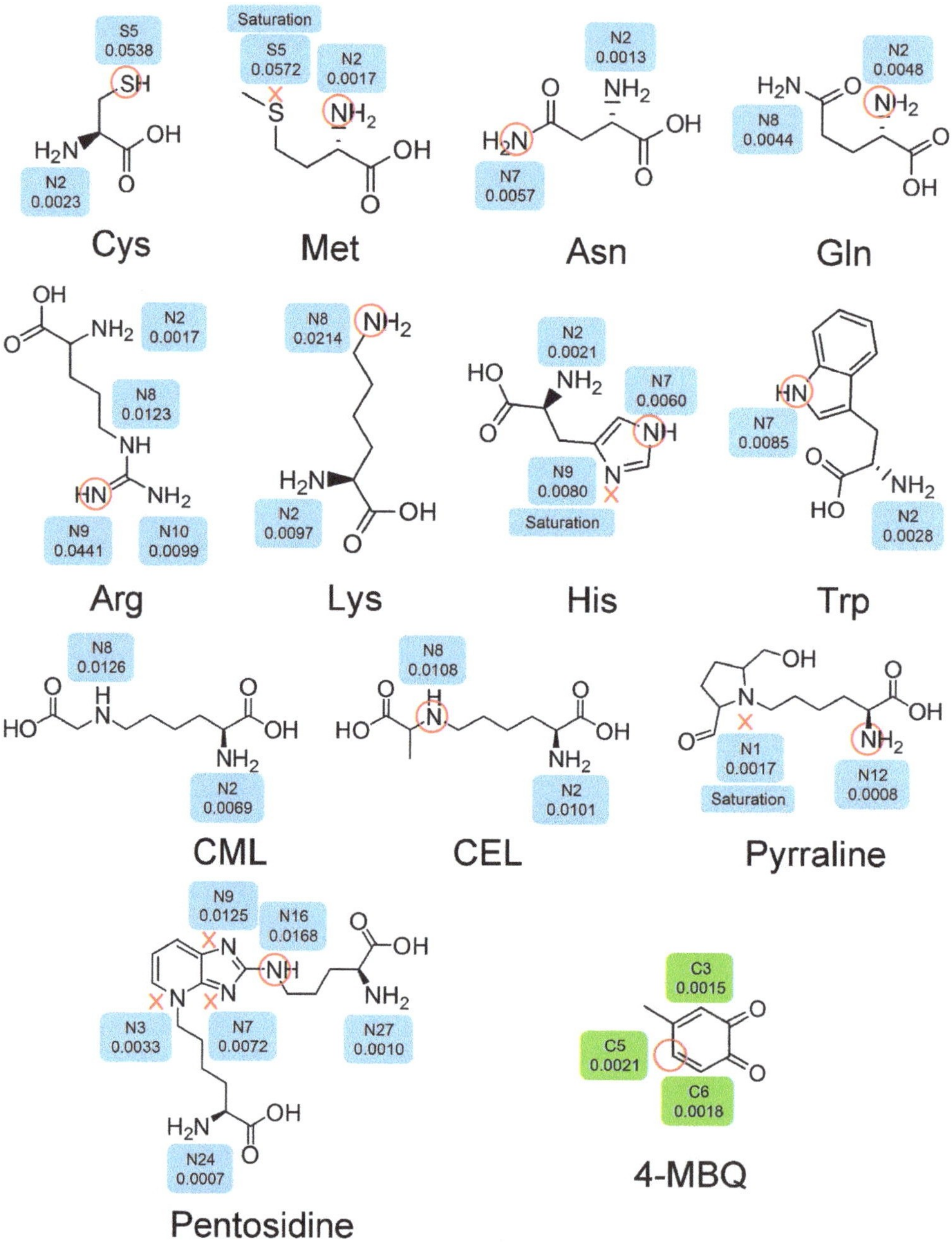

Figure 3. Local nucleophilic index (blue box) of key sites of nucleophiles containing multiple nucleophilic sites and local electrophilic index (green box) of key sites of 4-methylbenzoquinone. The red cross indicates that the site is saturated and cannot undergo nucleophilic addition reaction.

Table 3. Global nucleophilic and electrophilic index calculation results.

	Global Nucleophilic Index	Global Electrophilic Index
Cys	0.0910	0.0249
Pro	0.0852	0.0209
His	0.1143	0.0177
CML	0.0639	0.0293
Val	0.0657	0.0258
Ile	0.0713	0.0237
Met	0.1103	0.0238
Arg	0.1297	0.0188
Asn	0.0546	0.0286
Lys	0.0841	0.0238
Thr	0.0700	0.0239
Gly	0.0624	0.0253
Phe	0.0720	0.0276
Asp	0.0545	0.0289
Ala	0.0647	0.0252
Trp	0.1261	0.0220
Glu	0.0553	0.0284
Leu	0.0664	0.0256
Ser	0.0665	0.0244
Gln	0.0545	0.0280
Tyr	0.1004	0.0241
CEL	0.0665	0.0366
Pentosidine	0.1510	0.0320
Pyrraline	0.0914	0.0346

The local nucleophilicity of amino acids containing multiple nucleophilic sites were also calculated with quantum chemistry (shown in Figure 3). These data cannot accurately compare the nucleophilicity and electrophilicity of different substances, but they can compare different sites in the same substance. According to these results, the local nucleophilic index for the S position of Cys is much higher than that for the N2 position, which once again proves that the main reaction site of Cys is its sulfhydryl group. The nucleophilic index of the ε-amino group in CML is much higher than that of the α-amino group, which further proves that the ε-amino group is the main reaction site. The local nucleophilic index of the nitrogen on the imine of Arg is significantly higher than that at the other position, indicating that the imine is the main nucleophilic site, which is consistent with expectations. The N7 position of His and the N7 position of Trp also have secondary amine structures. The local nucleophilic indexes are much higher than that of a primary amine site, but the nucleophilic index of His N9 is also very high. However, His N9 is saturated and cannot undergo nucleophilic addition reactions. Other amino acids with multiple amino groups also have different local nucleophilic indexes, and these dates are in good agreement with the structure predicted by mass spectrometry.

Some typical AGEs have also been studied by using quantum chemistry to predict their reactivity with *o*-benzoquinone (shown in Table 3 and Figure 3). Due to the similar structures of CML and CEL, their nucleophilic and electrophilic indexes are relatively close. The global nucleophilic index of pentosidine is much higher than that of other AGEs since pentosidine has 6 amino sites, including a secondary amine structure that has the highest local nucleophilic index. Therefore, it is predicted that pentosidine is more likely to react with *o*-benzoquinone, while CML and CEL show similar reactivities. It should be noted that this conclusion is simply deduced from calculation, which can only be used as a reference, and the actual response ability needs to be further proved. In addition, the local electrophilic index of each site in 4MC was also calculated (shown in Figure 3). The C5 position has a higher electrophilic index, so this site should be the primary reactive site. Therefore, the inferred binding sites of *o*-benzoquinone in some studies may be inaccurate [10,16,17]. In general, quantum chemistry can predict the efficiency and site of reactions of *o*-benzoquinone and nucleophiles to a certain extent, which could be helpful for experimentalists. However, it is necessary to comprehensively analyze the global and local site indexes, and the actual reaction efficiency is also restricted by other aspects.

4. Conclusions

In conclusion, CV experiments confirmed that CML showed good reactivity compared to most amino acids, and the secondary amine structure greatly enhanced the reactivities of nucleophiles with *o*-benzoquinone. Coexistence experiments showed that even in complex systems, *o*-benzoquinone still captured CML. The structures of nucleophile-4-MBQ adducts were predicted from the MS and MS/MS results, and these predicted structures are in good agreement with the results calculated by quantum chemistry. In general, a series of experiments proved that *o*-benzoquinone could capture CML in single system or complex system, thereby showing potential to reduce the harmfulness of AGEs. Researchers can use this feature to investigate the actual impact of capture behavior on health.

Supplementary Materials: The following supporting information can be downloaded at: https://www.mdpi.com/article/10.3390/foods11101421/s1, Figure S1. Cyclic voltammograms and fitted curve of 4MC (1 mM) in the presence of different concentrations of nucleophiles (G: Tyr; H: his; I: Asn; J: Asp; K: Gln; L: Glu) at the surface of glass carbon electrode in 0.2 M phosphate buffer (pH 7.4). Scan rate of 10 mV/s. Temperature of 25 ± 1 °C; Figure S2. Cyclic voltammograms and fitted curve of 4MC (1 mM) in the presence of different concentrations of nucleophiles (M: Gly; N: Ala; O: Phe; P: Ser; Q: Thr; R: Trp) at the surface of glass carbon electrode in 0.2 M phosphate buffer (pH 7.4). Scan rate of 10 mV/s. Temperature of 25 ± 1 °C; Figure S3. Cyclic voltammograms and fitted curve of 4MC (1 mM) in the presence of different concentrations of nucleophiles (S: Val; T: Leu; U: Ile) at the surface of glass carbon electrode in 0.2 M phosphate buffer (pH 7.4). Scan rate of 10 mV/s. Temperature of 25 ± 1 °C; Figures S4–S24. Mass spectra, proposed structures and corresponding fragmentation pathways of adducts.

Author Contributions: Conceptualization, B.L., L.L. and Z.Z.; methodology, Z.Z. and Y.L.; validation, Z.Z. and Y.W.; formal analysis, Z.Z.; investigation, Z.Z.; resources, L.L. and B.L.; data curation, Z.Z. and Y.L.; writing—original draft preparation, Z.Z.; writing—review and editing, Y.W., H.Q. and B.L.; supervision, X.Z. and B.L.; project administration, X.Z. and B.L.; funding acquisition, L.L. and B.L. All authors have read and agreed to the published version of the manuscript.

Funding: This research was funded by Key Project of Guangzhou S&T Program (No. 201904020005) and the National Key R & D Program of China (No. 2017YFC1600401).

Institutional Review Board Statement: Not applicable.

Informed Consent Statement: Not applicable.

Data Availability Statement: The data presented in this study are available on request from the corresponding author.

Conflicts of Interest: The authors declare no conflict of interest.

References

1. Coelho, O.G.L.; Ribeiro, P.V.M.; Alfenas, R.C.G. Can grape polyphenols affect glycation markers? A systematic review. *Crit. Rev. Food Sci. Nutr.* **2021**, 1–11. [CrossRef] [PubMed]
2. Lavilla, C.J.; Billacura, M.P.; Hanna, K.; Boocock, D.J.; Coveney, C.; Miles, A.K.; Foulds, G.A.; Murphy, A.; Tan, A.; Jackisch, L.; et al. Carnosine protects stimulus-secretion coupling through prevention of protein carbonyl adduction events in cells under metabolic stress. *Free Radic. Biol. Med.* **2021**, *175*, 65–79. [CrossRef] [PubMed]
3. Sun, H.; Chen, J.; Sun, L.; Shi, B.; Li, J. Role of thioredoxin-interacting protein in diabetic fatty kidney induced by advanced glycation end-products. *J. Agric. Food Chem.* **2021**, *69*, 11982–11991. [CrossRef] [PubMed]
4. Omofuma, O.O.; Peterson, L.L.; Turner, D.P.; Merchant, A.T.; Zhang, J.; Thomson, C.A.; Neuhouser, M.L.; Snetselaar, L.G.; Caan, B.J.; Shadyab, A.H.; et al. Dietary advanced glycation end-products and mortality after breast cancer in the women's health initiative. *Cancer Epidemiol. Biomark. Prev.* **2021**, *30*, 2217–2226. [CrossRef]
5. Qin, C.; Li, Y.; Zhang, Y.; Liu, L.; Wu, Z.; Weng, P. Insights into oat polyphenols constituent against advanced glycation end products mechanism by spectroscopy and molecular interaction. *Food Biosci.* **2021**, *43*, 101313. [CrossRef]

6. Lund, M.N.; Ray, C.A. Control of maillard reactions in foods: Strategies and chemical mechanisms. *J. Agric. Food Chem.* **2017**, *65*, 4537–4552. [CrossRef]
7. Rowan, S.; Bejarano, E.; Taylor, A. Mechanistic targeting of advanced glycation end-products in age-related diseases. *Biochim. Biophys. Acta Mol. Basis Dis.* **2018**, *1864*, 3631–3643. [CrossRef]
8. Ito, S.; Sugumaran, M.; Wakamatsu, K. Chemical reactivities of ortho-quinones produced in living organisms: Fate of quinonoid products formed by tyrosinase and phenoloxidase action on phenols and catechols. *Int. J. Mol. Sci.* **2020**, *21*, 6080. [CrossRef]
9. Bittner, S. When quinones meet amino acids: Chemical, physical and biological consequences. *Amino Acids* **2006**, *30*, 205–224. [CrossRef]
10. Yin, J.; Hedegaard, R.V.; Skibsted, L.H.; Andersen, M.L. Epicatechin and epigallocatechin gallate inhibit formation of intermediary radicals during heating of lysine and glucose. *Food Chem.* **2014**, *146*, 48–55. [CrossRef]
11. Li, Y.; Li, L.; Lund, M.N.; Li, B.; Hu, Y.; Zhang, X. Kinetic investigation of the trapping of nε-(carboxymethyl)lysine by 4-methylbenzoquinone: A new mechanism to control nε-(carboxymethyl)lysine levels in foods. *Food Chem.* **2017**, *244*, 25–28. [CrossRef] [PubMed]
12. Land, E.J.; Ramsden, C.A.; Riley, P.A. Tyrosinase autoactivation and the chemistry of ortho-quinone amines. *Acc. Chem. Res.* **2003**, *36*, 300–308. [CrossRef] [PubMed]
13. Li, Y.; Jongberg, S.; Andersen, M.L.; Davies, M.J.; Lund, M.N. Quinone-induced protein modifications: Kinetic preference for reaction of 1,2-benzoquinones with thiol groups in proteins. *Free. Radic. Biol. Med.* **2016**, *97*, 148–157. [CrossRef] [PubMed]
14. Krayz, G.T.; Bittner, S.; Dhiman, A.; Becker, J.Y. Electrochemistry of quinones with respect to their role in biomedical chemistry. *Chem. Rec.* **2021**, *21*, 2332–2343. [CrossRef]
15. Oliveira, C.M.; Barros, A.S.; Ferreira, A.C.; Silva, A.M. Study of quinones reactions with wine nucleophiles by cyclic voltammetry. *Food Chem.* **2016**, *211*, 1–7. [CrossRef]
16. Li, Y.; Qi, H.; Fan, M.; Zhu, Z.; Zhan, S.; Li, L.; Li, B.; Zhang, X.; Zhao, X.; Ma, J.; et al. Quantifying the efficiency of o-benzoquinones reaction with amino acids and related nucleophiles by cyclic voltammetry. *Food Chem.* **2020**, *317*, 126454. [CrossRef]
17. Li, Y.; Huang, Q.; Yu, X.; Liu, Y.; Ma, J. Study of reactions of nε-(carboxymethyl) lysine with o-benzoquinones by cyclic voltammetry. *Food Chem.* **2019**, *307*, 125554. [CrossRef]
18. Parr, R.G.; Szentpály, L.v.; Liu, S. Electrophilicity index. *J. Am. Chem. Soc.* **1999**, *121*, 1922–1924. [CrossRef]
19. Domingo, L.R.; Chamorro, E.; Pérez, P. Understanding the reactivity of captodative ethylenes in polar cycloaddition reactions. A theoretical study. *J. Org. Chem.* **2008**, *73*, 4615–4624. [CrossRef]
20. Parr, R.G.; Yang, W. Density functional approach to the frontier-electron theory of chemical reactivity. *J. Am. Chem. Soc.* **1984**, *106*, 4049–4050. [CrossRef]
21. Geerlings, P.; De Proft, F.; Langenaeker, W. Conceptual density functional theory. *Chem. Rev.* **2003**, *103*, 1793–1873. [CrossRef] [PubMed]
22. Yang, W.; Mortier, W.J. The use of global and local molecular parameters for the analysis of the gas-phase basicity of amines. *J. Am. Chem. Soc.* **1986**, *108*, 5708–5711. [CrossRef] [PubMed]
23. Oláh, J.; Van Alsenoy, C.; Sannigrahi, A. Condensed fukui functions derived from stockholder charges: Assessment of their performance as local reactivity descriptors. *J. Phys. Chem. A* **2002**, *106*, 3885–3890. [CrossRef]
24. Le Floch, J.P.; Escuyer, P.; Baudin, E.; Baudon, D.; Perlemuter, L. Blood glucose area under the curve: Methodological aspects. *Diabetes Care* **1990**, *13*, 172–175. [CrossRef] [PubMed]
25. Fekedulegn, D.B.; Andrew, M.E.; Burchfiel, C.M.; Violanti, J.M.; Hartley, T.A.; Charles, L.E.; Miller, D.B. Area under the curve and other summary indicators of repeated waking cortisol measurements. *Psychosom. Med.* **2007**, *69*, 651–659. [CrossRef] [PubMed]
26. Myerson, J.; Green, L.; Warusawitharana, M. Area under the curve as a measure of discounting. *J. Exp. Anal. Behav.* **2001**, *76*, 235–243. [CrossRef] [PubMed]
27. Brotzel, F.; Chu, Y.C.; Mayr, H. Nucleophilicities of primary and secondary amines in water. *J. Org. Chem.* **2007**, *72*, 3679–3688. [CrossRef] [PubMed]
28. Yang, K.; Prasse, C.; Greenberg, M.M. Effect of histone lysine methylation on DNA lesion reactivity in nucleosome core particles. *Chem. Res. Toxicol.* **2019**, *32*, 910–916. [CrossRef]
29. Shannon, D.A.; Weerapana, E. Covalent protein modification: The current landscape of residue-specific electrophiles. *Curr. Opin. Chem. Biol.* **2015**, *24*, 18–26. [CrossRef]
30. Poole, L.B. The basics of thiols and cysteines in redox biology and chemistry. *Free Radic. Biol. Med.* **2015**, *80*, 148–157. [CrossRef]
31. Xie, R.; Tu, M.; Carvin, J.; Wu, Y. Detoxification of biomass hydrolysates with nucleophilic amino acids enhances alcoholic fermentation. *Bioresour. Technol.* **2015**, *186*, 106–113. [CrossRef]
32. Croom, E. Chapter three-metabolism of xenobiotics of human environments. In *Progress in Molecular Biology and Translational Science*; Hodgson, E., Ed.; Academic Press: Oxford, UK, 2012; pp. 31–88.
33. Dwyer, D.S. Electronic properties of the amino acid side chains contribute to the structural preferences in protein folding. *J. Biomol. Struct. Dyn.* **2001**, *18*, 881–892. [CrossRef] [PubMed]
34. Dwyer, D.S. Electronic properties of amino acid side chains: Quantum mechanics calculation of substituent effects. *BMC Chem. Biol.* **2005**, *5*, 2. [CrossRef] [PubMed]
35. Menger, F.M.; Mandell, L. *Electronic Interpretation of Organic Chemistry*; Springer: Berlin/Heidelberg, Germany, 1980.

36. Cameli, N.; Mariano, M.; Ardigò, M.; Corato, C.; Paoli, G.D.; Berardesca, E. Comparative instrumental evaluation of efficacy and safety between a binary and a ternary system in chemexfoliation. *J. Cosmet. Dermatol.* **2017**, *17*, 788–796. [CrossRef] [PubMed]
37. Bentley, T.W. Nucleophilicity parameters for amines, amino acids and peptides in water. Variations in selectivities for quinone methides. *Org. Biomol. Chem.* **2011**, *9*, 6685–6690. [CrossRef] [PubMed]
38. Christensen, H.N. New insights into intravenous peptide metabolism. *Nutr. Rev.* **1995**, *53*, 100–102. [CrossRef] [PubMed]
39. Papazi, A.; Kotzabasis, K. Inductive and resonance effects of substituents adjust the microalgal biodegradation of toxical phenolic compounds. *J. Biotechnol.* **2008**, *135*, 366–373. [CrossRef]